MANAGING HEALTH AND SAFETY

MIKE BATEMAN

MANAGING HEALTH AND SAFETY

MIKE BATEMAN

ICSA PUBLISHING

Published by ICSA Publishing Ltd
16 Park Crescent
London W1B 1AH

Typeset by ICSA Publishing and printed in Great Britain by
TJI Digital, Padstow, Cornwall

British Library Cataloguing in Publication Data

A catalogue record for this book is available from the British Library

ISBN – 186072–297–0

Contents

Preface

Since the Health and Safety at Work Act of 1974, employers have been required to take a pro-active approach to the identification and control of risks in the workplace. In recent years a specific requirement for risk assessment has been included in many sets of regulations. A steady flow of new regulations, many resulting from European Directives, constantly poses new challenges for the busy manager, a particular problem for those who only deal with health and safety on a part-time basis.

Managing Health and Safety provides an explanation of the law in plain English, together with an easily understood guide to the key elements of health and safety management. A major section of the book describes in practical terms how to carry out risk assessments – both general risk assessments and those required to comply with specific sets of regulations such as COSHH, manual handling and display screen equipment.

Several common health and safety topics such as work equipment, electricity and occupational health are dealt with in some detail, whilst there are also section dealing with subjects which have come into closer focus in recent times, for example managing asbestos in premises, occupational stress and work-related driving.

Managing Health and Safety has been written particularly with the non-health and safety specialist in mind, but it is hoped that specialists also will find much of value within it. Not only does it state what the law requires, but it also demonstrates what should be done to comply with it, in terms which can be easily understood by the busy manager.

Using this Book

 Case notes. These serve as cautionary tales, by providing real-life scenarios of the consequences of failure to comply with health and safety legislation.

 Sample documents appear throughout this work. These are included on the accompanying CD-ROM and can be adapted with the minimum of effort to suit the needs of an organisation. Included are forms such as those required for carrying out a risk assessment; health and safety inspections and reporting on accident investigations; and sample contents of a company's health and safety policy. Many of these are accompanied by completed examples in order to demonstrate their use and applicability.

Further reading

Details of additional information to which readers might wish to refer are included throughout the text. Precise details of the publications mentioned also appear at the end of individual chapters.

The tendency for readers to dip into books in order to find quick answers is reflected in the structure of this work. It is divided into logically sequenced sections and, within each chapter, further 'signposts' are provided by way of numbered headings which clearly flag up the subject. The sections contained in this work are:

- Health and safety law
- Managing health and safety
- Risk assessment requirements
- Other key areas of health and safety
- Sources of information

List of Abbreviations

ACM	asbestos-containing material
ACOP	Approved Code of Practice
APAU	Accident Prevention Advisory Unit
BS	British Standard
CCTV	closed circuit television
CDM	Construction (Design and Management) Regulations
CHIP	Chemicals (Hazard Information and Packaging for Supply) Regulations
CHSW	Construction (Health, Safety and Welfare) Regulations
CIEH	Chartered Institute of Environmental Health
COMAH	Control of Major Accident Hazards Regulations
COSHH	Control of Substances Hazardous to Health Regulations
CPS	Crown Prosecution Service
DDA	Disability Discrimination Act 1995
DETR	Department of the Environment, Transport and the Regions
DSE	display screen equipment
DSEAR	Dangerous Substances and Explosive Atmospheres Regulations
ELI	employers liability insurance
EMAS	Employment Medical Advisory Service
EU	European Union
FOPS	falling object protective structure
HASAWA	Health and Safety at Work etc. Act 1974
HSC	Health and Safety Commission
HSE	Health and Safety Executive
IEE	Institute of Electrical Engineers
IOSH	Institution of Occupational Safety and Health
LEV	local exhaust ventilation
LOLER	Lifting Operations and Lifting Equipment Regulations
MEL	maximum exposure limit
MEWP	mobile elevating work platform
NEBOSH	National Examination Board in Occupational Safety and Health
NTO	National Training Organisation
NVQ	National Vocational Qualification
OES	occupational exposure standard
OSRP	Offices, Shop and Railway Premises Act
PLI	public liability insurance
PPE	personal protective equipment
PTW	permit to work
PUWER	Provision and Use of Work Equipment Regulations
RIDDOR	Reporting of Injuries, Diseases and Dangerous Occurrences Regulations
ROPS	roll-over protection system
RoSPA	Royal Society for the Prevention of Accidents
RPE	respiratory protective equipment
SRSC	Safety Representatives and Safety Committees Regulations
SWL	safe working load
TUC	Trades Union Congress

Health and Safety Law

Introduction

AI.I   Introduction

It is not my intention, in a book of this type, to go into too much detail on legal concepts and processes. However, in order to understand their obligations in relation to the management of health and safety at work, it is essential that employers have a basic knowledge of how the law is structured in general terms and how health and safety laws are created and enforced.

AI.2   Criminal law

Standards of behaviour within UK society are regulated by law that is derived from Parliament. Some laws are in the form of Acts, such as the Health and Safety at Work etc. Act 1974 (HASAWA). However, many acts (including the HASAWA) empower the government, usually through the relevant secretary of state, to make regulations and orders that have the full force of the law. In fact the majority of health and safety law is in the form of detailed regulations, although the regulations are underpinned by important general requirements of the HASAWA – these are examined in greater detail in Chapter A2.

Breaches of the law are crimes (murder, assault, theft, etc.) and those responsible for such breaches are liable to prosecution. In most cases prosecutions are brought by the Crown Prosecution Service (CPS), but health and safety offences are prosecuted by the relevant enforcing authority either the Health and Safety Executive (HSE) or the local authority (usually the Environmental Health Department). Their roles and powers are described further in Chapter A2.

There may be exceptions to this where death occurs as a result of work activities. Consideration will be given to whether manslaughter charges are appropriate and the CPS will have a major role in that process. At the present time there ia a likelihood of a new offence of corporate killing being introduced – this will be dealt with in more detail in Chapter A2.

Prosecutions under health and safety law often result from accidents or cases of occupational ill health but this is not always the case. Many are also brought where the relevant enforcing authority is unhappy about the standards found in the workplace – unsatisfactory premises, facilities, work equipment or working practices.

Most health and safety prosecutions are heard in local magistrates' courts but more serious offences can go directly to the crown court or may be referred there by the magistrates. Appeals can go to either the Queen's Bench Division of the High Court or the Criminal Division of the Court of Appeal and ultimately to the House of Lords.

The process by which criminal charges are pursued and its interrelationship with the process of civil law (see AI.3 below) is described schematically in Fig. A1.1 while the structure of the courts system is shown in Fig. A1.2. Employment tribunals also play a part in the process of criminal law through their hearing of appeals against improvement or prohibition notices issued by the enforcing authorities (see A2.18).

A1.3   Civil law

Civil law is the process by which disputes between individuals and/or organisations are settled. Individuals who have suffered an injury or ill health as a result of work activities can claim damages under what is called the law of tort. A tort is a wrong or fault, and injury claims are brought under the tort of negligence, i.e. the injury is a result of a negligent failure to comply with obligations under statutory law or what is known as the 'common law'. Common law obligations and claims under civil law are explained further in Chapter A3.

Personal injury claims against employers under the civil law are most commonly brought by employees. However, employers also have both statutory law and common law obligations to many others such as customers, visitors, contractors' employees, tenants, neighbours, passers-by and even trespassers, and claims may result from each of these sources.

A1.4   The burden of proof

Under criminal law the prosecution must prove its case 'beyond all reasonable doubt' this applies in respect of breaches of health and safety law just as it does in respect of charges of murder or motoring offences. However, in civil cases the court must decide, on the basis of the evidence presented to it, where the truth lies on a 'balance of probabilities'.

Consequently even when criminal charges are dismissed by a magistrates' court or crown court, or the enforcing authority (the HSE or local authority) decides that the evidence is insufficiently strong to place before a court, a civil claim by the injured person may still prove successful.

A1.5   A brief history of health and safety law

The first health and safety legislation in the UK was the Health and Morals of Apprentices Act of 1802 which restricted hours of work and also made limited provision in respect of cleanliness and ventilation. However, this and three subsequent Acts were largely ineffectual, partly because of their limited scope and partly due to an absence of effective enforcement.

The Factory Act of 1833 (which applied only to textile mills) split the British Isles into four divisions, each with its own factory inspector and supported by eight superintendents. This gave a total strength of twelve to inspect 3,000 to 4,000 mills the HSE would probably claim that it has been understaffed ever since! It was not until the Factories Act of 1844 that any statutory safety provisions were introduced.

Despite some resistance in the 1850s, the legislation was steadily extended, in both its content and the range of workplaces to which it applied, until by 1878 almost all manufacturing industries were included. In 1881 the Employer's Liability Act was passed, making employers potentially liable in civil law for injuries to their employees. The Factory Act of 1891 gave the Home Secretary the power to make special regulations for dangerous trades.

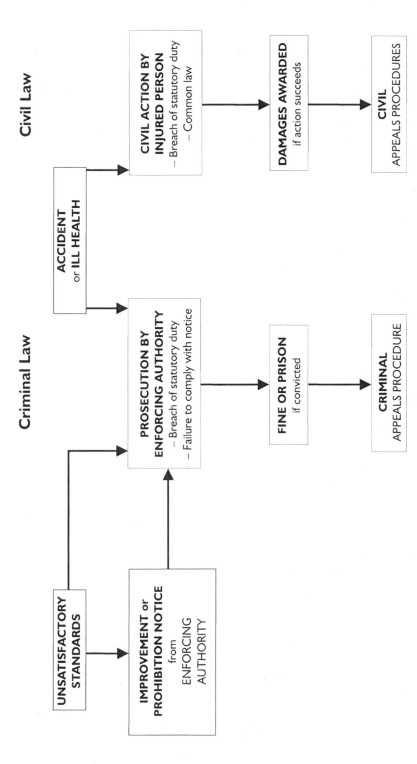

Figure A1.1 Process of Criminal and Civil Law

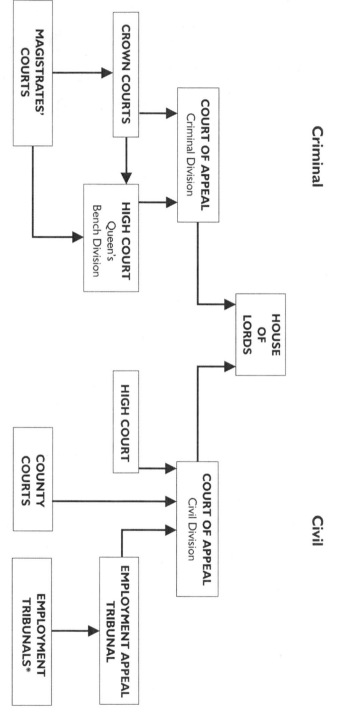

Criminal

Civil

MAGISTRATES' COURTS

CROWN COURTS

HIGH COURT
Queen's Bench Division

COURT OF APPEAL
Criminal Division

HOUSE OF LORDS

HIGH COURT

COUNTY COURTS

COURT OF APPEAL
Civil Division

EMPLOYMENT APPEAL TRIBUNAL

EMPLOYMENT TRIBUNALS*

Figure A1.2 The Courts Structure

* Employment tribunals deal with appeals against Improvement and Prohibition Notices

The twentieth century saw regulations introduced for several higher risk activities (docks, shipyards, building sites, etc.) while the Factories Act of 1937 removed a previously artificial distinction between factories where mechanical power was used and those where it was not. The same Act extended the safety provisions into areas such as means of access and the maintenance of floors and stairs, as well as expanding requirements in respect of mechanical safety. Further Acts were passed in 1948 and 1959 (the latter including increased requirements on fire precautions) before the Factories Act 1961.

That Act provided a model for the Offices, Shops and Railway Premises Act of 1963 which further extended the scope of health and safety legislation. Most of the 1961 and 1963 Acts have now been replaced by more modern statutory requirements resulting from the Health and Safety at Work etc. Act 1974 but some fragments still remain, including requirements to notify the relevant enforcing authority of the occupation or use of premises as a factory, office or shop.

A1.6   The Robens Report

Despite the progressive expansion of health and safety law described in the previous section, there were still many areas of work activity not covered by any legislation and employers had virtually no statutory responsibilities towards persons other than their employees. Regulations had developed piecemeal, often covering very specific trades or substances, and many had become very much out of date. In the early 1970s factories were still operating under Electricity Regulations passed in 1908 and the Wool, Goat Hair and Camel Hair Regulations of 1905 were also still in force.

Lord Alf Robens (previously a trade union leader, Labour MP and chairman of the National Coal Board) led a commission which addressed some of these problems. The Commission reported in 1972, recommending major changes. Most of their recommendations were incorporated into the Health and Safety at Work etc. Act 1974 which came fully into operation in April 1975.

Among the changes introduced in HASAWA were the following:

- The Act applied to all 'at work', including the self-employed wherever the work activity was being carried out.
- Both employers and employees had duties to protect others others at work and members of the public.
- Manufacturers, designers, suppliers and importers of work equipment and materials also had duties under the Act.

Responsibilities under the Act overlapped rather than being mutually exclusive.

- The Act contained general obligations rather than specific requirements.
- The Health and Safety Commission (HSC) and Health and Safety Executive (HSE) were established to coordinate policy and enforcement.
- The Act paved the way for future regulations supported by approved codes of practice, which could be kept up to date more easily.

The requirements of HASAWA and the role of the HSC and HSE are explored in more detail in Chapter A2.

A1.7 The European influence

The Single European Act of 1986 contained some important requirements in respect of health and safety. Article 118A made a specific commitment to improved health and safety at work on the part of the European Union (EU) while article 100A provided for the harmonisation of technical standards within the EU, including those relating to health and safety.

Although the EU can issue its own regulations which have direct application in member states, in relation to health and safety it mainly operates through issuing directives. These directives require member states to introduce their own legislation to implement the requirements of the directive, within a finite timescale. Introduction of the so-called 'six pack' of regulations in the UK, where six sets of regulations came into operation on 1 January 1993, was an illustration of this process.

The 'framework directive' adopted by the EU in 1989 contained many broad duties, including the requirement to assess risks associated with work activities and to introduce appropriate measures to prevent or control such risks. This matched regulations already in existence in the UK which required assessment, prevention and control of risks associated with asbestos, lead, hazardous substances and noise. The directive resulted in the general requirement for risk assessment contained in the Management of Health and Safety at Work Regulations (see Chapters B2 and C1).

Many other directives have been adopted or proposed since, including some that have resulted directly in new UK regulations (workplace health and safety, use of work equipment, use of personal protective equipment, manual handling of loads, display screen equipment, safety signs) or in changes to existing regulations (e.g. the COSHH Regulations). Additionally, many directives relating to technical standards and safety requirements for individual products have been adopted.

The specific requirements contained in many directives run counter to the general objective-setting approach of the Robens Report and HASAWA. There are also many concerns about the varying standards of implementation and enforcement within the member states of the EU and additionally about abuse of the CE marking – this is supposed to indicate conformance with technical standards.

Statutory Law
and its Enforcement

A2.1 Introduction

This chapter sets out the key requirements of the Health and Safety at Work Act 1974 (HASAWA) and summarises the content of many important sets of regulations – the detailed requirements of many of these regulations are dealt with in other chapters. The chapter also provides interpretation of many of the key terms used in health and safety law (reasonably practicable, competent person, etc.) and explains how the law is enforced. Additionally it looks ahead toward future trends in both legislation and its enforcement.

A2.2 HASAWA: the key provisions

HASAWA applies to everyone who is 'at work': employers, the self-employed and employees, with the exception of domestic servants in private households. The Act also protects the general public who may be affected by work activities. Paragraphs A2.3 to A2.10 of this chapter identify the main duty holders and the duties they hold under the Act. Some other sections of HASAWA, particularly those dealing with the enforcement of health and safety legislation, are covered later in this chapter.

A2.3 HASAWA, s. 2: duties of employers

Section 2 places extensive duties upon employers *to their employees*. Section 2(1) states: 'It shall be the duty of every employer to ensure, so far as is reasonably practicable, the health, safety and welfare at work of all his employees.' This sub-section is qualified by the term 'reasonably practicable' (which is interpreted in paragraph A2.11 of this chapter).

Section 2(2) goes on to detail more specific requirements relating to the following:

- provision and maintenance of plant and systems of work
- use, handling, storage and transport of articles and substances
- provision of information, instruction, training and supervision
- places of work and means of access and egress
- the working environment, facilities and welfare arrangements.

These are also qualified by the term 'reasonably practicable'.

Under Section 2(3) employers with five or more employees must prepare a written statement of their health and safety policy, together with the organisation and arrangements for carrying it out, and bring this to the notice of employees (see Chapter B1 for further guidance on health and safety policies).

Sub-sections 4, 6 and 7 of section 2 contain requirements in respect of the appointment of and consultation with safety representatives and the establishment of safety committees these subjects are dealt with in more detail in Chapter B6. (Sub-section 5 has been repealed.)

Section 2(1) in effect underpins all other legislation relating to the health, safety and welfare of employees, and the enforcing authorities regularly fall back on its all-embracing requirements to deal with situations that are not covered by other more specific legislation. Some examples of prosecutions brought under this section are provided in the case histories in chapters A2.3 and A2.4.

A range of examples of prosecutions under section 3(1) is provided in the case histories below.

 Prosecutions under HASAWA, s. 2 (1): case histories

- A local authority was fined £150,000 after two of its employees died when overcome by gas in an inspection chamber at a sewage pumping station. (This incident preceded the coming into operation of the Confined Spaces Regulations.)
- A diving company was fined £6,000 when one of its divers lost most of two fingers when his hand was trapped between two large concrete pipes being joined together in very dirty water with zero visibility.
- A transport company was fined £19,000 when an employee walking across its yard was knocked down and dragged along by a heavy goods vehicle.
- A company manufacturing offshore gas rigs and a firm of specialist sub-contractors were fined £75,000 and £150,000 respectively after a major explosion resulting from high pressure gas testing which killed three men and injured four others. (The companies were also fined identical amounts under s. 3(1) of HASAWA for endangering each other's employees – see below.)
- An oil refinery was fined £750,000 as a result of a fire in a catalytic converter which caused five workers to run for their lives (no one was hurt). This was the joint third highest ever fine for a breach of health and safety legislation.
- A laundry company was fined £325,000 after a worker died of heat exhaustion due to being trapped for nearly three hours in an industrial washing machine.

A2.4 HASAWA, s. 3: duties of employers and self-employed to persons other than their employees

Section 3(1) places similar wide-ranging duties on employers in relation to persons other than their employees. It states: 'It shall be the duty of every employer to conduct his undertaking in such a way as to ensure, so far as is reasonably practicable, that persons not in his employment who may be affected thereby are not exposed to risks to their health or safety.'

Section 3(2) puts self-employed persons under a similar duty in respect of other persons and also themselves. (Employees of self-employed persons are covered by section 2.)

Persons who might be protected by this section include the following:

- contractors and their employees
- agency staff
- volunteer workers

- visitors and delivery personnel
- customers, including passengers of transport undertakings
- lients in social or residential care
- patients within health services
- students at educational establishments
- members of the emergency services
- passers-by and occupants of neighbouring property
- trespassers.

However, it must be noted that these duties also are qualified by the phrase 'so far as is reasonably practicable' (see paragraph A2.11 below). This qualification would be of particular importance in relation to the extent of the duty to safeguard trespassers.

 Prosecutions under HASAWA, s. 3(1): case histories

- A construction company was fined £50,000 when a sub-contracted carpenter was run over by a delivery van reversing down a site access road. (The company was fined the same amount under section 2(1) and the van driver fined under section 7 of HASAWA – see paragraph A2.7 of this chapter.)
- A test laboratory was fined £25,000 for failing to provide a safe system of work for a staff member supplied by an agency. The worker suffered burns when she lost hold of a beaker of hot acid which ignited spontaneously in contact with her cotton laboratory coat.
- A health trust was fined £15,000 after a ten-month-old baby visitor pricked her finger with a used hypodermic syringe in a hospital.
- A ski centre was fined a total of £13,500 after an incident in which a 13-year-old user of the facility died when he hit a barrier. The boy, who was skiing for the first time, was allowed to go straight onto the main slope rather than the nursery slope.
- An employment agency was fined £175,000 and a rail maintenance company £150,000, after a young agency worker was hit and killed by a train. His experience and safety training were minimal and his immediate colleague at the time had previously been suspended for incorrect working practices.
- A health trust was fined £38,000 following the death of a patient who was injected with air instead of the appropriate fluid during a routine cardiac angiography.
- An animal feed mill and the owners of an HGV (an associated company) were fined a total of £50,000 as a result of an accident in which a 12-year-old passerby was killed by the vehicle reversing out of the mill. (The HGV driver was also fined under section 7 of HASAWA see paragraph A2.7 of this chapter.)
- A firm of railway engineering contractors left a lifting beam standing on soft ground in a public car park. A group of passing schoolchildren stopped in the car park for a picnic. A 16-year-old special needs child sat on the beam which toppled over, trapping both his legs against the ground. The contractors were fined £15,000.
- A technical college was fined £10,000 after a student lost the tips of two fingers on an incorrectly guarded guillotine.
- A rail operating company was fined £50,000 and a train operator £75,000 after a 12-year-old trespasser was killed on railway sidings. He had been able to climb through a poorly maintained fence.
- A timber merchant business was fined £100,000 when a forklift truck leaving its premises collided with a car travelling on the public road outside. The car driver was fatally injured by the raised forks of the truck.

- A health care company was fined £40,000 after a resident in one of its homes drank caustic soda from a jug which had been left unattended in an office which was accessible to residents.
- An earthmoving company was fined £16,000 for allowing mud to get onto a public road. A motorist subsequently lost control of his car and was killed in the resultant accident. The joint principal contractors for the project also each received fines.
- The principal contractor was fined £25,000 and a sub-contractor £12,500 after a pedestrian strayed into an area where roadworks were being carried out and was killed by a reversing lorry.

A2.5 HASAWA, s. 4: duties of persons concerned with premises to persons other than their employees

This section imposes duties on persons in relation to those who:

- are not their employees; but
- use non-domestic premises made available to them as a place of work.

Those who have duties under this section are persons who have:

- 'to any extent, control of premises' or
- means of access to or egress from such premises
- plant or substances in such premises

including those persons who have obligations by virtue of any contract or tenancy.
The section requires each duty holder:

> to take such measures as it is reasonable for a person in his position to take to ensure, so far as is reasonably practicable, that the premises, all means of access thereto or egress therefrom available for use by persons using the premises, and any plant or substances in the premises or, as the case may be, provided for use there, is or are safe and without risks to health.

The principal impact of this section is upon owners and managing agents who have duties in relation to premises under their control, even if they do not have a regular presence on the premises. For example, in a multi-occupancy office and retail complex, there will be duties in relation to common parts and equipment, for example:

- corridors and staircases
- the shopping concourse
- lifts and escalators
- access roads and car parks.

The duties will fall on whoever is deemed to have control of the relevant aspect normally the owner or managing agent, depending upon the circumstances.

Common parts of blocks of flats and equipment within them (e.g. lifts and electrical installations) have been determined to be 'non-domestic premises' available as a place of work to those who carry out repair and maintenance activities.

It should be noted that the duties under this section only relate to those who use the premises as a place of work, although cases relating to members of the public would probably be dealt with using section 3 of HASAWA.

A2.6 HASAWA, s. 6: duties of manufacturers, etc., as regards articles and substances for use at work

This section places duties on those who design, manufacture, import or supply articles for use at work (or articles of fairground equipment) and also on those who manufacture, import or supply substances. The duties relate to the following:

- the design and construction of articles, so that they are safe and without risks to health
- testing and examination of articles and substances
- the provision of adequate information about articles and substances (updated when new serious risks become known).

Additional duties in relation to the supply of work equipment and substances are contained in other Acts and Regulations. Some of these are explored in more detail later see paragraphs C3.5 (noise) and D1.3 (PUWER Reg. 10).

A2.7 HASAWA, s. 7: duties of employees at work

This section states that:
It shall be the duty of every employee while at work:

a)   to take reasonable care for the health and safety of himself and of other persons who may be affected by his acts or omissions at work; and

b)   as regards any duty or requirement imposed on his employer or any other person by or under any of the relevant statutory provisions, to co-operate with him so far as is necessary to enable that duty or requirement to be complied with.

(Please note that here, as in other parts of the law, the use of masculine pronouns is intended to include both men and women.)

Employees have a duty not only to take 'reasonable care' of themselves and their colleagues but also in relation to others who may be affected – the same range of persons to whom the employer has duties under section 3 of HASAWA. Employees' and employers' duties are not mutually exclusive – it is quite possible for an employer and one or more employees to be prosecuted as a result of the same set of circumstances. Two of the case histories referred to at para A2.4 (prosecutions under HASAWA, s. 3(1)) also resulted in the prosecution of individual employees, and such instances are not uncommon. Directors and senior managers have additional duties under section 37 of HASAWA.

A2.8 HASAWA, s. 8: duty not to interfere with or misuse things provided pursuant to certain provisions

The section states: 'No person shall intentionally or recklessly interfere with or misuse anything provided in the interests of health, safety or welfare in pursuance of any of the relevant statutory provisions.' This requirement is placed upon 'persons' rather than employees, although it is not known whether any prosecutions have been brought against persons other than those at work, i.e. members of the public.

It is intended to cover intentional or reckless actions such as the following:

- malicious activation of fire alarms or fire extinguishers
- damage to personal protective equipment

- deliberate bypassing of interlocking mechanisms
- vandalism or sabotage.

A2.9 HASAWA, s. 9: duty not to charge employees for things done or provided pursuant to certain specific requirements

Section 9 states: 'No employer shall levy or permit to be levied on any employee of his any charge in respect of anything done or provided in pursuance of any specific requirement of the relevant statutory provisions.' As a result of this section employers are not allowed to charge employees for anything provided or done as a result of health and safety legislation. Such practices have largely disappeared but still linger on in some areas of work activity, particularly in relation to the provision of personal protective equipment. (This is also covered in more detail in Chapter C6.)

A2.10 HASAWA, s. 37: offences by bodies corporate

Paragraph (1) of the section states: 'Where an offence under any of the relevant statutory provisions committed by a body corporate is proved to have been committed with the consent or connivance of, or to have been attributable to any neglect on the part of, any director, manager, secretary or other similar officer of the body corporate or a person who was purporting to act in any such capacity, he as well as the body corporate shall be guilty of that offence and shall be liable to be proceeded against and punished accordingly.'

Consequently where such cases of consent, connivance and neglect arise, the senior individuals concerned may be subject to fines as well as their company or public body. Some examples of prosecution of directors and senior managers are provided in the case histories below.

 Prosecutions of directors and senior managers: case histories

- Two directors of a firework company were personally fined a total of £46,250 for a catalogue of offences. As a result of one of these offences a man was killed during the unsafe disposal of condemned fireworks. The company had gone into bankruptcy and was only given a nominal fine.
- The managing director of a bakery was fined £20,000 and the production director £1,000 following an incident in which two men died from heat exposure during repair work inside a bread oven. (The bakery companies involved were fined a total of £350,000.)
- The company secretary and operations manager of a breakfast cereal manufacturer were fined £3,000 each after a worker was struck in the face by the lid of a pressure cooker, which was not properly secured.
- A director of a construction firm instructed an untrained employee to drive a forklift truck. A few days later a truck driven by the employee knocked over and killed a labourer. The director was fined £20,000 and the company £40,000.

A2.11 Levels of duty

Although the phrase 'so far as is reasonably practicable' is used to qualify some of the main sections of HASAWA and many regulations, some requirements must be complied with 'so far as is practicable' and many other requirements are 'absolute' duties.

Absolute requirements

An absolute duty is usually characterised by the use of the word 'shall', without qualification by 'practicable' or 'reasonably practicable'. In principle this means that the duty must be complied with, whatever the costs or practical difficulties involved. However, in practice many absolute duties are qualified by use of terms such as 'suitable', 'sufficient', 'adequate', 'appropriate', etc.

'Practicable'

Where requirements must be carried out 'so far as is practicable', this imposes a more onerous standard than 'reasonably practicable'. Once a precaution becomes 'practicable' in the light of current knowledge and invention, it must be taken, even though to take it may be expensive or inconvenient. However, a distinction is drawn between what is 'practicable' and what is 'physically possible'.

It is impracticable to take precautions against a danger that cannot yet be known to exist or to take precautions that have not yet been invented (*Jayne* v. *National Coal Board* (1963)). Consequently what is 'practicable' involves taking into account what is practice at the time. However, once a danger is recognised (e.g. following an accident), 'practicable' precautions must be taken.

'Reasonably practicable'

One of the most widely used definitions of 'reasonably practicable' is that provided by Lord Justice Asquith in his judgment on the case of *Edwards* v. *National Coal Board* (1949) in which he stated:

> 'Reasonably practicable' is a narrower term than 'physically possible' and seems to me to imply that a computation must be made by the owner in which the quantum of risk placed on one scale and the sacrifice involved in the measures necessary for averting risk (whether in money, time or trouble) is placed in the other, and that, if it be shown that there is a gross disproportion between them – the risk being insignificant in relation to the sacrifice – the defendants discharge the onus on them. Moreover, this computation falls to be made by the owner at a point in time anterior to the accident.

Use of the term 'reasonably practicable' means that employers (and others with duties) must weigh the level of risk against the costs of precautions. If the risk is insignificant compared to the money, time and trouble to avert it, the precautions need not be taken. In effect 'reasonably practicable' has always required a type of risk assessment to be carried out. Requirements for risk assessment have now become much more explicit, as described in Chapters C1 to C7.

However, account should not be taken of the duty holder's abilities to meet the costs of precautions what is 'reasonably practicable' is what is good value for money (or time or trouble) in health and safety terms, not what can be afforded by the duty holder.

The burden of proof

Section 40 of HASAWA states that in legal proceedings relating to duties qualified by the terms 'practicable' or 'reasonably practicable', it is the accused who must prove that it was not 'practicable' or 'reasonably practicable' to do more than was done to comply with the duty. This reverses the normal burden of proof in criminal cases although the defendants need only establish that they complied with the duty on a 'balance of probabilities' rather than 'beyond all reasonable doubt' (see paragraph A1.4 on burden of proof).

A2.12 Definitions and interpretations

A number of terms used in health and safety legislation are defined in sections 52 and 53 of HASAWA while definitions of others have been provided over the years by the courts (as in the cases of 'practicable' and 'reasonably practicable'). Definitions of some of the more common terms are given below.

Definitions given in s. 52
Section 52(1) of HASAWA states that:

   a) 'work' means as an employee or self-employed person;
   b) an employee is at work throughout the time when he is in the course of his employment, but not otherwise; and
   c) a self-employed person is at work throughout such time as he devotes to work as a self-employed person.

Sub-sections (2) and (3) provide for this definition to be extended by regulations. Several such extensions have been made, including the Health and Safety (Training for Employment) Regulations 1990 which include persons on certain types of training programme.

Definitions given in s. 53
Section 53 of HASAWA provides many definitions including the following:

- 'Employee' means an individual who works under a contract of employment, and related expressions shall be construed accordingly.
- 'Self-employed person' means an individual who works for gain or reward otherwise than under a contract of employment, whether or not he himself employs others.
- 'Plant' includes any machinery, equipment or appliance. (The term 'work equipment' is more commonly used nowadays see the Provision and Use of Work Equipment Regulations 1998.)
- 'Premises' includes any place and, in particular, includes: any vehicle, vessel, aircraft or hovercraft; any installation on land (including the foreshore and other land intermittently covered by water), any offshore installation and any other installation (whether floating or resting on the seabed or the subsoil thereof, or resting on other land covered with water or the subsoil thereof); and any tent or movable structure.

Competent person
Statutory definitions of this commonly used term 'competent person' are relatively rare and are not always helpful. Some guidance has been provided by the courts which has referred to a combination of theoretical knowledge and practical experience. The position has been improved by regulation 7 of the Management of Health and Safety at Work Regulations 1999 which requires employers to have competent health and safety assistance (see paragraph B2.6).

Paragraph (5) of the regulation states: 'a person shall be regarded as competent ... where he has sufficient training and experience or knowledge and other qualities to enable him properly to assist ...'. The HSE guidance accompanying the regulation states that competence 'does not necessarily depend on the possession of particular skills or qualifications'. Simple situations may require only the following:

- an understanding of relevant or current best practice;
- an awareness of the limitations of one's own experience and knowledge; and
- the willingness and ability to supplement existing experience and knowledge when necessary by obtaining external help and advice.

This definition and guidance provide a useful general basis for defining competence. An awareness of one's own limitations is particularly important – a person may be competent for one particular purpose or activity but not for another one.

A2.13 Enforcement of the law

In the UK health and safety is overseen by the Health and Safety Commission (HSC) which is appointed by the relevant secretary of state. The law is enforced in the field by the Health and Safety Executive (HSE), together with local authority staff, the latter mainly dealing with lower-risk work activities. The respective roles of the HSC, HSE and local authorities are examined below, together with the powers at the disposal of inspectors and the penalties that can be imposed by the courts for breaches of the law.

A2.14 Health and Safety Commission

The HSC was established by HASAWA and its composition, functions and powers are set down in sections 10–14 of the Act. Its members are appointed by the relevant secretary of state – responsibility currently rests with the Department for Work and Pensions.

The Commission consists of the following:

- a chairman
- three members appointed after consultations with employers' organisations
- three members appointed after consultations with employees' organisations (trade unions)
- up to three further members appointed after consultations with organisations representing local authorities, relevant professional bodies, etc.

As well as providing the linkage between government and the HSE, the HSC plays an important role in the preparation of new health and safety legislation, particularly in the light of European Directives. It issues codes of practice and guidance on how legislation is implemented (see later in this chapter) and can appoint individuals or committees to provide it with advice on specific topics or work activities. With the consent of the secretary of state it may establish investigations and inquiries into accidents and incidents. These powers were particularly well used in relation to a series of railway accidents during the 1990s but have also been used in respect of industrial and construction incidents.

A2.15 Health and Safety Executive

The HSE is the operational arm of the HSC. The Executive consists of a director, appointed by the HSC with the approval of the secretary of state, and two other members. These two are also appointed in the same way, but after consultation with the director.

The Field Operations Directorate of the HSE covers many employment sectors including manufacturing, construction, agriculture, education and local and central government (including the fire and police services), whilst there are specialist sections dealing with nuclear safety and hazardous installations (including mining, offshore installations and some parts of the chemical industry).

During 2004 the government announced that responsibility for enforcing health and safety legislation within the rail industry was being transferred from the HSE to the Office of Rail Regulation. This move drew criticism from the TUC and some individual trade unions as well as from the HSC and HSE themselves. The process of transfer was likely to take some time.

While the HSE has a vital role in enforcing the law (and its powers and policy for doing this are described below), it is also an important source of information on health and safety matters. Its role in fulfilling this function is described in more detail in Chapter E1.

A2.16 Local authorities

Local authorities are responsible for enforcing health and safety legislation in lower-risk activities such as the retail sector, some warehouses, hotels and catering, sport and leisure, consumer services, places of worship and most offices. (The HSE deals with offices supporting industrial and similar activities at the same location, e.g. offices at factories or on construction sites.)

Enforcement is normally through the Environmental Health Department which has to deal with many other functions including food safety, pest control, noise nuisance, etc., and not all environmental health staff will necessarily be health and safety specialists. However, local authority inspectors can still have the same powers as HSE inspectors (see below), although in practice authorities may limit these powers where individuals do not have appropriate knowledge and experience.

Concerns have been expressed in a number of quarters about the standards of health and safety enforcement by local authorities. In 2004 a Parliamentary Select Committee recommended an audit of local authority performance with action against authorities failing in their enforcement duties.

A2.17 Powers of inspectors

The powers of inspectors are detailed within section 20 of HASAWA. In some respects (e.g. powers of entry, power to take statements) their powers are greater than those of the police. Their powers include the following:

- to enter premises at any reasonable time (or at any time if they think a situation is, or may be, dangerous)
- to carry out examinations and investigations
- to take measurements, photographs or recordings
- to take samples of articles or substances (including atmospheric samples)
- to take possession of articles or substances (for examination purposes, in order to make them safe or for use as evidence)
- to require information and take statements from persons (in relation to examinations or investigations)

- to inspect and take copies of any books and documents (both statutory records and other records)
- to require persons to provide facilities and assistance within their control (e.g. use of copying facilities, access to a private room for interviews).

Persons being interviewed or giving statements are allowed to nominate a person to accompany them. Their answers are not admissible in evidence against themselves or their husband or wife. In practice, where inspectors are interviewing someone against whom legal proceedings are likely, they will use the standard form of caution used by the police.

Inspectors may be accompanied onto premises by any person duly authorised by their enforcing authority. Where they have reasonable cause to expect serious obstruction in exercising their duties they are empowered to take a police constable with them. Intentional obstruction of an inspector in exercising his powers is an offence under section 33 of HASAWA.

It is not unknown for persons to impersonate inspectors or at least give a misleading impression that they are an inspector. Section 19 of HASAWA, which relates to the appointment of inspectors, also requires them to produce evidence of their appointment when required. Both HSE and local authority inspectors would normally carry their accreditation with them at all times while they are carrying out their duties.

Under HASAWA, inspectors also have the powers to serve improvement and prohibition notices (HASAWA, ss. 21–4) and to institute and prosecute criminal proceedings (HASAWA, ss. 38 and 39). Both of these powers are dealt with below. Inspectors are restricted in respect of disclosure of information that has been acquired using their powers (HASAWA, s. 28).

In 2004 the HSC issued a warning about organisations falsely claiming to regulate health and safety legislation and demanding payment for documentation or registration they claim is required by law. Employers should stay alert for such approaches and, if necessary, seek guidance from the HSE. Both HSE and local authority inspectors visiting in person to enforce health and safety requirements must carry an identification warrant – persons should be requested to show this if there is doubt as to their identity.

A2.18 Improvement and prohibition notices

The power to issue improvement and prohibition notices was introduced under HASAWA. The intention was to provide an alternative to the use of potentially time-consuming prosecutions and to give a quick and efficient method for dealing with situations presenting serious danger.

Improvement notices

Improvement notices are generally used where inspectors are not confident that employers (or other duty holders) will take appropriate action within an acceptable time scale. They are often used where there has not been a satisfactory response to an oral recommendation or to a letter, or in order to focus the employer's mind on the importance of taking action.

The inspector must be of the opinion that the law has been contravened and that the contravention will continue or be repeated. In serving an improvement notice the inspector must state the following:

- in his opinion, what statutory provision has been contravened
- the reasons that he is of that opinion
- the period within which the contravention must be remedied (this must be at least as long as the period for lodging an appeal).

Appeals against notices must be made within twenty-one days of service and are heard by employment tribunals. The notice is then suspended until the appeal is either heard or withdrawn.

Grounds for an appeal may be that the person on whom the notice has been served believes that it is wrong in law (e.g. the inspector's interpretation of what is 'reasonably practicable' differs from the employer's) or that the notice allows insufficient time to remedy the contravention. The tribunal may either cancel or affirm the notice or make modifications to it.

Prohibition notices

Inspectors may serve prohibition notices where they are of the opinion that activities involve (or will involve) 'a risk of serious personal injury'. Such activities may be identified during the investigation of an accident or during routine inspection work. The prohibition notice must state the following:

- the matters giving rise to the risk of serious personal injury
- any statutory provisions that, in the inspector's opinion, are being contravened, giving the reasons.

The activities to which the notice relates must not be carried on until appropriate remedial action has been taken. The notice may come into operation at the end of a period specified within it, or immediately, should the inspector think fit. While appeals against prohibition notices may also be made, such appeals cannot be lodged as a delaying tactic.

The prohibition notice remains in operation until such time as an industrial tribunal either hears the appeal, or directs that the notice be suspended pending the hearing of the appeal or until the withdrawal of the notice. The prohibition notice is a particularly potent tool for inspectors to use, particularly as contravention of a notice is one of the offences that can be punished by up to two years' imprisonment (see paragraph A2.19 below).

A2.19 Penalties

The penalties imposed under section 33 of HASAWA for breaches of health and safety legislation vary, depending on the nature of the offence and whether it involves summary conviction (at the magistrates' court) or conviction on indictment (at the crown court).

In both summary conviction and conviction on indictment, a period of imprisonment may be imposed as well as a fine. There is a six-month time limit for commencement of prosecution in the magistrates' court but there are no time limits in respect of indictable offences. Details of a prosecution for a breach of asbestos licensing requirements, which resulted in a three-month prison sentence for the owner of a demolition firm, are provided on page 20.

Summary conviction

Breaches of sections 2 to 6 of HASAWA and contraventions of notices or breaches of court remedy orders (see below) can currently attract a fine of up to £20,000, while breaches of other sections of HASAWA and health and safety regulations are subject to a fine of up to £5,000. It should be noted that the enforcing authorities may prosecute for more than one offence, in which case multiple fines of up to £20,000 (or £5,000) may be imposed.

Magistrates' courts can also impose penalties of imprisonment for up to six months for contravention of a requirement or prohibition imposed by an improvement or prohibition notice; or failing to comply with a court remedy order (imposed under section 42 of HASAWA).

Conviction on indictment

Where cases are heard on indictment at the crown court, conviction can result in 'a fine' with no maximum figure specified. The largest fine for a single breach of health and safety legislation increased in 2004 to £2 million, following the prosecution of a train operating company resulting from the tragic Ladbroke Grove rail crash.

Crown courts can also impose penalties of imprisonment for up to two years for contraventions of improvement and prohibition notices or breaches of court remedy orders (see above under 'Summary conviction'). Up to two years' imprisonment can also be imposed for several offences including breaches of licensing requirements (e.g. those relating to asbestos) and explosives-related offences.

A2.20 Manslaughter

Where death results from work activities, those considered responsible may face charges of manslaughter. Work-related deaths are normally investigated jointly by the HSE, the police and the Crown Prosecution Service (CPS). The CPS would make the final decision on whether to institute manslaughter charges, but the HSE may also bring additional charges under health and safety legislation. Where the CPS decides not to prosecute for manslaughter, the HSE may still pursue health and safety charges.

Manslaughter charges for work-related accidents are relatively rare and it is particularly difficult to obtain convictions in respect of corporate manslaughter. Nevertheless there have been a number of successful manslaughter prosecutions and details of some of these are provided below.

In large organisations the directors and others who make important decisions affecting health and safety issues are often remote from the situations that actually result in fatal accidents, and this introduces technical difficulties in achieving manslaughter convictions against them. In recent years a significant lobby has built up to make such people much more responsible for deaths resulting from their decisions or lack of effective control.

Recommendations were made by the Law Commission in 1996 that a new offence of 'corporate killing' be introduced and this was followed in May 2000 by a Home Office consultation document on the subject.

However, progress since then has been painfully slow, drawing criticism from many quarters, particularly in the light of the deaths of the Morecambe Bay cockle-pickers. In the Queen's Speech in November 2004 the government announced its intention of publishing a draft bill introducing a new offence of corporate manslaughter. This would then be subject to pre-legislative scrutiny by a joint

Parliamentary Select Committee which would receive both written and oral evidence. With an election likely to take place during 2005, the eventual progress of an actual bill would be dependent upon the attitude of the new government.

 Prosecutions resulting in prison sentences: case histories

- A small demolition company set about ripping apart an old factory building containing asbestos, using an excavator and loading the debris onto a lorry. No precautions were taken to protect either the workers or the surrounding community from asbestos dust. The firm's owner was convicted on five charges under the Asbestos (Licensing) Regulations 1983 and the Control of Asbestos Regulations 1987. He was sent to prison for three months and ordered to pay £4,000 costs.

- An activity centre involved in a canoeing accident in which four young people died, and its managing director, were both convicted of manslaughter. Evidence was put forward regarding lack of experience of the staff supervising the canoeists, lack of appropriate equipment and inadequate communication arrangements. The managing director was jailed for three years and the company fined £60,000 (a relatively rare case of a corporate manslaughter conviction).

- A managing director of a haulage company was sentenced to twelve months' imprisonment for manslaughter after the death of a 21-year-old employee who was sprayed in the face with a toxic chemical. The managing director was also fined under HASAWA while the company was fined £7,000 under s. 2(1) of HASAWA and £15,000 for manslaughter (another rare corporate manslaughter conviction).

- A building contractor was jailed for eighteen months for the manslaughter of two men who were killed when a tunnel kiln they were demolishing collapsed on top of them. The company for whom the work was being carried out was also fined £125,000 for offences under the Construction (Design and Management) Regulations 1994.

- An untrained 16-year-old agricultural student was killed when the JCB farm loader he was driving was struck by a lorry at an unlit road junction (the boom of the loader was protruding onto a major road). The father and son who ran the farm received suspended prison sentences of twelve and fifteen months respectively for his manslaughter. This case, which reached court in early 2002, brought the reported total of people receiving immediate or suspended prison sentences for work-related health and safety offences to twenty-four.

- A ten-year-old boy, accompanying his mother who was assisting on a school trip, drowned in an activity involving jumping into the cold and turbulent water of a rock pool. Several other older children and the boy's mother required hospital treatment for hypothermia. The teacher in charge of the trip pleaded guilty to two charges, one of manslaughter and one under s. 7 of HASAWA. He was sentenced to a year's imprisonment on the manslaughter conviction. No separate penalty was imposed for the HASAWA offence.

- A landlord and a workman were jailed for five and three years respectively following manslaughter convictions resulting from the deaths of two teenagers from carbon monoxide poisoning. The landlord engaged the workman (who had no gas-fitting qualifications) to carry out work on the flues of gas fires in the flat where the two teenagers lived.

- A cleaning contractor removed a safety cage and other safety devices from a ride-on machine used for cleaning out a chicken shed. As a result, an employee was crushed between the lifting arms and the body of the machine. The contrac-

tor (who had previously been fined following another employee's death) was jailed for a year for the operative's manslaughter.
- The managing director of a heating and ventilation firm was jailed for a year following the death of an apprentice working in a confined space in a boatyard. A build-up of solvent fumes resulted in an explosion. His company was also convicted of manslaughter, as well as several health and safety offences and fined a total of £90,000.

A2.21 Other important statutory requirements

This section looks at additional statutory requirements that managers need to consider.

The Factories Act 1961 and the Offices, Shop and Railway Premises Act 1963 (OSRP Act) are progressively being replaced by more modern legislation and are expected to be repealed entirely in the near future. The Acts (and Regulations passed under them) only apply to factories, offices, shops and railway premises as defined. Both still contain requirements to notify the relevant enforcing authority of the occupation or use of premises as a factory, office or shop.

A limited number of workplaces are subject to the Fire Precautions Act 1971 – mainly larger factories, offices, shops and hotels. Chapter C7 makes further reference to the requirements of the Act.

Many sets of regulations have been introduced under HASAWA. Some of the more important regulations and those of more general application are listed below.

- Confined Spaces Regulations 1997. The regulations require work in confined spaces to be either avoided or carried out in accordance with a safe system of work, developed as a result of a risk assessment. Suitable arrangements for emergencies must also be made. Further reference to the regulations is made in Chapter B3 which deals with emergency procedures.
- Construction (Design and Management) Regulations 1994. The broad definition of 'construction work' used in the regulations means that they apply to many medium-sized engineering and maintenance projects as well as traditional construction activities and all demolition work. Specific duties are given to the 'client' (who must appoint a 'planning supervisor') and to the 'principal contractor' (who may in some cases also be the client). Key requirements of the regulations are for the development and implementation of a formal 'health and safety plan' and the creation of a 'health and safety file' for the project. Further details are contained in Chapter B7.
- Construction (Health, Safety and Welfare) Regulations 1996 (CHSW). The CHSW regulations replaced most of the older regulations affecting construction. The regulations include requirements relating to work at heights, use of access equipment, dangers from falling objects or collapses, excavations, etc. Chapter D2 provides further detail of their requirements.
- Control of Asbestos at Work Regulations 2002. These regulations continue previous requirements relating to work involving asbestos and to the storage, distribution and labelling of raw asbestos and asbestos waste and also the supply of products containing asbestos. However, the 2002 regulations contain an important new 'duty to manage asbestos in non-domestic premises'. This requires 'dutyholders' (as defined in the regulations) to

carry out an assessment of the presence and condition of possible asbestos-containing material (ACM). They must then prepare a written plan to manage the risks from ACM, particularly in relation to occupants of the premises and those carrying out maintenance or construction work. Further details are available in paragraph D2.10.

- Control of Substances Hazardous to Health Regulations 2002 (COSHH). The COSHH regulations were introduced in 1988 but have been amended several times since. The regulations are based on an assessment of the risks created by hazardous substances. Further details are contained in Chapter C2.
- Dangerous Substances and Explosive Atmospheres Regulations 2002 (DSEAR). The regulations apply to substances with the potential to create risks from fire, explosions, exothermic reactions, etc. They require employers to carry out an assessment of activities involving dangerous substances and to take appropriate measures to control risks. Several types of control measures are specified in the regulations. Paragraph C7.4 provides further details.
- Electricity at Work Regulations 1989. These regulations deal with the standards of electrical equipment, its maintenance and the carrying out of electrical work see Chapter D3 for further information.
- Fire Precautions (Workplace) Regulations 1997. The regulations contain basic requirements regarding fire precautions for all workplaces, including those that are also subject to the Fire Precautions Act (see Chapter C7).
- Health and Safety (Consultation with Employees) Regulations 1996. These regulations extend the previous requirements (contained in the Safety Representatives and Safety Committees Regulations see below) to also require employers to consult with workers not covered by trade union safety representatives. Chapter B6 contains further details.
- Health and Safety (Display Screen Equipment) Regulations 1992. Where there is significant use of display screen equipment (DSE), employers must carry out assessments of DSE workstations and offer 'users' eye and eyesight tests (which may necessitate provision of spectacles for DSE work). Further information is contained in Chapter C5.
- Health and Safety (First Aid) Regulations 1981. At least basic first aid equipment, controlled by an 'appointed person', must be provided in all workplaces. Higher-risk activities or larger numbers of employees may require additional equipment and fully trained first aiders. Further details are contained in Chapter D4.
- Health and Safety (Signs and Signals) Regulations 1996. The regulations require safety signs to be provided, where appropriate, for risks that cannot adequately be controlled by other means. Signs must be of standard designs and colours. Reference to the regulations is contained in paragraph B8.5.
- Lifting Operations and Lifting Equipment Regulations 1998 (LOLER). The regulations replaced previous requirements in the Factories Act, OSRP Act and various sector-based regulations (e.g. construction) relating to lifting equipment, including hoists and lifts. Further details are contained in Chapter D1.

- Management of Health and Safety at Work Regulations 1999. Employers and the self-employed are required by the regulations to manage the health and safety aspects of their business in a systematic and responsible way. They must also carry out risk assessments of their activities. Following several amendments the regulations were revised in 1999. Chapter B2 summarises the general requirements of the regulations and Chapter B3 deals with several of their more important implications, while Chapter C1 covers risk assessment.
- Manual Handling Operations Regulations 1992. Manual handling operations involving risk of injury must either be avoided or must be assessed by the employer, with steps taken to reduce the risk, so far as is reasonably practicable. Chapter C4 contains further details of the requirements.
- Noise at Work Regulations 1989. These regulations require an assessment to be made of noise risks in the workplace in order to identify appropriate control measures. Chapter C3 contains further details.
- Personal Protective Equipment at Work Regulations 1992. Employers are required by the regulations to assess the personal protective equipment (PPE) needs of their work activities, provide the necessary PPE, and take reasonable steps to ensure its use. Chapter C6 provides further details.
- Provision and Use of Work Equipment Regulations 1998 (PUWER). These regulations (which were revised and extended in 1998) deal with equipment safety, including the guarding of machinery. The definition of 'work equipment' also includes hand tools, vehicles, laboratory apparatus, etc. See Chapter D1 for further details.
- Reporting of Injuries, Diseases and Dangerous Occurrences Regulations 1995 (RIDDOR). Fatal accidents, major injuries (as defined) and dangerous occurrences (as defined) must be reported immediately to the enforcing authority. Other accidents involving four or more days' absence must be reported in writing within seven days. Chapter B4 provides further details of the requirements.
- Safety Representatives and Safety Committees Regulations 1977. Where trade unions are formally recognised by employers, the unions may appoint safety representatives, who have rights which are set out in the regulations. The representatives may also require the employer to establish a Safety Committee. See Chapter B6 for more details.
- Workplace (Health, Safety and Welfare) Regulations 1992. Physical working conditions, safe access and welfare provisions are the subject of these regulations. More information is contained in Chapter D2.

A2.22 Approved Codes of Practice and HSE guidance

The Robens Committee (see paragraph A1.6) wished to see health and safety legislation simplified and to be kept up to date more easily. Section 16 of HASAWA empowered the Health and Safety Commission (HSC) to approve and issue codes of practice to provide practical guidance on the requirements of sections 2 to 7 of HASAWA and health and safety regulations.

Section 17 of HASAWA made these Approved Codes of Practice (ACOPs) admissible in criminal proceedings and a proven breach of a Code proof of a breach of the relevant statutory requirements, unless the defendant could prove compliance with the requirement by other means. This switched the onus of proof onto the defendant, rather as is the case for proving what is not 'practicable' or 'reasonably practicable' (see paragraph A2.11).

Almost all new health and safety regulations and significant amendments to regulations are accompanied by an ACOP and, as Robens intended, existing ACOPs are updated periodically. Some ACOPs are also issued on topics that may not be subject to specific health and safety regulations, but in order to provide employers with guidance on meeting their general obligations under HASAWA (e.g. on safety, health and welfare standards in zoos).

Where ACOPs relate to regulations, they are often published in a single booklet together with the relevant regulations. These booklets also include guidance on compliance with the law. While guidance does not have the specific legal status of the ACOP it nevertheless provides much practical advice on how compliance can be achieved.

The HSC/HSE produce many separate guidance booklets and leaflets on a wide range of topics. These are available from HSE Books and much guidance is also available on the HSE's website (see also Chapter E1 which covers sources of information and advice).

A2.23 Future trends

Many of the government's future intentions in respect of health and safety legislation and its enforcement were signalled in a strategy statement entitled 'Revitalising Health and Safety', published (in June 2000) by the Department of the Environment, Transport and the Regions (DETR), then the parent department of the Health and Safety Commission.

The forty-four action points included in Annex C to the document include actions relating to the following:

- a fundamental review of incident reporting regulations (RIDDOR)
- extension of the £20,000 maximum fine in magistrates' courts and their powers of imprisonment to a much wider range of offences
- the 'naming and shaming' of convicted companies and individuals (already implemented on the HSE's website)
- consideration of allowing private prosecutions for health and safety offences
- the removal of Crown immunity from enforcement action.

Some of the action points and the other intentions included in the strategy statement related more to leadership and encouragement, including the following:

- businesses reporting publicly on their health and safety standards
- motivation of employers via insurance companies
- the government and other public bodies leading by example
- greater contact with small firms on health and safety issues
- development of a new occupational health and safety strategy
- better education on health and safety matters (from children at school up to 'safety-critical' professionals such as architects and engineers).

However, the government has been slow to implement many of these good intentions, just as it has in respect of the introduction of 'corporate killing' offences (see paragraph A2.20 above). Whether this is the result of deliberate delay or concentrating on other priorities is not clear. Certainly the position has not been helped by the restructuring of government departments – the HSC/HSE had two changes in parent department in two years after the publication of 'Revitalising Health and Safety'.

The HSC was forced to admit that their 2003/04 statistics for work-related deaths, major injuries and over three day injuries showed no clear evidence of change towards meeting interim targets set out in the original 'Revitalising Health and Safety' document.

The government's sincerity has not been shown in a good light by the budgetary constraints it has placed on the HSE. A report in 2004 by a Parliamentary Work and Pensions Select Committee made 30 recommendations in a variety of areas including doubling the number of inspectors in the HSE's Field Operations Directorate and reversing an earlier decision to switch resources from inspection and enforcement towards information and education.

Nevertheless, the government rejected the majority of the Select Committee's recommendations and as a result drew criticism from many quarters including the TUC, the HSE inspectors' own union and the Institution of Occupational Safety and Health (IOSH).

The pace of new health and safety directives from the EU has slackened somewhat in recent years but there is always pressure from a variety of quarters for new regulations – work-related violence and work-related stress are two topical examples although many health and safety professionals feel that these two issues can be dealt with adequately under the general obligations of HASAWA, together with relevant guidance from the HSE.

Source materials

1.   HSE (1998) 'What to expect when a health and safety inspector calls', HSC 14.
2.   HSE (2002) 'HSC's enforcement policy statement', HSC 15.

CHAPTER

A3

Civil Law

A3.1 Introduction

Chapter A1 explained the differences between criminal and civil law and outlined the court structures involved in both types of legal proceedings. This chapter examines how employers may become liable to pay damages under the civil law, the processes by which civil claims are made and decided, and some of the terminology used in civil proceedings. While the chapter concentrates on potential industrial injury claims by employees it will also make reference to possible claims from other sources, e.g. contractors' employees, visitors, tenants, customers and members of the public.

A3.2 The basis for civil claims

Civil claims are brought under the law of tort (meaning a wrong or fault) and particularly the tort of negligence. The claim would be made against a person (the defendant) who has negligently failed to comply with their statutory obligations or their common law 'duty of care' (see below) and as a result another person (the plaintiff) has suffered injury or loss. For negligence at common law the plaintiff must prove the following:

- that the defendant owed them a duty of care
- that the defendant did not fulfil that duty
- that, as a result of that breach, the plaintiff has suffered injury or loss.

A3.3 The 'duty of care'

All UK citizens have a duty of care to each other. This means taking reasonable care to avoid acts or omissions, which can be reasonably foreseen to be likely to injure one's neighbour. A neighbour is anyone who ought reasonably to be kept in mind. In a domestic situation the duty would be quite literally to the occupants of neighbouring properties as well as visitors, delivery workers (e.g. postal staff, milkmen) and service providers (e.g. window cleaners). On the roads drivers have duties to other road users – other drivers, pedestrians, cyclists, horse riders, etc.

Employers will have a duty of care to a wide range of people – their employees, contractors and their employees, visitors, delivery staff, tenants, customers, service users, occupants of neighbouring property, passers-by, the emergency services and even trespassers (see paragraph A3.5 below). This matches their general duties under HASAWA, ss. 2 & 3 (see paragraphs A2.3 and A2.4).

In respect of employees, employers have a common law duty to provide them with the following:

- a safe place of work
- a safe system of work (see Chapter B8)

- safe plant, equipment and tools
- safe fellow workers.

As can be seen, these duties closely match those imposed by HASAWA, s. 2(2), many of which have been extended by other statutory duties such as the Workplace (Health, Safety and Welfare) Regulations 1992 and the Provision and Use of Work Equipment Regulations 1998.

The duty to provide safe fellow workers and the concept of 'vicarious liability' are explored below.

A3.4   Safe fellow workers

In order to fulfil the duty of care to provide safe fellow workers, employers must ensure that their employees:

- are provided with appropriate levels of information and training
- are adequately experienced and competent to perform the task in hand
- have appropriate levels of competent supervision
- have a suitable attitude and behavioural characteristics (employers have been held liable for the actions of practical jokers and bullies).

Employers are liable to persons (both employees and non-employees) who are injured by negligent acts of their employees, committed in the course of their employment. This concept is known as 'vicarious liability'. Once negligence by the employee is proven, the employer is strictly liable and no fault on the part of the employer has to be demonstrated.

Thus if an untrained employee drove a forklift truck and injured a fellow employee or a customer due to negligent driving, their employer could be liable, even though the employee was driving without the employer's knowledge or permission.

An important requirement is that the employee causing the injury must be acting within the course of their employment. Courts have decided that the following are within the course of employment and therefore the scope of vicarious liability:

- travel for work purposes (whether using private or employers' vehicles)
- activities incidental to work, e.g. going to the toilet or canteen
- deliberately disobeying clear instructions (smoking while unloading petrol).

An employer may also be vicariously liable for the negligence of contractors and their employees, where they have in effect become part of that employer's own workforce, e.g. labour-only contractors or agency staff.

A3.5   Duties of occupiers to visitors and trespassers

The Occupiers' Liability Act of 1957 established the liability of occupiers under civil law to exercise a duty of care in respect of lawful visitors to their property. Property could include occupied, unoccupied or derelict buildings as well as open land, including derelict land. The occupier's duty of care was extended by the Occupiers' Liability Act of 1984 to protect trespassers, in respect of any injury suffered on the premises, either because of any danger due to the state of the premises or things done, or omitted to be done.

27

The following three criteria must be satisfied in respect of trespassers:

1) The occupier must be aware of the danger, or have reasonable grounds to believe that it exists.

2) The occupier must know, or have reasonable grounds to believe, that a trespasser is in the vicinity of the danger concerned, or that a trespasser may come into the vicinity of the danger.

3) The risk must be one against which, in all the circumstances of the case, the occupier may reasonably be expected to offer some protection.

Any case brought under the 1984 Act would be decided on its merits but there are two important factors to consider.

The standard of care required

A greater duty of care exists in relation to major risks (e.g. high voltage electrical equipment, radioactive substances, high scaffolding, open shafts or deep water) as opposed to minor risks (e.g. tripping hazards, uneven land or shallow streams). The practicality of taking and maintaining precautions would also have to be taken into account. It may not be reasonable to expect to fence off a large and relatively low-risk construction site in its entirety, especially in an area where vandalism was prevalent. However, it may be reasonable to prevent access to individual high-risk areas within the site, e.g. scaffolding and equipment stores.

The type of trespasser

Case law has already established that a greater duty of care exists in respect of child trespassers, although parents are expected to assume the prime responsibility for controlling the whereabouts of very young children. Occupiers must also be aware of the presence of items on property or sites that may attract children, e.g. unattended vehicles or derelict machinery.

The duty of care owed to adults will also vary. It will be much greater to those simply walking through property (especially if straying accidentally from a public footpath or sheltering from the rain) than to those indulging in illicit activities such as theft, vandalism or fly tipping. However, even these latter categories of trespasser cannot be completely ignored.

A3.6   Breach of statutory duty

Section 47 of HASAWA makes breaches of duties imposed by health and safety regulations actionable at civil law, unless the regulations themselves state otherwise. However, the same section excludes civil actions for breaches of sections 2 to 8 of HASAWA. This is to a certain extent academic, since many of these duties are matched by the employer's common law 'duty of care' (see paragraph A3.3 above) or the employer's vicarious liability for negligent acts of employees (see paragraph A3.4 above).

As a result of changes to legislation in October 2003, civil actions can now be brought as a result of employers being in breach of the Management of Health and Safety at Work Regulations and the Fire Precautions (Workplace) Regulations 1997. Both of these sets of regulations were previously excluded from civil actions.

These changes may have some practical significance. For example, it could be difficult for an employer to claim that a particular risk was not foreseeable when no risk assessment of the activity had been carried out at all. A failure to apply the management cycle (see paragraph B 3.3) in order to ensure control measures were implemented could also be important.

A3.7 Pursuing civil claims

Most civil claims are initiated following the injured person seeking advice from a solicitor. In the case of injured employees the solicitor may be a specialist engaged by the employee's trade union. However, in recent years many firms of solicitors and claims agencies have begun to advertise heavily in the press and on TV, offering their services to victims of accidents and occupational diseases, often on a 'no win, no fee' basis.

The first stage would normally be a letter from the injured person's solicitor to their employer (or the person they consider responsible, e.g. a contractor), stating the nature of the injury, the circumstances of the accident or occupational disease and the basis of the claim. The prudent employer should ensure that such letters of claim are passed on promptly to their insurers (see paragraph A3.12). This will usually be followed by a period of investigation of the circumstances by the insurance company's representatives and probably some correspondence with the claimant's solicitors. At this stage the claim may be withdrawn or a settlement may be agreed mutually between the parties.

Where settlements cannot be agreed in this way, formal legal proceedings will normally be instituted by the claimant (who now becomes the plaintiff), usually against their employer (who becomes the defendant). Some solicitors issue formal proceedings at a relatively early stage of the case, whereas others prefer to negotiate first. Formal proceedings must be initiated within three years of the time of knowledge of the cause of action.

For a traumatic injury such as a fracture or amputation this would be the time of the related accident. However, where the onset of symptoms is more gradual, e.g. the development of back problems, it may be some time before the cause is diagnosed as being related to work activities, such as poor manual handling practices. Some occupational diseases, particularly those relating to asbestos, take many years to develop – the three years would only commence once a definite diagnosis of the disease and its likely causation were made.

Once court proceedings have been initiated, the exchange of information between the parties becomes much more formalised, but the negotiations for an out-of-court settlement often continue, with the majority of cases being settled without a court hearing. It is not unusual for agreements to be made in the court immediately before the hearing commences, as both parties are conscious of the high legal costs associated with the proceedings themselves. Employers must bear in mind that although they may be the defendant in the case, their insurance company holds the purse strings and will make the final decision on whether an out-of-court settlement should be agreed. However, the employer should still point out relevant considerations to their insurers, e.g. that if a particular claim is conceded, it may well open the floodgates for a series of similar claims.

Where a defendant has offered what he believes to be a reasonable out-of-court settlement but this has been rejected by the plaintiff, it is open to the defendant to make what is called a payment into court. If, when the case is heard, the judge (who is unaware of the amount of the payment into court) makes an award that is lower than the payment in, the plaintiff becomes responsible for the costs of both parties from the point at which the payment into court is made. A payment into court can often concentrate the minds of the plaintiff and his advisers since they see the potential for a significant part of their damages to be taken away in legal costs. In normal circumstances (and where the eventual award exceeds the payment into court) the success-ful plaintiff would expect all the costs to be paid by the defendant.

Lower value civil claims are heard within the county court system whereas higher value ones would go to the High Court. In the High Court both parties are normally represented by barristers, thus increasing the cost of proceedings. Cases are usually heard by a single judge who must decide on whether the defendant has been negligent and is liable to pay damages (liability) and the amount of the damages (quantum). In some cases either liability or quantum may have been agreed between the parties and the judge has only to decide on the outstanding issue. The basis on which damages are assessed is summarised below (see paragraph A3.8). Damages may also be reduced because of contrib-utory negligence by the plaintiff (see paragraph A3.9 below).

The court would hear witnesses of fact from both sides and there will often also be expert witnesses on medical and technical matters. Previously each side would engage its own separate expert witnesses, but this was changed by new rules which came into operation from 1999. These provided for one medical and one non-medical expert to be used jointly by both sides. The same rules (the Civil Procedure Rules 1998) introduced a pre-action protocol which was intended to limit delay and also a fast-track system for dealing with claims up to a value of £15,000.

Once the case has been decided by either the county court or the High Court, it is open to the losing party to seek leave to appeal to the Court of Appeal. Further appeals can be made from there to the House of Lords (see Figure A1.2 on page 4). Relatively few claims go to appeal, particularly because of the considerable additional expense involved. Decisions on points of law made by the Court of Appeal and the House of Lords create legal precedents which must subsequently be followed by the lower courts in deciding similar cases.

A3.8   Damages

The civil courts take into account a variety of factors in determining the level of damages in each case. These include the following:

- the plaintiff's loss of earnings prior to the hearing (including normal overtime earnings, in addition to basic wages)
- possible future loss of earnings (due to future absences or being placed in a disadvantageous position in the labour market)
- pain and suffering
- disfigurement
- the loss of ability to pursue hobbies, sports and other social activities
- medical or nursing expenses
- the possible need for adaptations to be made to housing (e.g. to accom-modate a wheelchair) or for special equipment.

While many of these will relate to the plaintiff and their individual circumstances, comparisons will be made to damages awards made by other courts in similar circumstances.

Some of these factors can be of major importance. The widow of a successful businessman who died from an asbestos-related disease (resulting from earlier manual work) was awarded £4.37 million in damages. A major part of this sum reflected his future earning potential.

Regulations were due to come into force in 2005 which would allow all NHS hospital and ambulance costs to be recoverable in all cases where personal injury compensation is paid. This would apply to claims made under both employer's liability and public liability insurance.

A3.9   Contributory negligence

It is quite common for the defendant to claim that the injury was caused wholly or partly by the negligence of the injured person, for example in disobeying instructions or failing to follow clear safety rules. A distinction must be drawn between a moment's thoughtlessness or an inadvertent act (possibly in a situation of fatigue, boredom or difficult working conditions) and an act of negligence. The latter is more likely to relate to a deliberate choice to do something dangerous, e.g. not wearing personal protection where rules clearly require it.

The percentage of contributory negligence in each individual case will vary with the facts. It is a matter for the court to determine, based on the evidence presented.

 Contributory negligence: a case history

- An employee was working in the vicinity of an activity involving the skimming of dross from a container of molten metal and was burnt in one eye by a stray hot particle. The court found that his employer did not have a 'safe system of work', in that non-essential workers should have been kept clear when this activity was taking place. However, the court also found that the employers had clear rules requiring eye protection to be worn in this area, that the injured man had not been wearing eye protection and that suitable eye protection was readily available and would have prevented the injury. (Evidence was produced of formal warnings given previously to the man for failing to wear eye protection.) The plaintiff was adjudged to be 50 per cent contributory negligent and the damages reduced accordingly.

A3.10 Other defences

Employers may also claim the following defences in the case of injuries to employees.

Foreseeability

Employers may claim that the injuries suffered by their employees (or others) were not reasonably foreseeable – often called an 'act of God', i.e. outside normal expectation or control. This is particularly important in relation to a number of occupational diseases where the risks may not have been apparent at the time that exposure to the cause of the disease took place.

In relation to noise-induced hearing loss, mesothelioma (an asbestos-related cancer) and vibration white finger, the courts have established dates after which employers should have been aware of the risks of the condition developing and of the precautions necessary to control those risks. Claims relating to exposure prior to these dates would be unsuccessful but subsequent claims would succeed, if the relevant precautions had not been taken.

Voluntary assumption of risk

If an employee freely consents to taking risks in connection with work activities then the employer may escape a common law negligence claim by using a defence known as *volenti non fit injuria* (meaning 'one who consents cannot complain'). However, the defence cannot be used against claims for breach of statutory duty – it is not possible to contract out of statutory obligations or statutory protection.

Not in the course of employment

Employers are not liable for injuries that are not sustained in the course of employment. This would mainly relate to times when employees were not at work, although incidences of work-related violence occurring outside working hours may be treated differently (see paragraph A3.11 below). There would be no liability for injuries sustained at work while performing unauthorised activities, e.g. carrying out private work on the employer's premises during working hours. However, a distinction must be drawn between the previous situation and one where the employee is carrying out authorised activities but in unauthorised ways.

A3.11 Social Security benefits

Industrial industries benefits are paid from government funds to employed earners who suffer personal injury arising out of and in the course of their employment. These benefits are paid on a 'no fault' basis, regardless of whether the employee or the employer was at fault in relation to the causes of the accident.

Most cases are straightforward but there may be situations where claims are rejected because the employee is not at work at the time in question (e.g. carrying out private work) or is deliberately doing something without authority or permission. However, a Benefits Agency officer who was attacked outside working hours by a person he had previously reported for fraud was deemed to be eligible for benefit, since the assault was held to have occurred in the course of his employment.

Benefits are also payable for a wide range of industrial diseases, although claimants must demonstrate the following:

- that they are suffering from a prescribed disease (often beyond a specified threshold, e.g. a specified percentage of disability or degree of hearing loss)
- that the disease is prescribed for their occupation (or they have been exposed to a specified substance)
- that their disease was due to the occupation or exposure (this may be presumed in the absence of evidence to the contrary).

In some cases the claimant must have been exposed to the risk for a specified number of years.

Where claimants are subsequently successful in making damages claims against their employers, the state is now able to claw back Social Security payments from damages awards made by the courts. These generally relate to compensation for lost earnings, for the cost of care and for loss of mobility. Damages for pain and suffering and for loss of amenity are not subject to this arrangement.

A3.12 Insurance against civil claims
Employers' liability insurance (ELI)
The Employers' Liability (Compulsory Insurance) Act 1969 requires most employers to take out insurance in respect of claims made by their employees. This insurance must now involve a minimum indemnity of £10 million.

A number of types of employer are exempt from this requirement, including government departments, nationalised industries and a variety of other public bodies. In practice some of these bodies still choose to carry ELI. Employers are also not required to insure employees who are their spouse, parent, child or other close relatives and employees working abroad, although again many employers still choose to take out insurance in respect of such persons.

Copies of ELI certificates must be displayed prominently at each place where employees covered by the insurance work, and certificates must be retained by employers for a period of forty years from their commencement date.

Public liability insurance (PLI)
While PLI is not compulsory, most prudent employers ensure that they have it. As well as providing cover for injuries to visitors and trespassers to whom the employer has obligations under the Occupiers' Liability Acts (see paragraph A3.5), PLI would also normally insure employers against accidental injury or damage involving other persons such as passers-by, neighbours, customers, etc. Employers who do not have such insurance could easily find themselves bankrupted by a successful claim.

Insurance brokers will often offer smaller employers a package deal involving both ELI and PLI, whereas larger organisations may choose to place these types of insurance separately. Organisations finding placements for modern apprentices, other types of trainees and work experience students will usually require evidence of PLI before making the placement. Companies engaging subcontractors will also often require the subcontractors to have a specified level of PLI (see Chapter B7).

Professional indemnity insurance
Persons offering professional services such as architects, engineers, etc., may be subject to civil claims in respect of incorrect advice or mistakes in calculations, specifications, etc. Although not compulsory, it is prudent for such persons to ensure that they are adequately insured against such claims. Many health and safety consultancies hold professional indemnity insurance, although hopefully the possibility of claims against such consultancies is rather less than for other types of professionals. Some clients will insist on evidence of insurance before engaging consultancies or others providing professional services.

There are several other types of insurance that may be appropriate, dependent upon the nature of the employer's business. These include products liability insurance (in respect of commodities, articles, etc., supplied by the business), employer's practices liability insurance (providing cover for claims of unfair dismissal or discrimination), insurance of property, equipment or goods (against theft, loss, damage, etc.), insurance in respect of travel and work overseas and of course motor vehicle insurance. Advice on appropriate insurance cover should be sought from a competent insurance broker.

Managing Health and Safety

The Health and Safety Policy

B1.1   Introduction

There is a legal requirement for all those employing five or more people to
have a written health and safety policy. This must contain a statement of general policy on health and safety at work and also the organisation and arrangements in place for putting the policy into practice.

However, preparing the policy should be much more than meeting a legal
obligation. The policy should set the scene for the effective management of
health and safety within the organisation by providing a clear statement of
intent, an important reference document for staff at all levels of what their
responsibilities are and guidance to where detailed information on health and
safety arrangements is available.

The policy will often be important in providing potential client or partner
organisations with evidence of the employer's health and safety standards.
Companies engaging contractors are increasingly requesting potential contractors to send copies of their health and safety policies and risk assessments,
together with other relevant documents (see Chapter B7).

B1.2   What HASAWA requires

The legal requirement for a health and safety policy stems from section 2(3) of
HASAWA which states:

> Except in such cases as may be prescribed, it shall be the duty of every
> employer to prepare and as often as may be appropriate revise a written
> statement of his general policy with respect to the health and safety of his
> employees and the organisation and arrangements for the time being in
> force for carrying out that policy, and to bring the statement and any revi-
> sion of it to the notice of his employees.

An exception from this requirement has been made in respect of undertakings
employing less than five employees (the Employers' Health and Safety Policy
Statements (Exception) Regulations 1975). A subsequent legal decision determined that this threshold of five employees only related to those employees
present on the premises at the same time.

B1.3   What the policy should contain

There should normally be three main sections in a health and safety policy:

1) *The policy statement.* This should include a commitment to achieving
   high standards of health and safety and will often also incorporate a reference to complying with legal obligations. Although there is only a legal
   duty for the policy to relate to employees' health and safety, most
   employers also make a commitment in respect of others who may be

affected by their activities, particularly if the policy is likely to be supplied to clients or partner organisations. Paragraph B1.4 provides further detail of what might be included in the statement while paragraph B1.5 contains a sample policy statement.

2) *Organisation*. The organisation section of the policy would normally set out who has what responsibilities for health and safety. For small employers this might only involve the owner or manager and possibly a supervisor or chargehand, while in larger companies reference might need to be made to several levels of managerial responsibility, together with a number of staff with specific roles. Paragraph B1.6 details who might be included in this section, while paragraph B1.7 provides an illustrative example of how responsibilities might be allocated.

3) *Arrangements*. Employers should have much of the detail of their health and safety arrangements contained within their risk assessment documents (see Chapters C1 to C7). The HSE leaflet 'Stating Your Business' suggests that reference might also be made in the 'arrangements' section of the policy to works' rules, safety checklists, training programmes, emergency instructions, etc. While all of the detail need not be contained within the policy document itself, the document should state where the detail can be found. Paragraph B1.8 provides a list of the type of health and safety arrangements that should be included in this section.

A health and safety policy that remains in the office filing cabinet, which is only accessed whenever a client wishes to see a copy of it, is of little value. The policy should be:

- communicated effectively to employees (see paragraph B1.10 below)
- implemented in practice (see paragraph B1.11)
- reviewed periodically to ensure that it is up to date (see paragraph B1.12).

The HSE leaflet 'Stating Your Business' provides guidance on preparing a health and safety policy document for small firms. The leaflet provides a one-page policy statement for the employer to sign up to, a page for allocating responsibilities to specific named individuals and several pages to record details of risk assessments and common types of health and safety arrangements. These pages provide prompts on key issues, alongside which employers can insert their own arrangements. For example:

_____ will check that new plant and equipment meets health and safety standards before it is purchased. The first aid box(es) is/are kept at _____.

This type of format is similar to that used in 'model' risk assessment formats (see Chapter C1).

It provides a convenient means for smaller employers to comply with their legal obligations but is unlikely to be suitable for larger organisations, although it may provide useful guidance on policy content.

B1.4   The policy statement
The HSE booklet 'Successful Health and Safety Management' states that 'Effective policies are not simply examples of management paying lip service

to improved health and safety performance but a genuine commitment to action.' It goes on to say: 'policy means the general intentions, approach and objectives – the vision – of an organisation and the criteria and principles upon which it bases its action'.

Several examples of health and safety philosophy are provided within the booklet, including the following:

> In the field of health and safety we seek to achieve the highest standards. We do not pursue this aim simply to achieve compliance with current legislation but because it is in our best interests.

> The company believes that excellence in the management of health and safety is an essential element within its overall business plan.

The content of the policy statement should reflect the genuine commitments that the company wishes to make and the nature of its activities (e.g. whether it is likely to have an impact on the health and safety of persons other than its employees). It might include references to the following:

- a general commitment to achieving high standards
- those in respect of whom it is making this commitment (employees and others such as visitors, contractors, clients, tenants, the public)
- awareness of and commitment to meeting legal obligations·
- legal obligations representing minimum standards
- the prevention of accidents and ill health
- consultation with employees
- competence and training of employees
- provision of information, instruction and supervision
- specific aspects of health and safety, e.g. work equipment, working conditions, systems of work.

The statement should be signed by a senior person in the organisation (e.g. owner, chairman, chief executive) and dated. Most statements are restricted to a single page, thus allowing them to be displayed prominently on notice boards and in reception areas. An example of a policy statement is provided in Figure B1.1.

B1.5   Sample policy statement

The policy statement in Figure B1.1 has been adapted from one produced by a medium-sized manufacturer and installer of specialist electronic equipment. The company has its own factory and carries out installation work in commercial and residential property using either its own installation staff or a network of approved installation subcontractors.

B1.6   Organisation and responsibilities

The policy document must set out the organisation for implementing all of the good intentions of the policy statement. This is usually done by setting out the responsibilities of people within the organisation. These may relate to those at different levels within the organisation (e.g. managers or supervisors) and also to individual posts carrying specific responsibilities (e.g. health and safety specialists).

However, the allocation of responsibilities must reflect the size of the organisation and the nature of its activities. In a small business the owner or

---

Figure B1.1   Sample policy statement

Acme Electronics Limited Health and Safety Policy

POLICY STATEMENT

We fully recognise our responsibilities under the Health and Safety at Work etc. Act 1974 and associated legislation for the health and safety of our employees and of others who may be affected by our activities. We are particularly conscious of our obligations towards clients, residents, members of the public and others who may be affected by installation work being carried out by our own staff or by contractors working under our control.

We regard the effective management of health and safety as an integral component of our general management responsibilities, ranking in importance with other management functions. Detailed responsibilities for health and safety management are set out elsewhere in this Policy.

We will actively seek the cooperation of our employees in the implementation of this Policy by bringing it to their attention during their initial induction into the company and also through regular consultation with them during team briefing sessions and through meetings of our health and safety committee.

This Policy will be reviewed annually and any changes to it brought to the attention of our employees through team briefing sessions or via notice boards.

Signed . . . .   A. King/Chief Executive
September 2005

---

manager will have the majority of the responsibilities, including many often allocated to relatively low levels in a larger company, such as monitoring the condition of equipment or enforcing PPE requirements.

Dependent upon the size and structure of the organisation, responsibilities may need to be allocated at the following levels:

- board of directors
- chief executive or managing director
- senior managers
- junior managers
- supervisors/team leaders/chargehands
- all employees.

Those with specialist responsibilities who may be separated out could include the following:

- health and safety manager or officer
- occupational health or occupational hygiene staff
- human resources specialists
- engineers or other technical staff.

In situations where health and safety advice is provided from outside sources (e.g. an external consultant or a specialist at a different location within the employer's business), reference should be made to how this person can be contacted.

An illustration is provided in paragraph B1.7 of how responsibilities might be allocated within a medium-sized organisation. While the responsibilities are typical of those that would need to be referred to in this kind of business activity, some responsibilities would be likely to be allocated to different levels of management within larger or smaller organisations.

B1.7 Sample allocation of organisational responsibilities

The example in Figure B1.2 is adapted from the responsibilities section of the health and safety policy document produced by the electronics company who provided the sample policy statement contained in B1.5 above. It reflects the company's dual activities – both manufacturing within their factory and installing equipment on site (the managing director having prime responsibility for the latter).

B1.8 Arrangements for implementing the policy

As well as having the legal duty under section 2(3) to record the arrangements in force for carrying out their health and safety policy, employers also have duties under regulations 3 and 5 of the Management of Health at Work Regulations 1999 to record the following:

- the results of their risk assessments
- their arrangements 'for the effective planning, organisation, control monitoring and review of the preventive and protective measures'.

(See Chapters B2 and C1.)

It is quite acceptable for employers to make cross-references to risk assessment records, rather than including all of the details in the health and safety policy. The HSE leaflet 'Stating Your Business' also refers to making cross-references to other documents such as works' rules, safety checklists, training programmes and emergency instructions, where appropriate.

'Stating Your Business' (see paragraph B1.3 above) contains sections for employers to insert details about their arrangements for the following:

- risk assessments
- consultation with employees
- safe plant and equipment (including maintenance and defect reporting)
- safe handling and use of substances
- information, instruction and supervision
- ensuring competency for tasks and the provision of training
- accidents, first aid and work-related ill health
- monitoring working conditions and practices (actively via inspections or audits and reactively by investigating accidents, incidents, etc.)
- emergency procedures – fire and evacuation.

Other arrangements to which reference might be appropriate, dependent upon the nature of the employer's business and the risks associated with it, include those for the following:

- assessment and control of risks to young persons and new or expectant mothers
- personal protective equipment (PPE) (establishment of standards and provision of PPE)

FIND THIS ON CD

Figure B1.2 Sample allocation of organisational responsibilities

Acme Electronics Limited

RESPONSIBILITIES FOR HEALTH AND SAFETY

1. Managing Director

The Managing Director has overall responsibility for the implementation of our Health and Safety Policy and a particular responsibility for installation activities. His specific responsibilities include:

1.1 Ensuring, in conjunction with the Board, that adequate resources are available for the Policy's implementation.
1.2 Delegating responsibility for specific aspects of health and safety to other members of the management team.
1.3 Monitoring their performance in carrying out these responsibilities, providing assistance and support where appropriate.
1.4 Overseeing the health and safety aspects of installation projects, in conjunction with other Acme Electronics staff and the designated installer. (This includes ensuring compliance with requirements of the CDM Regulations.)
1.5 Overseeing the health and safety standards of subcontractors (who must have been approved to carry out installation work on the company's behalf).
1.6 Ensuring that health and safety is discussed regularly at meetings of the management team.
1.7 Reviewing the Health and Safety Policy annually, and arranging to make any revisions found to be necessary.

2. Factory Manager

The Factory Manager has principal responsibility for health and safety within the factory.
He also holds the post of Health and Safety Officer (see also section 5).
His specific responsibilities include:

2.1 Ensuring the provision of safe and suitable plant together with satisfactory arrangements for its maintenance.
2.2 Ensuring the provision and maintenance of a satisfactory working environment and a safe place of work.
2.3 Ensuring that safe systems of work are established for activities within the factory.
2.4 Ensuring that staff within the factory are provided with adequate information, instruction, training and supervision.
2.5 Ensuring that the company's risk assessments are kept up to date and that necessary control measures are implemented.
2.6 Delegating specific health and safety responsibilities to others and monitoring their effectiveness in carrying out those responsibilities.
2.7 Conducting regular inspections of health and safety standards within the factory.
2.8 Investigating accidents and incidents within the factory and ensuring that the Company complies with its obligations under RIDDOR 95 in relation to activities both within the factory and elsewhere.

2.9 Providing an in-house source of health and safety advice and identifying the need for external advice when appropriate.

2.10 Chairing meetings of the health and safety committee.

3. Other Managers and Supervisors

All managers and supervisors have prime responsibility for health and safety within their own sphere of management responsibility. In particular this includes:

3.1 Monitoring the standard and condition of plant, equipment and the working environment and initiating corrective action where appropriate.

3.2 Establishing and maintaining safe systems of work.

3.3 Monitoring the performance and behaviour of employees, including their compliance with personal protective equipment (PPE) requirements and taking corrective action where appropriate.

3.4 Identifying new or unusual risks associated with their work and initiating appropriate action to assess those risks and implement necessary control measures.

3.5 Ensuring that accidents and incidents are reported promptly to the Factory Manager.

3.6 Identifying training needs among their staff and communicating these to the Factory Manager.

The Contracts Manager also has a responsibility for monitoring the health and safety standards of approved installation subcontractors and taking any appropriate remedial action (including withdrawal of their approval).

4. All Employees

All employees have legal obligations to take reasonable care of their own health and safety and for that of others who may be affected by their acts or omissions. As well as their colleagues this duty also extends to clients and their employees, residents, members of the public, contractors etc. Specific responsibilities of employees include:

4.1 Behaving in a responsible manner.

4.2 Following established safe working practices and procedures.

4.3 Complying with reasonable instructions and conforming with PPE requirements.

4.4 Reporting significant or unusual risks or problems to an appropriate member of management.

4.5 Reporting accidents and potentially serious incidents to an appropriate member of management.

4.6 Suggesting where improvements to health and safety arrangements might be made.

5. Health and Safety Advice

5.1 The Factory Manager provides an in-house source of health and safety advice.

5.2 A Health and Safety Consultant has been appointed as an external source of health and safety advice to the Company. Contact with him should normally be made via the Managing Director or Factory Manager.

5.3 Other sources of advice or expertise will be sought as appropriate. These might include the Company's solicitors, employers' organisations or relevant consultancies.

FIND THIS ON CD

Figure B1.3   Sample 'arrangements' section of policy

Acme Electronics Limited

ARRANGEMENTS FOR HEALTH AND SAFETY

Many of the arrangements for implementing the Health and Safety Policy are set out in the Company's Risk Assessment documentation. Other important arrangements are listed below.

1. **Consultation with employees**
   The Company has an agreement with the recognised Trade Unions regarding the appointment of Safety Representatives. It has a well-established Health and Safety Committee, chaired by the Factory Manager. This Committee has a formal constitution, agreed with the Safety Representatives.

2. **Health and Safety Training**
   All new staff and internal transferees receive a health and safety induction within their new department using a series of standard checklists. Full details of health and safety training standards and arrangements for training delivery are contained in the Company Training Manual.

3. **First aid**
   Details of the locations of first aid equipment and trained first aiders are posted on notice boards around the premises. The PA to the Factory Manager is responsible for arranging training for first aiders and keeping the notices up to date.

4. **Accident and incident investigation**
   The Company has a formal procedure for investigation, using a standard report form. The Factory Manager is responsible for submitting any RIDDOR reports.

5. **Monitoring Health and Safety standards**
   The Company has a formal procedure for carrying out health and safety inspections. Arrangements for these are overseen by the Health and Safety Committee. An annual health and safety audit is carried out by an independent consultant, appointed by the Factory Manager.

6. **Young Persons and New or Expectant Mothers**
   Checklists for use in assessing risks to these persons are available, when required. Responsibility for the assessment is primarily with the manager of the department in which they work.

7. **Personal protective equipment (PPE)**
   Detailed requirements for PPE are incorporated into the risk assessment documents. Each department has a summary of PPE requirements displayed prominently. Instruction about PPE requirements is included in health and safety inductions.

8. **Statutory examinations and inspections**
   The Maintenance Manager is responsible for arranging all necessary examinations and inspections. Records are kept in the Maintenance Department.

9. **Selection and Management of Contractors**
   The Company has a formal procedure for selecting and managing contractors. Responsibility for implementing the procedure rests with:
   The Maintenance Manager – for contractors working on Company premises
   The Projects Manager – for sub-contractors carrying out installation work.

10. **Visitors**
    The Company requires visitors (and contractors) to sign in and out of the premises. They are issued with a visitor's pass which contains relevant health

> and safety information. The staff member hosting visitors (and contractors) has primary responsibility for their health and safety, including ensuring that they read the information on their pass.
>
> 11. **Driving**
>   The Company policy on work-related driving is contained within the Employees' Handbook.
> 12. **Smoking**
>   A Company policy on smoking has been agreed by the Health and Safety Committee. This also is contained within the Employees' Handbook.

- permit to work systems
- occupational health – screening and monitoring
- statutory examinations and inspections of equipment
- selection and management of contractors
- controlling visitors
- homeworkers or outstationed staff
- employees working abroad
- work-related driving (inc. use of mobile phones)
- smoking.

Many employers now keep risk assessment records, health and safety-related procedures, etc., in electronic form and it is acceptable for the policy document to simply refer to where these documents can be found. Although HASAWA refers to 'a written statement' of health and safety policy, there is no reason that the policy itself cannot be kept electronically, providing that hard copies are also available.

B1.9 Sample 'arrangements' section of a policy

The example provided in Figure B1.3 relates to the electronics company used as the basis for the sample policy statement and allocation of organisational responsibilities earlier in the chapter.

It demonstrates how the 'arrangements' section can contain brief summaries of the health and safety arrangements which are in place, together with cross-references to the main risk assessment and other relevant documents. – Obviously the complexity and content of this section would vary according to the size of each organisation and the nature of its business activities.

B1.10 Communication to employees

HASAWA requires employers to bring the policy document (and any revision of it) to the notice of their employees. The most appropriate time for doing this is at the time that new employees join the company, as part of their induction programme. This provides an opportunity to make all new employees aware that the organisation that they are joining places a high priority on health and safety and to remind them of their own health and safety responsibilities. Young people especially are likely to need to be told about their responsibilities and also of any restrictions on their activities.

Copies of the policy document (particularly the statement of intent) can be displayed prominently on noticeboards, etc. Many employers provide each employee with their own copy of the policy, sometimes contained within a general staff handbook. Keeping the policy document relatively brief (especially the arrangements section – see paragraph B1.9 above) can often be an aid to its effective communication.

B1.11 Implementation of the policy

Reference was made earlier in this chapter to the importance of the policy representing 'a genuine commitment to action' rather than mere 'lip service'. Employees and others will justifiably become cynical about an organisation that does not show a serious commitment towards implementing standards that it has set for itself. Regulation 5 of the Management Regulations establishes a 'management cycle':

PLAN

      ORGANISE

             CONTROL

                 MONITOR

                     REVIEW

for ensuring the effective implementation of 'preventive and protective measures' and this is explored in more detail in Chapter B3. However, a failure to implement the health and safety policy effectively is not only a breach of the employer's legal obligations; it also means that injury and ill health are more likely to arise from work activities.

B1.12 Review of the policy

HASAWA requires the policy to be revised 'as often as may be appropriate'. Any need for revision would have to be established by conducting a periodic review. The HSE leaflet 'Stating Your Business' suggests carrying out an annual review – including the policy within a document control system should also ensure that a regular review is carried out. A review could be conducted by a suitable individual, such as a health and safety specialist or the owner of the business, or alternatively it could be carried out by the health and safety committee.

Reviews may identify the need for the policy to be updated to take account of the following:

- changes in the management structure, requiring a redistribution of responsibilities

- changes in business activities, necessitating alterations to the responsibilities or arrangements section or both

- new or altered health and safety arrangements, e.g. a new permit-to-work procedure or changes to the health and safety inspection system.

An out-of-date health and safety policy is likely to create a poor impression with new employees, health and safety specialists reviewing it on behalf of potential clients and the enforcing authorities, all of whom will draw their own conclusions as to the employer's commitment to health and safety.

Source materials

1.   HSE (2000) 'Stating Your Business', INDG 324.
2.   HSE (1997) 'Successful Health and Safety Management', HSG 65.

Requirements of the Management Regulations

B2.1   Introduction

The Management of Health and Safety at Work Regulations (Management Regulations) were first introduced as part of the so-called 'six pack' of new regulations in 1992. Many would consider them to be the most significant health and safety legislation since HASAWA in 1974, containing not only a general requirement for risk assessment but many other duties of fundamental importance to the way that health and safety is managed.

This chapter only contains a summary of the regulations themselves. However, more information on some of their more important requirements – the application of the 'management cycle', the appointment of competent health and safety assistance, the establishment of emergency procedures and the provision of health and safety training – is contained in Chapter B3. Risk assessment and a number of other important topics relating to the Management Regulations are dealt with elsewhere in the book.

The 1999 version of the Management Regulations incorporated a number of amendments that had been made to the original 1992 Regulations. These changes mainly involved risk assessments and other issues relating to young persons, new or expectant mothers and fire precautions. The Regulations, together with an Approved Code of Practice and guidance, are contained in a single HSE booklet.

B2.2   Regulation 3: risk assessment

Paragraph (1) of Regulation 3 states:

Every employer shall make a suitable and sufficient assessment of:

a) the risks to the health and safety of his employees to which they are exposed while they are at work; and

b) the risks to the health and safety of persons not in his employment arising out of or in connection with the conduct by him of his undertaking, for the purpose of identifying the measures he needs to take to comply with the requirements or prohibitions imposed upon him by or under the relevant statutory provisions and by Part II of the Fire Precautions (Workplace) Regulations 1997.

Paragraph (2) imposes similar requirements on self-employed persons, and paragraph (3) requires a risk assessment to be reviewed if:

- there is reason to suspect that it is no longer valid; or
- there has been a significant change in the matters to which it relates.

Paragraph (4) contains a requirement that a risk assessment must be made or reviewed before an employer employs a young person, while paragraph (5) identifies a number of specific issues that must be taken into account in respect of young persons (particularly their inexperience, lack of awareness of risks and immaturity). (Additional requirements relating to young persons are contained in Regulation 19, while Regulation 16 contains specific requirements on risk assessments in respect of new or expectant mothers.)

Paragraph (6) states that employers with five or more employees must record the following:

- the significant findings of their risk assessments; and
- any group of employees identified as being especially at risk.

While Regulation 3(1) refers to the assessment of risks, it also states that the purpose of risk assessment is to identify the measures needed 'to comply with ... the relevant statutory provisions'. Risk assessment involves identification of both the risks and the necessary precautions, including in many cases determining what is 'practicable' and 'reasonably practicable'.

Chapter C1 provides further guidance on risk assessment, both in respect of what the law requires and how assessments should be carried out. This also includes the special requirements relating to young persons and new or expectant mothers.

B2.3   Regulation 4: principles of prevention to be applied

This regulation states: 'Where an employer implements any preventive and pro-tective measures he shall do so on the basis of the principles specified in Schedule 1 to these Regulations.' Schedule 1 specifies the general principles of prevention set out in an EC directive (Article 6(2) of Council Directive 89/391/EEC):

(a) avoiding risks;
(b) evaluating the risks which cannot be avoided;
(c) combating the risks at source;
(d) adapting the work to the individual, especially as regards the design of workplaces, the choice of work equipment and the choice of working and production methods, with a view, in particular, to alleviating monot-onous work and work at a predetermined work-rate and to reducing their effect on health;
(e) adapting to technical progress;
(f) replacing the dangerous by the non-dangerous or the less dangerous;
(g) developing a coherent overall prevention policy which covers technolo-gy, organisation of work, working conditions, social relationships and the influence of factors relating to the working environment;
(h) giving collective protective measures priority over individual protective measures; and
(i) giving appropriate instructions to employees.

These principles are ones that have been adopted in the UK for many years and are contained in specific regulations such as COSHH and PUWER. Avoiding risks or combating them at source are clearly preferable to protecting employ-ees simply by giving them appropriate instructions. However, in the cases of both COSHH and PUWER, the relevant regulations require a hierarchical approach to be taken, i.e. prevention of exposure is preferable to control at source which in turn is preferable to the use of personal protective equipment (see Chapters C2 and D1).

Regulation 4 does not contain such a requirement for a hierarchical approach, thus making it somewhat academic – a useful checklist of principles to be applied but with no element of compulsion to consider the more effective precautions first.

B2.4   Regulation 5: health and safety arrangements

Regulation 5 states:

(1) Every employer shall make and give effect to such arrangements as are appropriate, having regard to the nature of his activities and the size of his undertaking, for the effective planning, organisation, control, monitoring and review of preventive and protective measures.

(2) Where the employer employs five or more employees, he shall record the arrangements referred to in paragraph (1).

This regulation requires all employers to take a systematic approach to their management of health and safety in order to ensure that precautions are implemented effectively. It refers to what is often called the management cycle:

PLAN
    ORGANISE
           CONTROL
                MONITOR
                    REVIEW

The management cycle is explained more fully in Chapter B3. It should also be noted that paragraph (2) requires these management arrangements to be recorded by employers with five or more employees. For smaller organisations simple references within the 'arrangements' section of their health and safety policy (see Chapter B1) or within risk assessment records should be sufficient. However, larger employers are likely to have detailed procedures and manuals setting out their management systems, e.g. for health and safety inspections, accident and incident investigation.

B2.5   Regulation 6: health surveillance

This regulation states: 'Every employer shall ensure that his employees are provided with such health surveillance as is appropriate having regard to the risks to their health and safety which are identified by the assessment.'

The need for health surveillance will often be identified during risk assessments carried out under more specific regulations. The COSHH Regulations contain their own requirements for health surveillance while noise assessments may reveal the need for periodic audiometry, and manual handling assessments may indicate that some staff involved in manual handling activities should be subject to periodic physical checks.

While most health surveillance will relate to risks to the employees themselves, risks to others must not be overlooked, e.g. the dangers that could be caused by train drivers or electricians suffering from colour blindness or other vision defects.

The ACOP accompanying Regulation 6 states that health surveillance should be introduced where:

- there is an identifiable disease or adverse health condition related to the work concerned; and
- valid techniques are available to detect indications of the disease or condition; and
- there is a reasonable likelihood that the disease or condition may occur under the particular conditions of work; and
- surveillance is likely to further the protection of the health and safety of the employees to be covered.

Health surveillance is covered in greater detail in Chapter D4.

B2.6   Regulation 7: health and safety assistance

Paragraph (1) of Regulation 7 states:

Every employer shall, subject to paragraphs (6) and (7), appoint one or more competent persons to assist him in undertaking the measures he needs to take to comply with the requirements and prohibitions imposed upon him by or under the relevant statutory provisions and by Part 2 of the Fire Precautions (Workplace) Regulations 1997.

A competent person is defined by paragraph (5) as having 'sufficient training and experience or knowledge and other qualities to enable him properly to assist in undertaking the measures referred to in paragraph (1)'.

Paragraphs (6) and (7) allow self-employed employers and partners to carry out this role themselves – providing they are competent as defined above.

The remaining paragraphs of the regulation contain the following requirements:

- employers must ensure adequate cooperation between persons they have appointed;
- the numbers of persons and the time and means at their disposal must be adequate in relation to the size of the undertaking and the risks present in it;
- employers must provide the persons appointed with relevant information;
- competent persons who are employed by the employers should be appointed in preference to outsiders (e.g. consultants).

Several of these aspects, together with the training available for health and safety specialists, are explored in greater detail in Chapter B3.

B2.7   Regulation 8: procedures for serious and imminent danger and for danger areas

Paragraph (1) of the regulation requires employers to:

- establish and give effect to appropriate procedures to be followed in the event of serious and imminent danger to persons at work;
- nominate sufficient competent persons to implement evacuation procedures;
- restrict access by employees to high-risk areas to those who have had adequate instruction.

Paragraphs (2) and (3) contain further detailed requirements in respect of evacuation procedures and competent persons. More detail on emergency procedures is contained in Chapter B3.

The requirement to restrict access on grounds of health and safety might relate to the presence in an area of bare live electrical conductors, hazardous substances or even potentially dangerous people (e.g. in prisons) or animals (e.g. on farms or in zoos). The risk may be present there at all times, or only temporarily (because of short-term activities, e.g. maintenance or repair) or occasionally (because of foreseeable breakdowns or accidents). Those who are permitted access must have received adequate instruction and training both to recognise the risks present and to take appropriate precautions.

B2.8   Regulation 9: contacts with external services

Regulation 9 states that: 'Every employer shall ensure that any necessary contacts with external services are arranged, particularly as regards first aid, emergency medical care and rescue work.'

In some cases this may only mean that staff on site have appropriate emergency telephone numbers but higher risk workplaces would also need to make the emergency services (and other relevant resources) aware of foreseeable emergencies that might arise and liaise with such bodies on how they would respond. This may extend as far as conducting training exercises or regular review meetings. This is referred to further in the sections of Chapter B3 dealing with emergency procedures.

B2.9   Regulation 10: information for employees

Paragraph (1) requires employers to provide employees with comprehensible and relevant information on the following:

- the risks to their health and safety identified by risk assessment
- the preventive and protective measures
- emergency procedures and competent persons nominated to implement them.

This requirement is closely related to the general requirements for training contained in Regulation 13 (see paragraph B2.12 below and also the sections of Chapter B3 dealing with employee training).

Paragraphs (2) and (3) of the regulation require employers to provide parents or guardians of children (younger than the school leaving age) with information about risks and precautions, before employing their children. In the case of work experience activities this would usually be dealt with by the school or work experience agency, but those employing children in part-time or holiday jobs would need to do this directly. (There are other restrictions affecting the timing, duration and nature of work carried out by children.)

B2.10 Regulation 11: cooperation and coordination

This regulation requires employers and self-employed persons sharing workplaces (whether on a temporary or a permanent basis) to:

- cooperate with each other in respect of health and safety requirements
- coordinate their precautionary measures
- take all reasonable steps to inform each other about health and safety risks.

Sharing of workplaces may be temporary (such as in short-term construction projects) or more permanent (as in a multi-occupancy building). Information may need to be provided to other occupants about the use of hazardous or highly flammable substances or radioactive materials or in respect of high-risk activities, particularly those that may affect the other occupants.

Coordination and cooperation will be particularly necessary in respect of fire and other emergency procedures but may be relevant for other reasons, e.g. traffic control. Landlords and managing agents will also have responsibilities for common parts of premises or common services such as fire alarms (through their duties under section 4 of HASAWA), even if they do not have any employees working on the premises. Principal contractors on construction projects are also given an important coordinating role through the requirements of the Construction (Design and Management) Regulations (CDM).

Additional requirements are placed on 'host' employers through Regulation 12 (see paragraph B2.11 below). Chapter B7 provides more guidance on compliance with Regulations 11 and 12 and the implications of the CDM Regulations in relation to the selection and management of contractors.

B2.11 Regulation 12: persons working in host employers' or self-employed persons' undertakings

Regulation 12 requires host employers (or self-employed persons) to provide other employers (or self-employed persons) whose employees are working in the host's undertaking with comprehensible information on the following:

- risks to those visiting employees' health and safety arising out of or in connection with the conduct of the host's undertaking
- precautions already taken by the host in relation to those visiting employees.

Additionally the host must: 'ensure that any person working in his undertaking who is not his employee ... is provided with appropriate instructions and comprehensible information regarding any risks to that person's health and safety which arise out of the conduct' of the host's undertaking. In both cases this must include information about any evacuation procedures.

The types of risks that visiting employers and their workers need to be informed about should have been identified during the risk assessment process. Examples of some such risks and various means of passing on relevant information are provided in Chapter B7.

B2.12 Regulation 13: capabilities and training

Paragraph (1) of Regulation 13 states: 'Every employer shall, in entrusting tasks to his employees, take into account their capabilities as regards health and safety.' The capabilities of employees relate not only to the training that they may have received but also to their abilities to put that training into practice. Some tasks may require a considerable degree of skill in order to be carried out safely. In some situations employees may be required to make a judgement about the risks involved and make an appropriate choice of the precautions to be used. This is often described as a dynamic risk assessment (see paragraph C1.21). The employer must identify such tasks through the risk assessment process and be satisfied that those staff performing them have the necessary skills or are capable of making appropriate judgements on risks and precautions.

Paragraph (2) requires employers to ensure that their employees are provided with adequate health and safety training:

(a) on their being recruited into the employer's undertaking; and
(b) on their being exposed to new or increased risks because of
   - being transferred or having a change of responsibilities
   - the introduction of new work equipment or changes in equipment already in use
   - the introduction of new technology
   - the introduction of a new system of work or changes to existing systems.

Paragraph (3) of the regulation states:
The training referred to in paragraph (2) shall –

(a) be repeated periodically where appropriate;
(b) be adapted to take account of new or changed risks to the health and safety of the employees concerned; and
(c) take place during working hours.

The need for periodic refresher training may be identified during the risk assessment process or alternatively through feedback from monitoring systems such as audits, inspections or accident and incident investigations. The ACOP to the regulations states that 'if it is necessary to arrange training outside an employee's normal hours, this should be treated as an extension of time at work'. This would normally involve time off in lieu or overtime payments. The ACOP emphasises that section 9 of HASAWA prohibits employers from requiring employees to pay for their own health and safety training. (The Safety Representatives and Safety Committee Regulations 1977 require employers to consult safety representatives about arrangements for health and safety training.)
   Training is dealt with more fully in Chapter B3.

B2.13 Regulation 14: employees' duties
This regulation states:

(1) Every employee shall use any machinery, equipment, dangerous substance, transport equipment, means of production or safety device provided to him by his employer in accordance both with any training in the use of the equipment concerned which has been received by him and the instructions respecting that use which have been provided to him by the said employer in compliance with the requirements and prohibitions imposed upon that employer by or under the relevant statutory provisions.
(2) Every employee shall inform his employer or any other employee of that employer with specific responsibility for the health and safety of his fellow employees –
   (a) of any work situation which a person with the first-mentioned employee's training and instruction would reasonably consider represented a serious and immediate danger to health and safety; and
   (b) of any matter which a person with the first-mentioned training and instruction would reasonably consider represented a shortcoming in the employer's protection arrangements for health and safety.

Insofar as that situation or matter either affects the health and safety of that first-mentioned employee or arises out of or in connection with his own activities at work, and has not previously been reported to his employer or to any other employee of that employer in accordance with this paragraph.

These duties on employees – to work in accordance with the training and instructions that they have received and to report serious and immediate dangers and shortcomings in the employer's health and safety precautions – overlap with duties placed on employees by section 7 of HASAWA.

B2.14 Remaining Management Regulations

The remaining parts of the Regulations either are of limited relevance to most employers or are dealt with elsewhere in this book.

- Reg. 15: temporary workers. This regulation places duties on employers to provide workers on fixed duration contracts or from employment agencies with information about safety-related qualifications or skills that are required and details of any statutory health surveillance. It also requires them to inform employment agencies (employment businesses) similarly about qualifications and skills and specific health and safety features of work activities (e.g. it involves work at heights or considerable manual handling).
- Reg. 16: risk assessment in respect of new or expectant mothers.
- Reg. 17: certificate from registered medical practitioner in respect of new or expectant mothers.
- Reg. 18: notification by new or expectant mothers. Issues relating to new or expectant mothers are examined in greater detail in Chapter C1.
- Reg. 19: protection of young persons. Protection of young persons is also covered in Chapter C1. This requirement overlaps with Regulation 3 (see paragraph B2.2 above).
- Reg. 20: exemption certificates. The Secretary of State for Defence may exempt home forces, visiting forces and their headquarters from most of the requirements of the regulations, in the interests of national security.
- Reg. 21: provisions as to liability. Any act or default of an employee or a person appointed to provide competent advice does not give employers a defence in law against a failure to comply with their own obligations. (In practice inspectors will take account of the circumstances of each individual case.)
- Reg. 22: a breach of a duty imposed on an employer under the regulations does not confer a right of action in civil proceedings in respect of persons not in his employment. (The previous exclusion of right of action by employees was removed in October 2003.)
- Reg. 23: extension outside Great Britain. The regulations apply to offshore installations, wells, pipelines, pipeline works and connected activities and provide workers with protection while they are off duty at such locations.
- Regs 24–30. These regulations contain minor amendments to other sets of regulations and deal with other technical matters.

Source materials

1. HSE (2000) 'Management of Health and Safety at Work' (Regulations, ACOP and Guidance), L 21.

Key Elements of Management

B3.1 Introduction

The previous chapter summarised the legal obligations placed by the Management Regulations on both employers and the self-employed. Within this chapter several of the key elements of effective safety management and the requirements of the Management Regulations are explored in greater detail. These are as follows:

- application of the management cycle (Reg. 5)
- competent health and safety assistance (Reg. 7)
- emergency procedures (Regs 8 and 9)
- health and safety training (Reg. 13).

Several other important aspects of safety management are covered in other chapters within this section of the book, while risk assessment has a separate section devoted to it.

B3.2 Reasons to manage health and safety well

The reasons for managing health and safety effectively are usually presented under three headings: legal, financial and humanitarian.

Legal obligations

The statutory requirements relating to health and safety are extensive (as was demonstrated in Chapter A2), while the Management Regulations (see Chapter B2) contain specific obligations in respect of health and safety management. Those at work, particularly employers and the self-employed, also have obligations under civil law both to employees and to others who may be affected by their work activities. The HSE calculated the total costs to the economy of workplace accidents and work-related ill health in 2001/2 as between £13.2 and £22.2 billion.

Financial considerations

While employers and the self-employed can insure against successful civil claims being made against them (see paragraph A3.12), in the long run their insurance premiums will reflect their claims history. Many insurers are now also proactively seeking evidence of the effective management of health and safety, from both existing policy holders and new business, and this too will have an impact on premiums.

However, there are many other costs imposed on businesses through accidents, ill health and non-injury incidents through staff absence, damaged equipment and materials and a variety of other disruptive effects. HSE research (ref. 1) has demonstrated that the ratio between the insured and uninsured costs of accidents can be as high as 1:36. The potential costs associated with accidents, etc., are explored further in Chapter B4.

Humanitarian aspects

It is almost impossible to be involved (however marginally) in a fatal accident, to see the long-term impact of a serious work injury or to meet someone in the later stages of a terminal occupational disease, without being profoundly affected by the experience. Although the statistics are improving, a typical year in the UK sees:

- 300 deaths of people at work, or resulting from work activities
- 30,000 major injuries (fractured limbs, amputations, etc.)
- 2,000 deaths directly attributed to occupational diseases
- 10,000 deaths in which occupational disease played a significant part.

(Deaths and injuries from work-related road accidents are not included.)

Many cases of occupational disease occurring now (particularly asbestos and other dust-related diseases) result from poor standards many years ago and, it is hoped, we will soon see a significant downward trend.

Apart from the humanitarian considerations, a business that regularly experiences serious accidents or is known to cause health problems for its workers is unlikely to have a well-motivated workforce or a good image within the wider community. This is likely to have a negative effect on the company's financial position.

B3.3   The management cycle

Regulation 5 of the Management Regulations (see paragraph B2.4) requires employers to have appropriate arrangements for the effective

PLANNING
      ORGANISATION
              CONTROL
                    MONITORING
                            REVIEW

of their preventive and protective measures, i.e. to have management systems in place which ensure that precautions are implemented effectively.

The HSE booklet 'Successful Health and Safety Management' provides a slightly different model to follow:

- POLICY: Establishment of an effective health and safety policy within the organisation.
- ORGANISING: An effective management structure and arrangements in place to deliver the policy (see Chapter B1).
- PLANNING AND IMPLEMENTATION: A planned and systematic approach to implementing the policy.
- MEASURING PERFORMANCE: Measurement of performance against agreed standards (through active and reactive monitoring methods).
- REVIEWING PERFORMANCE: Learning from monitoring and experience and applying the lessons learned through a feedback loop into an appropriate earlier stage of the cycle.

All stages of this process can be subjected to auditing.

There are no major differences between the two models and various other health and safety management systems adopt a similar approach. Application of these management principles within a small organisation should be relatively simple, whereas larger organisations will need rather more detailed procedures and arrangements. Several of the more common types of health and safety management arrangements are described in this section of the book. Even where smaller employers feel that the degree of formality suggested is not appropriate to their situation, they are still recommended to follow the same principles.

Each of the five stages of the management cycle contained in the Management Regulations is explained in greater detail below and Figures B3.1–B3.3 on pages 58–59 provide further illustrations of how the cycle can be applied in practice.

B3.4   Planning

Health and safety policy
The health and safety policy (see Chapter B1) is a key component in effective management. Not only does the statement of intent set the tone for the company's approach but the allocation of responsibilities for its implementation to appropriate levels and individuals in the organisation is essential for its eventual success.

Health and safety procedures
An important prerequisite for the effective implementation of health and safety arrangements throughout larger organisations is often the establishment of a formal procedure. Some of the topics dealt with in this section of the book are likely to benefit from the use of formal procedures, for example:

- accident and incident investigations
- health and safety inspections
- selection and management of contractors.

Annual health and safety plans
Continuous improvement should be sought in health and safety just as in other aspects of management. In some cases action may be necessary to cope with new legislation, changes in ACOPs or HSE guidance. Other actions may be required because of changes in equipment, materials, processes or personnel. As with any plan, those responsible for implementing it should be identified, together with the intended time scales and necessary resources.

B3.5   Organisation

'Successful Health and Safety Management', published by the HSE, refers to the four Cs of effective organisation of health and safety:

1. Control: dealt with at the next stage of this management cycle (see paragraph B3.6).
2. Cooperation: between individuals and groups.
3. Communication: throughout the organisation.
4. Competence: of individuals.

Cooperation
While employees have a legal obligation to cooperate with their employers on health and safety matters, legislation alone is unlikely to achieve effective cooperation. Employees often have valuable knowledge and experience that can be tapped into and they are more likely to commit themselves wholeheartedly to those health and safety activities of which they feel a degree of ownership. Such ownership is often achieved through another 'C' consultation. This may be car-

ried out formally via safety committees or team meetings or informally through contact with individual employees or safety representatives (see Chapter B6).

Cooperation will also be much more forthcoming from a workforce that believes that its management is genuinely committed to health and safety improvements, rather than paying lip service. Employers often get the degree of cooperation that they deserve.

Communication

Policies and procedures will only be implemented if they have been communicated effectively to all the people who are affected by them. Communication is much more than mere distribution of documents or circulation of e-mails, and explanation is often vital to success – not just of a procedure itself, but also of the reasons. Two-way communication will help this considerably – again safety committees and team meetings have an important role, as do informal forms of communication.

Competence

Individuals who are allocated responsibilities, whether within the health and safety policy or not, for implementing of a procedure or for carrying out an action in an annual health and safety plan, must have the necessary degree of competence. Training is an important element in acquiring the knowledge necessary for achieving competence, but so also is experience and the ability to put the knowledge and experience to good use. Health and safety-related training is covered in more detail later in this chapter (see paragraph B3.22 onwards).

B3.6   Control

Control of health and safety can be achieved through a variety of means.

Health and safety procedures

While the establishment of procedures for health and safety is part of the 'planning' stage of the management cycle, the daily use of procedures also provides an important reference point for the day-to-day control of activities within the workplace.

Supervision

Good supervision is an essential element in achieving control of what happens in the workplace on a day-to-day basis. All those with supervisory responsibilities have an important role to play in ensuring that procedures are implemented, PPE and other standards are met and that those carrying out tasks have received appropriate training. In many cases supervisors will be involved in delivering training to staff under their control.

Activities of health and safety specialists

Those with specialist roles in health and safety also play a key part in maintaining control of health and safety matters. This may be through providing training in specialist areas (e.g. accident investigation, permit-to-work systems, duties of fire wardens) or by giving guidance on interpretation and implementation of legislative requirements or health and safety procedures. The role of the health and safety specialist is described more fully later in the chapter (see B3.9 onwards).

B3.7   Monitoring

The monitoring of the progress of planned activities and of actual conditions on the ground is vital in all aspects of business management, and health and safety is no exception. Monitoring techniques can be divided into proactive and reactive methods.

Proactive monitoring

This type of monitoring activity provides feedback on performance without an accident or some other undesirable incident having occurred. Among proactive methods are the following:

- routine health and safety inspections
- task observations
- health and safety auditing systems
- procedure compliance checks
- progress reports (on annual health and safety plans, etc.)
- occupational hygiene and health surveys.

A number of proactive techniques are described more fully in Chapter B5 while Chapter D4 deals with occupational health and hygiene.

Reactive monitoring

Accident statistics alone are a poor indicator of health and safety performance. A low accident rate, particularly in a small organisation, does not necessarily mean that risks are being controlled effectively and health and safety procedures are being followed. While a high accident rate may indicate health and safety problems, it may also be partially a result of high absenteeism generally or poor morale.

However, investigation of individual accidents, incidents or cases of ill health provides an opportunity to identify the causes, particularly underlying causes such as lack of training, inadequate supervision and poor attitudes to health and safety. Chapter B4 provides further detail on the investigation of accidents and incidents.

B3.8   Review

Where monitoring activities identify shortcomings in health and safety arrangements, the reasons for these shortcomings need to be identified and an appropriate course of action agreed upon. Implementation of these actions would then need to be planned, with the whole management cycle being applied once again. Important means of carrying out the review stage of the cycle are as follows:

- health and safety committees (dealt with in more detail in Chapter B6)
- management of health and safety meetings (reviewing the progress of action plans and preparing new ones)
- activities of health and safety specialists (determining the reasons for poor performance and recommending possible courses of remedial action).

B3.9   Competent health and safety assistance

Regulation 7 of the Management Regulations (see paragraph B2.6) requires all employers to appoint competent health and safety assistance. Paragraph (5) of the regulation refers to such a person having 'sufficient training and experience or knowledge and other qualities'. These attributes of the competent person are examined in further detail below. Other aspects referred to by Regulation 7 are as follows:

FIND THIS ON CD	Figure B3.1	Application of the management cycle to fire evacuation arrangements

PLAN	Ensure that the fire alarm system is adequate (e.g. sufficient call points)
	Identify suitable assembly points
	Decide how the roll call is to be structured
	Determine how to deal with non-employees (visitors, customers)
	Identify who will be responsible for testing alarms, initiating evacuation drills, etc.
ORGANISE	Indicate assembly points with suitable signs
	Communicate the procedure to all staff
	• by prominent notices
	• via e-mails
	• include in the induction of new staff
	Identify and train
	• sufficient fire wardens
	• persons to test the alarms
	Decide and communicate the times at which alarms will be tested
CONTROL	Replace fire wardens (or alarm testers) who leave or transfer
MONITOR	Conduct regular evacuation drills
	Check the alarm test records periodically
	Check signs and notices (during routine inspections)
REVIEW	Shortcomings identified during drills
	Failures to test alarms regularly
	(At Health and Safety Committee meetings or directly with relevant staff.)

FIND THIS ON CD	Figure B3.2	Application of the management cycle to health and safety induction training

PLAN	Make a statement that induction training will be carried out (in the health and safety policy, staff handbook, etc.)
	Determine who will be responsible for carrying out the induction (line supervisors, training section, the department)
	Decide the intended content of the induction
ORGANISE	Circulate a draft induction checklist for comment
	Communicate the final version to all responsible for conducting inductions
	Decide where completed checklists should be returned to
CONTROL	Ensure that a blank checklist is provided for all new starters (and transferees)
MONITOR	Ensure that checklists are returned and have been completed properly

	Make spot checks on sample new starters
	• Did they receive their induction when they should have?
	• Have they acquired the knowledge that they should have done?
	Discuss the process and checklist content with those carrying out inductions
REVIEW	Reasons that inductions may be late or incomplete, or have not taken place
	(At Health and Safety Committee meetings, management meetings or directly with relevant staff.)

FIND THIS ON CD   **Figure B3.3**   **Application of the management cycle to accident and incident investigation**

PLAN	Draft an accident and incident investigation procedure, identifying
	• what should be investigated (e.g. accidents, incidents, ill health)
	• who should investigate (e.g. line supervisor, review by manager)
	• how quickly investigations should be carried out and reports submitted
	• whether safety representatives will be involved
	• RIDDOR reporting responsibility
	Draft an associated report form
ORGANISE	Incorporate relevant suggestions into the final procedure and report form
	Communicate the procedure and form to all staff
	Train relevant staff in investigation techniques
	Include references to the procedure in
	• the staff handbook
	• the induction of new staff
CONTROL	Maintain the supply of report forms (hard copies or available in electronic form)
	Train new supervisors (and managers) in investigation techniques
MONITOR	Check report forms for
	• adequacy of investigation
	• promptness of investigation and reporting
	• completion of remedial actions
	Cross-check between accident report forms and first aid treatment records
	Compare the numbers of reports for accidents and non-injury incidents
REVIEW	The suitability of the form
	The clarity and completeness of the procedure
	Shortcomings in investigation or reporting (including undue delay)
	(At Health and Safety Committee meetings or directly with relevant staff.)

- employers ensuring cooperation between competent persons
- the scale of competent assistance required
- provision of relevant information to competent persons
- a preference for employees to be appointed to this role, rather than consultants
- self-employed employers or partners being permitted to appoint themselves.

The HSE's ACOP and guidance booklet on the Management Regulations provides further interpretation on what is required as does the HSE booklet 'Successful Health and Safety Management'.

B3.10 The scale of assistance required

In smaller businesses that involve a relatively low level of risk, employers may appoint either themselves or an employee as the source of competent health and safety assistance, even though they have no formal qualifications. The guidance document refers to the following requirements:

- an understanding of relevant current best practice
- an awareness of the limitations of one's own experience and knowledge
- the willingness and ability to supplement existing experience and knowledge, when necessary, by obtaining external help and advice.

The awareness of one's own limitations is important for anyone involved in health and safety. Even the most experienced practitioners will regularly come across situations where their experience and knowledge are lacking. External sources of help and advice may be the HSE and its extensive range of published material, other publications (in printed or electronic form) or health and safety consultants. (See Chapter E1 for details of sources of information.)

Small businesses whose activities involve a higher level of risk and medium-sized organisations are more likely to need the assistance of someone with a formal qualification in health and safety. This may be an external consultant (see paragraph B3.13 below), or the employer may choose to train up a member of the workforce (see paragraph B3.12). Even with a trained person in-house, there may still be a need for assistance from external sources.

Large organisations may require the services not just of more than one health and safety practitioner, but also of occupational health specialists (medical practitioners, nurses, ergonomists, etc.) and occupational hygienists. The roles of occupational health and hygiene specialists are explained more fully in Chapter D4.

B3.11 The role of the health and safety specialist

The precise role of the specialist will vary, depending on the size of the business and the nature and extent of the risks involved in its activities. Not only must the specialist be competent to provide the level of assistance required but he or she must also have an appropriate status within the organisation. Health and safety specialists most commonly are located within human resources departments but they may also find themselves assigned within technical or engineering sections or reporting directly to a senior member of management. The job description of the specialist is likely to involve the following:

1. Policy and procedures:
   (a) developing new policies and procedures relating to health and safety
   (b) reviewing and revising existing policies and procedures
   (c) preparing short-term and long-term health and safety plans.
2. Implementation:
   (a) organising the putting of policies and procedures into practice
   (b) delivering health and safety training
   (c) investigating accidents and incidents
   (d) carrying out occupational hygiene surveys
   (e) organising the health and safety committee.
3. Guidance and prompting:
   (a) maintaining relevant reference material and information systems
   (b) providing advice on request
   (c) prompting others from whom action is required
   (d) identifying future issues.
4. Communication and promotion:
   (a) through health and safety committees
   (b) directly with safety representatives
   (c) maintaining a high profile in the workplace
   (d) through newsletters, poster campaigns, etc.
5. Monitoring:
   (a) the progress of health and safety plans
   (b) auditing health and safety management systems
   (c) conducting workplace inspections, etc.
   (d) operational procedures and training standards.
6. Management:
   (a) managing the work of other health and safety specialists.
7. External liaison:
   (a) with the HSE and other enforcing authorities (local authorities, fire service)
   (b) other specialists within a group of companies or similar work activities
   (c) insurance companies and legal representatives
   (d) contractors and their health and safety specialists
   (e) clients and customers.

As can be seen, the specialist's role involves all of the stages of the management cycle (see paragraph B3.3 above). Reference is made in Regulation 7 of the Management Regulations to the 'other qualities' of the source of competent health and safety assistance. These are separate from whatever professional qualifications this person should have, and should include the following:

- the ability to put theory into practice
- good interpersonal skills
- an ability to overcome difficulties and setbacks
- high levels of integrity and independence.

B3.12 Specialist training and qualifications
There are several bodies in the UK involved in the delivery of health and safety specialist training and the administration of related qualifications. Among the most prominent are the following.

NEBOSH (National Examination Board in Occupational Safety and Health)

- NEBOSH General Certificate: The NEBOSH Certificate is one of the main entry routes into health and safety specialism. It involves a programme of study of between 80 and 100 hours, two written examination papers (Identifying and Controlling Hazards and Management of Safety and Health) and the practical assessment of a workplace, including the preparation of a written report.
- NEBOSH Construction Certificate: This is similar to the General Certificate, with the syllabus and examinations slanted towards construction issues.
- NEBOSH Diploma Part 1: The Diploma programme is much more demanding than the Certificate and mainly attracts candidates with previous knowledge and experience of health and safety matters. The course consists of five main modules (Management of Risk; Legal and Organisational Factors; The Workplace; Work Equipment; and Agents – mainly covering occupational health issues), together with shorter modules on communication and training skills. An assignment must be carried out on each of the main modules – these can often be conducted in the candidate's own workplace. The assignments represent 50 per cent of the total marks with two examinations accounting for the remainder. Study time is stated as 172 hours of tuition and 92 hours of private study but in practice many candidates spend longer on study and assignments.
- NEBOSH Diploma Part 2: The structure and assessment of Part 2 is similar to Part 1, with the modules having the same titles but going into some specific aspects in much more detail. NEBOSH specify 194 hours of tuition and 100 hours of private study for Part 2.

NEBOSH programmes are delivered in a variety of ways – day release, block courses, distance learning, etc. All centres are subject to a NEBOSH approval process and quality control checks, which include moderation of locally marked practical assessment reports (Certificate) and assignments (Diploma). Many colleges and private training providers are approved to deliver the Certificate but approved centres for the Diploma are harder to find. The NEBOSH telephone number is 0116 263 4700 and NEBOSH has a website at www.nebosh.org.uk.

IOSH (Institution of Occupational Safety and Health)
IOSH is the principal professional body in the UK for health and safety practitioners. At one time NEBOSH was part of IOSH but now the two bodies have an arm's-length relationship. IOSH now offers its own training programmes. The programme most of relevance to health and safety specialists is IOSH's Managing Safely.

The programme consists of eight modules, each of which requires three or four hours of directed study. One module, dealing with 'specialised hazards', can be tailored toward the needs of particular industries or occupational activities.

Assessment is by one-hour test (multiple choice or short-answer questions) together with a practical 'risk assessment', preferably carried out in the candidate's own workplace. Centre accreditation and moderation of locally marked work is carried out on a similar basis to NEBOSH. IOSH also offers other shorter courses, but these are unlikely to be suitable for health and safety specialists. IOSH's telephone number is 0116 257 3100 and the website is at www.iosh.co.uk.

The British Safety Council
The British Safety Council's Diploma in Safety Management involves similar topics to the NEBOSH General Certificate. It is assessed by multiple choice tests and reports.

Universities
An increasing number of universities offer first degree courses with a significant health and safety content. Similarly postgraduate courses offering Master's degrees or postgraduate certificates or diplomas are also expanding. Such programmes also often include environmental modules.

National Vocational Qualifications
National Vocational Qualifications (NVQs) are based not on examinations but on the assessment of competence within the workplace. In August 2002 new vocational standards in occupational health and safety were launched by the leaders of IOSH, the British Safety Council, the Royal Society for the Prevention of Accidents (RoSPA) and NEBOSH. These standards were developed with the National Training Organisation for Employment (Employment NTO) and include two qualifications at Level 5 in the NVQ structure – for those with a board or senior management role in delivering health and safety in the workplace. Further details are available from the Employment NTO (tel. 0116 251 7979).

Other sources of training
Many other organisations offer training that may be beneficial to the health and safety specialist. These include employers' organisations (Engineering Employers Federation, Construction Industry Training Board), trade unions and the TUC, RoSPA and the Chartered Institute of Environmental Health (CIEH).

B3.13 Health and safety consultants
Many employers use the services of health and safety consultants and other health and safety specialists. In some cases consultants are engaged to carry out a specific programme of work while in others they may be retained on an annual fee to provide a guaranteed minimum level of service. While some larger consultancies have significant numbers of consultants (employees and associates) and have national or regional prominence, many consultants (including the author) work independently for a limited range of clients.

Means of identifying suitable consultants include the following:

- recommendations from business contacts, trade associations, etc.
- their possession of relevant qualifications
- membership of professional organisations (e.g. IOSH)
- relevant experience of the type of work involved
- the availability of references from satisfied clients.

IOSH (see paragraph B3.12 above) and several other professional bodies keep registers of appropriately qualified consultants. Before formally engaging a consultant to carry out a programme of work, it is important to define what the consultant is required to do and, of course, the fees that will be charged. Where it is not possible to be precise about costs, the consultant will often agree a range of likely charges or an overall maximum for the project. Many consultants will make a preliminary visit to discuss a project free of charge.

63

However, lengthier visits to define the scope of a complex programme of work may incur a charge (although this may be offset against future fees if the consultant's services are engaged for the main project).

Further guidance on engaging consultants and other specialists is contained in a free HSE leaflet, 'Need Help on Health and Safety?'. This also contains a listing of a number of professional bodies and other useful sources of advice.

B3.14 Emergency procedures

There are several statutory obligations that require employers to establish emergency procedures or arrangements.

Management of Health and Safety at Work Regulations 1999

Regulation 8 requires employers to 'establish and, where necessary, give effect to appropriate procedures to be followed in the event of serious and imminent danger' (see paragraph B2.7). Regulation 9 requires employers to arrange necessary contacts with external services as regards first aid, emergency medical care and rescue work (see paragraph B2.8).

Fire Precautions Act 1971

Fire certificates issued under the Act may impose requirements in respect of 'appropriate instructions or training on what to do in case of fire', as well as for fire fighting equipment, fire alarms, etc. Chapter C7 provides more detail of the requirements of the Act and its implications.

Fire Precautions (Workplace) Regulations 1997

Emergency actions to safeguard employees, such as fire fighting, rescue, etc., are required by Regulation 4. The requirements of these regulations are explored further in Chapter C7.

Confined Spaces Regulations 1997

Regulation 5 requires 'suitable and sufficient arrangements for the rescue of persons in the event of an emergency' in a confined space. The emergency may relate to risks in the confined space (e.g. fire, gases or vapours, lack of oxygen) or to the need for rescue because of a fall or general illness.

Construction (Health, Safety and Welfare) Regulations 1996

Regulation 20 (as amended) requires arrangements to be made for dealing with emergencies, including procedures for evacuating the site, persons to implement emergency arrangements and contacts with emergency services.

Control of Asbestos at Work Regulations 2002

Regulation 14 requires arrangements to be in place to deal with accidents, incidents or emergencies related to the use of asbestos in a work process or to the removal or repair of asbestos-containing materials. Steps must be taken to deal with the unplanned release of asbestos, including the exclusion of persons without appropriate RPE and protective clothing from the affected area.

Control of Major Accident Hazards Regulations 1999 (COMAH)

Control of Major Accident Hazards Regulations 1999 (COMAH) only applies to sites that contain certain quantities of specified dangerous substances. The regulations require both on-site and off-site emergency plans to be prepared with the following objectives:

- containing and controlling incidents to minimise effects and damage
- implementing necessary measures to protect persons and the environment from the effects of major accidents
- communicating information to the public, the emergency services and relevant authorities
- arranging necessary restoration and clean-up measures.

See page 74 for further reading on COMAH requirements.

Control of Substances Hazardous to Health Regulations 2002 (COSHH)
The 2002 version of the COSHH Regulations included a new regulation 13 dealing with 'Arrangements to deal with accidents, incidents and emergencies'. Whilst the regulation contained a number of detailed requirements, with further detail in the accompanying ACOP and guidance, the principles – to be followed are those set out in paragraphs B3.15 to B3.21 of this chapter.

B3.15 Types of emergency to consider
The identification of foreseeable situations of 'serious and imminent danger' should be part of the overall process of risk assessment. What is foreseeable will be based on the experience of employers and their staff, the experience of others carrying out similar activities and guidance from the HSE and other authoritative sources. Such situations might include the following:

- fire
- bomb threats or suspect packages
- leaks or discharges of hazardous substances
- major process problems, e.g. runaway chemical reactions
- power failures
- severe weather conditions
- violence to employees
- work in confined spaces (see paragraph B3.14 above)
- dangerous animals, e.g. in zoos, farms, veterinary practices
- crowds out of control.

B3.16 Components of emergency procedures
Where the risks are sufficient to justify the preparation of an emergency procedure, then that procedure is likely to need to contain a number of different components. These can be summarised as follows:

- discovery, alert, investigation and initial control
- external liaison
- evacuation
- control and rescue
- information and training.

Each of these is dealt with in greater detail below. Specific guidance on fire evacuation procedures is provided in Chapter C7.

B3.17 Discovery, alert, investigation and initial control
Any person discovering what they believe to be a situation of serious and imminent danger must be given guidance on what action to take. Depending on the circumstances the first action may be:

- to alert others to the emergency situation; or
- to investigate the situation further.

Factors such as the nature of the emergency, the availability of suitable equipment and the capabilities of the individual must be taken into account. In cases of doubt, emphasis should be placed upon alerting others. Some procedures may need to include detailed arrangements for initial control of the situation, e.g. closing down processes or equipment that could make the situation worse, isolating containers of hazardous or flammable substances, or using fire fighting equipment.

B3.18 External liaison

Emergencies are likely to require prompt contact with a variety of organisations and people outside the workplace, including the following:

- emergency services (fire brigade, ambulance, police, local hospitals)
- public bodies (HSE, Environmental Agency, local authorities)
- neighbours (they may be directly affected by the emergency)
- senior management (e.g. to control the emergency or deal with the media)
- specialist services (e.g. waste disposal services, specialist contractors).

Relevant contact telephone numbers should be included within the procedure and posted prominently in suitable locations, e.g. switchboard, security office, medical centre. Consideration must be given as to who will make such contacts during night or weekend work or outside normal working hours.

B3.19 Evacuation

In some emergencies evacuation of part or all of the premises may follow immediately upon the alert being sounded, while in others the premises might only be evacuated after an investigation has been carried out. In all cases the evacuation must take place

- to a place of safety
- via a safe route.

While some places of safety (e.g. fire assembly points) may be predetermined, others may need to be chosen to match the circumstances of the emergency, such as:

- upwind of a chemical leak
- in the case of a bomb alert, away from plate glass and areas where a bomb might have been concealed, e.g. cars, skips.

Procedures need to take account of others who may be on the premises, such as visitors, contractors, etc., and also of persons with disabilities. Particularly where members of the public are present, someone may need to be appointed to 'sweep' the premises to ensure that they are clear.

A roll call should be conducted at the assembly point to ensure that everyone can be accounted for and this information then conveyed to the emergency services. Administrative staff are often the best people to carry out this task since they may well have the best idea of who was on the premises at the time the emergency arose. Once again consideration might need to be given to evacuation at night, over the weekend or outside normal working hours.

B3.20 Control and rescue

The nature of the emergency may require ongoing activity to control the situation, and possibly to rescue anyone who may have been trapped. Close liaison with the emergency services and other specialists will be an important part of this process. Points to include in this section of an emergency procedure are as follows:

- identification of an emergency controller (and a deputy)
- a defined emergency control point (possibly with an alternative)
- emergency teams (e.g. for rescue or to control leaks)*
- emergency equipment (relevant to possible emergencies)*
- restricting access (preventing unauthorised access to affected areas)
- call-in of staff or additional resources
- contacting relatives (to be carried out sensitively)
- responsibility for liaison with the media.

* The ACOP to the Confined Spaces Regulations 1997 lays particular emphasis on the availability of appropriate rescue and resuscitation equipment and of people trained in actions necessary in the event of an emergency. It also refers to the importance of safeguarding the rescuers – an important issue in other types of emergency arrangements.

Provision will also have to be included for determining when the emergency situation has ended, and when and how normal activities may resume. This decision is likely to be made by the emergency controller in conjunction with the emergency services.

B3.21 Information and training on emergency arrangements

All employees and others affected will need to be informed about the emergency procedure and particularly their role in it. Those allocated specific roles in the procedure, e.g. emergency controllers, fire wardens, receptionists, first aiders, medical staff, members of emergency teams and security personnel, are likely to need more specific training, some of which is likely to involve hands-on practice.

Full-scale drills may be impracticable or undesirable, but it may be possible to carry out partial practices supported by role play or 'table top' exercises to simulate other parts of the procedure. For an emergency procedure to have any value, it must work when it is needed and there must be some confidence that this will be the case.

The instruction and training of staff in relation to fire procedures is covered in the HSE booklet 'Fire Safety: An Employer's Guide'. This booklet also provides guidance on other aspects of fire procedures. Figure B3.4 contains an example of a typical notice containing basic instructions on what to do in case of fire. (Various other aspects of fire safety are dealt with in Chapter C7.)

B3.22 Health and safety-related training

Earlier in this chapter (see paragraph B3.5) reference was made to the 'organisation' step in the 'management cycle' and to the importance of both competence and communication in managing health and safety effectively. The provision of appropriate training and instruction is an essential part of establishing competence at all levels within an organisation, while training events also aid in the process of communication.

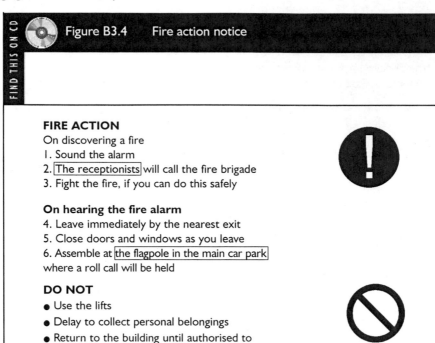

Figure B3.4   Fire action notice

FIRE ACTION
On discovering a fire
1. Sound the alarm
2. The receptionists will call the fire brigade
3. Fight the fire, if you can do this safely

On hearing the fire alarm
4. Leave immediately by the nearest exit
5. Close doors and windows as you leave
6. Assemble at the flagpole in the main car park
where a roll call will be held

DO NOT
● Use the lifts
● Delay to collect personal belongings
● Return to the building until authorised to

Under section 2(2)(c) of the Health and Safety at Work Act every employer must provide 'such information, instruction, training and supervision as is necessary to ensure, so far as is reasonably practicable, the health and safety at work of his employees'. Regulation 13 of the Management Regulations (see paragraph B2.12) also requires employers:

- to take into account employees' capabilities as regards health and safety
- to provide adequate health and safety training for employees (on recruitment, transfer, change of responsibilities, or changes to work equipment, technology, systems of work, etc.).

Many specific training requirements are contained in a variety of codes of regulations – see paragraph B3.26 and Figure B3.6 respectively.

B3.23 Health and safety training needs
Health and safety training needs should be established for all levels in the organisation:

- directors and senior managers
- middle and junior managers
- supervisors and team leaders
- health and safety specialists
- technical specialists
- maintenance staff
- other employees
- new recruits.

The content and style of delivery of the training are likely to vary considerably between these groups. Topics likely to be appropriate include the following:

- basic health and safety legislation
- new legislation and its implications
- principles of health and safety management
- company health and safety policy and responsibilities under it
- investigation of accidents and incidents
- health and safety auditing and inspection techniques
- fire and other emergency procedures
- isolation and permit to work procedures
- task procedures (both formal and informal)
- changes to equipment, processes, procedures, etc.

B3.24 Integration of health and safety into other training

While some forms of health and safety training will need to take place separately, other types can be integrated with training on other topics. These latter types include the following.

Induction of new recruits

The training of new recruits on health and safety matters can be carried out at the same time as they are briefed on other aspects of their new employment, e.g. pay and conditions, disciplinary procedures, general company rules, trade union membership. Figure B3.5 contains a sample checklist of items that might be included in a health and safety induction at a manufacturing plant.

Training in task procedures

Health and safety should be fully integrated into the training of employees to carry out any operational or maintenance task, whether or not a formal procedure exists for the task. The trainee should learn how to carry out the task safely as well as learning about relevant quality and environmental issues. Figure B8.1 on page 148 provides an example of how relevant health and safety matters can be integrated into a task procedure – in effect the procedure provides a risk assessment for the task in question.

Training of newly appointed managers and supervisors

Managers and supervisors will acquire new roles and responsibilities on their appointment. They need to be made aware of these and equipped to carry them out effectively. Topics of relevance include: their responsibilities under the health and safety policy; accident and incident investigation techniques; their role under any emergency procedures. Such training could accompany training in other important issues such as in implementing disciplinary procedures, ordering equipment and materials, recording and reporting staff attendance, holidays, etc.

Good standards are not achieved by simply administering a one-off inoculation of health and safety training but by integrating health and safety considerations into the way the business is run.

B3.25 Organisation of health and safety training

Within businesses that are managing health and safety systematically, training is likely to take place on two levels. Some training will take place routinely – induction of new recruits, training of staff making significant job changes, operator training, training of first aiders. Other training will be focused on shorter-term needs, e.g. changes in legislation, local changes, staff development needs, etc.

Figure B3.5   Sample health and safety introduction checklist

Associated Timber Products
Health and Safety Induction

Name of employee _____   Date commenced _____

	INDUCTION TOPIC		Employee signature
1	Copy of health and safety policy supplied and own responsibilities explained		
2	Fire evacuation routes		
3	Fire evacuation assembly point (north side of car park)		
4	Location of first aid equipment and accident book		
5	Introduction to first aiders and appointed persons		
6	Location of toilets, washing facilities, lunchroom		
7	Personal responsibilities for housekeeping standards emphasised		
8	Emergency arrangements for lifts explained		
9	DSE eye test required	Yes / No	
10	DSE workstation self-assessment completed	Yes / No	
11	Manual handling techniques / use of manual handling aids explained		
12	Manual handling video seen	Yes / No	
13	This employee is permitted to work on the equipment listed below:		

14	PPE ISSUED		TO BE WORN
	Safety footwear	Yes / No	At all times in factory and yard
	Hearing protection	Yes / No	When using nail guns and saws
	Safety specs/goggles	Yes / No	When using powered tools (inc. nail guns) and timber treatment chemicals
	Gloves	Yes / No	When handling timber treatment chemicals or for rough manual handling work
15	Any other topics?		

I confirm that this induction was completed on _____ and that I consider that the above-named employee has been adequately trained to operate the equipment identified in section 13.

Signed _____   Position _____

Items 1 to 8 and 15 should be covered for *all* employees;

Items 9 and 10 for office-based employees;

Items 11 to 14 for production and maintenance employees.

Induction should normally be completed on the first day of employment.

Figure B3.6   Specific competence and training requirements

Regulations	Competence	Training	Notes
Construction (Design and Management) Regulations 1994 (CDM)	Reg. 8	Reg. 17	Planning supervisor, designer, principal contractor
Construction (Health, Safety and Welfare) Regulations 1996 (CHSW)	Regs 6, 9, 10, 12, 13, 14 & 29	Reg. 28	Various activities
Control of Asbestos at Work Regulations 2002		Reg. 9	
Control of Substances Hazardous to Health Regulations 2002 (COSHH)		Reg. 12	
Electricity at Work Regulations 1989	Reg. 16[1]		
Fire Precautions (Workplace) Regulations 1997		Reg. 4(2)	Training to implement fire fighting measures
Health and Safety (Consultation with Employees) Regulations 1996		Reg. 7	Employee rep. training
Health and Safety (Display Screen Equipment) Regulations 1992		Reg. 6	DSE 'users'
Health and Safety (First Aid) Regulations 1981		Reg. 3(2)	First aiders
Health and Safety (Safety Signs and Signals) Regulations 1996		Reg. 5	

Regulation			
Ionising Radiations Regulations 1999	Reg.13 + ACOP	Regs. 14 & 17[2]	Radiation protection adviser
Lifting Operations and Lifting Equipment Regulations 1998 (LOLER)	Reg. 8		Planning lifting operations
Manual Handling Operations Regulations 1992		Guidance[3]	
Noise at Work Regulations 1989	Reg. 4(2)	Reg. 11	Competent to make a noise assessment
Personal Protective Equipment at Work Regulations 1992		Reg. 9	
Pressure Systems Safety Regulations 2000		Reg. 8	Drawing up a scheme of examination
Provision and Use of Work Equipment Regulations 1998 (PUWER)		Reg. 9[4]	Users, supervisors and managers
Safety Representatives and Safety Committees Regulations 1977		Reg. 4(2)	Safety rep. training
Work in Compressed Air Regulations 1996		Reg. 15	

1. Technical knowledge or experience necessary to prevent danger and injury, e.g. in live working.
2. Reg. 17 deals with appointment of radiation protection supervisors (training is referred to in guidance).
3. Reference to manual handling training is included because of its importance (many other guidance documents refer to training).
4. There are also specific requirements for training chainsaw operators (PUWER ACOP – ref. 9) and drivers of rider-operated lift trucks (a separate ACOP – ref. 10).

Some training needs may also be identified as corrective actions following audits, inspections or investigations, e.g. the need for refresher training.

Management need to establish that they are adequately resourced to carry out all of the routine training required, much of which is likely to be undertaken in-house. Short-term health and safety training needs are more likely to be dealt with on a reactive basis or as part of an annual training plan. While many of these needs may also be met in-house, e.g. through delivery by a health and safety specialist, there is also likely to be use of external training courses or consultants (see paragraph B3.13 above).

B3.26 Specific training requirements
Specific requirements in respect of competence (to perform certain tasks or carrying out defined roles) and training are contained in many sets of regulations. Requirements for training are often accompanied by requirements for information, instruction or supervision and sometimes by all three – as in HASAWA, s. 2(2)(c).

Figure B3.6 on page 72 provides a table showing where requirements for competence and/or training are contained in many more important sets of regulations. However, this list is not comprehensive – some regulations of limited application have been omitted for brevity. There are also many other references to the need for training contained within HSE guidance documents.

B3.27 Auditing of training
Health and safety-related training should be included within auditing programmes. Aspects that can be monitored include the following:

- adherence to training performance standards or training plans
- quality and effectiveness of training delivery
- adequacy of testing and assessment standards
- training records.

Chapter B5 provides further information on the design and implementation of health and safety auditing programmes.

Source materials
1. HSE (1997) 'Successful Health and Safety Management', HSG 65.
2. HSE (2000) 'Management of Health and Safety at Work: Management of Health and Safety at Work Regulations 1999', ACOP, L 21.
3. HSE (2000) 'Need Help on Health and Safety?' (free leaflet), INDG 322.
4. HSE (1999) 'A Guide to the Control of Major Accident Hazards Regulations 1999', L 111.
5. HSE (1999) 'Emergency Planning for Major Accidents', HSG 191.
6. HSE (2002) 'Control of Substances Hazardous to Health' ACOP and Guidance, L5.
7. HSE (1997) 'Safe Work in Confined Spaces: Confined Spaces Regulations 1997', ACOP, L 101.
8. HSE (1999) 'Fire Safety: An Employer's Guide'.
9. HSE (1998) 'Safe Use of Work Equipment: Provision and Use of Work Equipment Regulations 1998', ACOP and guidance, L 22 (pp. 37, 38).
10. HSE (1999) 'Rider-operated Lift Trucks: Operator Training', ACOP and guidance, L 117.

Investigating Accidents and Incidents

B4.1 Introduction

There are several important aspects to the investigation of accidents and near-miss incidents. First, there is the need to establish the facts associated with a particular incident – to find out *what* happened. Reports may need to be submitted under the Reporting of Injuries, Diseases and Dangerous Occurrences Regulations 1995 (RIDDOR) and details may also have to be submitted to insurers because of the possibility of a compensation claim (see Chapter A3).

However, it is also important to determine the reasons that the incident occurred. Some of the causes may be fairly obvious (e.g. poor technique, untidiness) but these immediate causes may indicate more basic and underlying shortcomings in the management of health and safety (e.g. lack of training, inadequate supervisory control). Investigation provides a reactive means of monitoring health and safety performance.

Once the causes are determined, appropriate *remedial action* should be taken to *prevent recurrence* of this type of incident in the future. Various studies (see paragraph B4.8) have demonstrated that near-miss incidents and instances of damage are much more common than minor accidents, which in turn occur much more frequently than serious accidents. Near-miss and damage incidents together with minor accidents should be used as opportunities to learn about and eliminate their causes, before the same set of circumstances have the opportunity to cause something more serious – shutting the stable door *before* the horse has bolted.

Reducing the numbers of accidents and incidents can also give considerable financial benefits – many employers are unaware of the true costs associated with the damage and disruption caused by incidents occurring within their workplaces, with many minor incidents not even coming to their attention at all. All of these aspects are examined within this chapter which also provides guidance in investigation technique, sets out the key elements to be included in an investigation procedure and contains a specimen accident/incident report form.

B4.2   RIDDOR requirements

The Reporting of Injuries, Diseases and Dangerous Occurrences Regulations (RIDDOR) were revised in 1995. They place a legal duty on the employer to report certain accidents and ill health to the enforcing authority – the HSE or the local authority. In 2001 a new national Incident Contact Centre for reporting all RIDDOR incidents was established (further details of this are provided in paragraph B4.7). Most accidents involving moving vehicles on public roads do not have to be reported under RIDDOR unless other work activities are involved, e.g. loading and unloading or nearby construction work.

Incidents must be reported as follows:

1. **Death*** or major injury† to an employee anywhere or to a self-employed person working on the employer's premises (including acts of violence) or if a member of the public is killed or taken to hospital:
   (a) notify the enforcing authority without delay, e.g. by telephone
   (b) send a completed form F2508 within ten days.
2. **'Over three day'**† injury to an employee anywhere or to a self-employed person working on the employer's premises (including acts of violence):
   (a) send a completed form F2508 within ten days.
3. **Diseases**† where a doctor notifies that an employee is suffering from a reportable work-related disease:
   (a) send a completed form F2508A immediately.
4. **Dangerous occurrence**† as defined within the regulations:
   (a) notify the enforcing authority immediately, e.g. by telephone
   (b) send a completed form F2508 within ten days.

* Deaths of employees as a result of an accident at work, occurring within one year of the accident, must be reported in writing, whether or not the accident has been reported previously.

† These terms are explained further in paragraphs B4.3 to B4.6.

The regulations are published in full, together with associated guidance, in an HSE booklet while a free leaflet, 'RIDDOR Explained', provides a summary of the regulations together with a blank copy of form F2508 (see Figure B4.1).

Deaths or injuries must be connected with work to be reportable under RIDDOR. An incident in which an employee or a member of the public suffered a heart attack would not be reportable, unless it was suspected to be due to a work-related cause, e.g. faulty electrical equipment. However, a slip, trip or fall in a workplace would be reportable, even if the exact cause could not be determined. In cases of doubt it is always wise to report the incident and let the enforcing authority make the final decision.

B4.3   Major injuries

Major injuries (which must be reported without delay) have been summarised by the HSE in its leaflet 'RIDDOR Explained' as:

- fracture other than to fingers, thumbs or toes
- amputation
- dislocation of the shoulder, hip, knee or spine
- loss of sight (temporary or permanent)
- chemical or hot metal burn to the eye or any penetrating injury to the eye
- injury resulting from an electric shock or electric burn leading to unconsciousness or requiring resuscitation or admittance to hospital for more than twenty-four hours
- any other injury: leading to hypothermia, heat-induced illness or unconsciousness; or requiring resuscitation; or requiring admittance to hospital for more than twenty-four hours
- unconsciousness caused by asphyxia or exposure to a harmful substance or biological agent

- acute illness requiring medical treatment, or loss of consciousness arising from absorption of any substance by inhalation, ingestion or through the skin
- acute illness requiring medical treatment where there is reason to believe that this resulted from exposure to a biological agent or its toxins or infected material.

Full details of the RIDDOR requirements are contained in Schedule 1 to the regulations with further guidance provided in their RIDDOR booklet.

B4.4 'Over three day' injuries

An 'over three day' injury is one that results in the person concerned being absent from work or unable to carry out the full range of their normal duties (e.g. put on 'light work') for more than three days, not counting the day of the injury itself. This includes days when they would not normally be expected to work (e.g. weekends, rest days in shift rosters or holidays).

The position is clear where a period of scheduled absence is bracketed by absence due to injury, i.e. an accident occurs on a Thursday, the person is absent on Friday and absent on Monday, following a scheduled weekend off. However, an accident occurring on a Wednesday, resulting in absence on Thursday and Friday but a return on Monday after a weekend off provides a more difficult problem. If the accident victim states that they could have worked over the weekend, if necessary, then the accident need not be reported. Some larger employers operating shift patterns where this problem occurs on a regular basis have reached local agreements with HSE inspectors on the approach to be taken. Once again, however, if in doubt it is best to report.

B4.5 Reportable diseases

The position regarding reportable diseases is even more complicated, although fortunately the vast majority of employers will never need to worry about it. First, the disease must be notified to the employer by a doctor – a mere claim by an employee that he or she has the disease is not sufficient. Second, although some diseases are relatively common, they must have occurred within a specified process or activity or be due to contact with a specified substance for them to be reportable.

The HSE leaflet 'RIDDOR Explained' summarises reportable diseases as including the following:

- certain poisonings
- some skin diseases such as occupational dermatitis, skin cancer, chrome ulcer, oil folliculitis/acne
- lung diseases including occupational asthma, farmer's lung, pneumoconiosis, asbestosis, mesothelioma
- infections such as leptospirosis, hepatitis, tuberculosis, anthrax, legionellosis and tetanus
- other conditions such as occupational cancer, certain musculoskeletal disorders, decompression illness and hand-arm vibration syndrome.

Whenever an employee reports any occupational disease it is prudent for the employer to investigate it (for the reasons described in paragraph B4.1). Once the nature of the disease and the circumstances of its causation are better

Figure B4.1 Form F2508

HSE
Health & Safety
Executive

Health and Safety at Work etc Act 1974 **?**
The Reporting of Injuries, Diseases and Dangerous Occurrences Regulations 1995

Click here for report guidance

Report of an injury or dangerous occurrence

Filling in this form
This form must be filled in by an employer or other responsible person.

Part A

About you

1 What is your full name?

2 What is your job title?

3 What is your telephone number?

About your organisation

4 What is the name of your organisation?

5 What is its address and postcode?

6 What type of work does the organisation do?

Part B

About the incident

1 On what date did the incident happen?

2 At what time did the incident happen?
(Please use the 24-hour clock eg 0600)

3 Did the incident happen at the above address?

Yes ☐ Go to question 4

No ☐ Where did the incident happen?

☐ elsewhere in your organisation – give the name, address and postcode

☐ at someone else's premises – give the name, address and postcode

☐ in a public place – give details of where it happened

If you do not know the postcode, what is the name of the local authority?

4 In which department, or where on the premises did the incident happen?

F2508 (05.00)

Part C

About the injured person

If you are reporting a dangerous occurrence, go to Part F. If more than one person was injured in the same incident, please attach the details asked for in Part C and Part D for each injured person.

1 What is their full name?

2 What is their home address and postcode?

3 What is their home phone number?

4 How old are they?

5 Are they male? ☐
female? ☐

6 What is their job title?

7 Was the injured person (tick only one box)

☐ one of your employees?

☐ on a training scheme? Give details:

☐ on work experience?

☐ employed by someone else? Give details of the employer:

☐ self-employed and at work?

☐ a member of the public?

Part D

About the injury

1 What was the injury? (eg fracture, laceration)

2 What part of the body was injured?

SAMPLE

Next Page

3 Was the injury (tick the one box that applies)

☐ a fatality?

☐ a major injury or condition? (see accompanying notes)

☐ an injury to an employee or self-employed person which prevented them doing their normal work for more than 3 days?

☐ an injury to a member of the public which meant they had to be taken from the scene of the accident to a hospital for treatment?

4 Did the injured person (tick all the boxes that apply)

☐ become unconscious?

☐ need resuscitation?

☐ remain in hospital for more than 24 hours?

☐ none of the above.

Part E

About the kind of accident

Please tick the one box that best describes what happened, then go to Part G.

☐ Contact with moving machinery or material being machined

☐ Hit by a moving, flying or falling object

☐ Hit by a moving vehicle

☐ Hit something fixed or stationary

☐ Injured while handling, lifting or carrying

☐ Slipped, tripped or fell on the same level

☐ Fell from a height

How high was the fall?

☐ _____ metres

☐ Trapped by something collapsing

☐ Drowned or asphyxiated

☐ Exposed to, or in contact with, a harmful substance

☐ Exposed to fire

☐ Exposed to an explosion

☐ Contact with electricity or an electrical discharge

☐ Injured by an animal

☐ Physically assaulted by a person

☐ Another kind of accident (describe it in Part G)

Part F

Dangerous occurrences

Enter the number of the dangerous occurrence you are reporting. (The numbers are given in the Regulations and in the notes which accompany this form)

☐ _____

Part G

Describing what happened

Give as much detail as you can. For instance

- the name of any substance involved
- the name and type of any machine involved
- the events that led to the incident
- the part played by any people.

If it was a personal injury, give details of what the person was doing. Describe any action that has since been taken to prevent a similar incident. Use a separate piece of paper if you need to.

Part H

Your signature

Signature

_____

Date

_____

If returning by post/fax, please ensure this form is signed, alternatively, if returning by E-Mail, please type your name in the signature box

Where to send the form

Incident Contact Centre, Caerphilly Business Centre, Caerphilly Business Park, Caerphilly, CF83 3GG, or email to riddor@natbrit.com or fax to 0845 300 99 24

Continue

For official use		
Client number	Location number	Event number
		☐ INV REP ☐ Y ☐ N

© HMSO

SAMPLE

known, reference may need to be made to Schedule 3 to RIDDOR and associated guidance to determine whether it is reportable.

B4.6 Dangerous occurrences

The full listing of dangerous occurrences that must be reported under RIDDOR is contained within Schedule 2 to the regulations and related guidance is provided in the HSE booklet. The summary of dangerous occurrences is reproduced in Figure B4.1.

Employers are recommended to look through this list and identify any dangerous occurrences that they anticipate may occur in their work activities, so that they are prepared to make a report should such an eventuality arise. Reference to the more detailed guidance mentioned above will be necessary in cases of doubt.

B4.7 Reporting responsibility and records

Responsibility for reporting events under RIDDOR is normally as follows:

- injury or disease of an employee at work – that person's employer
- dangerous occurrences – the person in control of the premises at the time (in connection with their carrying on any trade, business or undertaking).

Consequently responsibility for reporting dangerous occurrences involving contractors would be with:

- the 'host' employer for dangerous occurrences on their premises
- the 'principal contractor' for dangerous occurrences on construction projects.

There are a number of exceptions to these general principles. These are summarised in Figure B4.3 but full details are contained in the HSE RIDDOR booklet.

Reports can be made to the National Incident Control Centre in a number of different ways:

Phone: 0845 300 9923 (8.30am–5.00pm)
Fax: 0845 300 9924 (any time)
Internet: www.riddor.gov.uk (any time)
E-mail: riddor@natbrit.com
Post: Incident Contact Centre, Caerphilly Business Park, Caerphilly CF83 3GG

Reports can also still be made directly to the enforcing authorities:

- HSE – addresses and telephone numbers of area offices are in the telephone directory (or from HSE's Infoline, tel. 08701 545500)
- local authorities – addresses and telephone numbers of the environmental health department should also be in the telephone directory.

Records of all RIDDOR reports must be kept for at least three years. This can be done by keeping copies of completed report forms or by recording the details on a computer or in written form. The record must include the following:

- the date and method of reporting
- the date, time and place of the event
- personal details of persons suffering injury or disease
- a brief description of the circumstances.

Figure B4.2   Dangerous occurences reportable under RIDDOR
(Source: HSE leaflet: 'RIDDOR Explained')

1   Collapse, overturning or failure of load-bearing parts of lifts and lifting equipment.

2   Explosion, collapse or bursting of any closed vessel or associated pipework.

3   Failure of any freight container in any of its load-bearing parts.

4   Plant or equipment coming into contact with overhead power lines.

5   Electric short circuit or overload causing fire or explosion.

6   Any unintentional explosion, misfire, failure of demolition to cause the intended collapse, projection of material beyond a site boundary, injury caused by an explosion.

7   Accidental release of a biological agent likely to cause severe human illness.

8   Failure of industrial radiography or irradiation equipment to de-energise or return to its safe position after the intended exposure period.

9   Malfunction of breathing apparatus while in use or during testing immediately before use.

10 Failure or endangering of diving equipment, the trapping of a diver, an explosion near a diver, or an uncontrolled ascent.

11 Collapse or partial collapse of a scaffold over five metres high, or erected near water where there could be a risk of drowning after a fall.

12 Unintended collision of a train with any vehicle.

13 Dangerous occurrence at a well (other than a water well).

14 Dangerous occurrence at a pipeline.

15 Failure of any load-bearing fairground equipment, or derailment or unintended collision of cars or trains.

16 A road tanker carrying a dangerous substance overturns, suffers serious damage, catches fire or the substance is released.

17 A dangerous substance being conveyed by road is involved in a fire or released.

The following dangerous occurrences are reportable except in relation to offshore workplaces:

18 Unintended collapse of: any building or structure under construction, alteration or demolition where over five tonnes of materials falls; a wall or floor in a place of work; any false-work.

19 Explosion or fire causing suspension of normal work for over twenty-four hours.

20 Sudden, uncontrolled release in a building of:
    - 100 kg or more of a flammable liquid;
    - 10 kg or more of a flammable liquid above its boiling point; or
    - 10 kg or more of a flammable gas; or
    - 500 kg of these substances if the release is in the open air.

21 Accidental release of any substance that may damage health.

Note: additional categories of dangerous occurrences apply to mines, quarries, relevant transport systems (railways, etc.) and offshore workplaces.

(Three years is also the normal time limit for initiating civil claims – see paragraph A3.7.)

Records of *all* work-related accidents (whether RIDDOR reportable or not) must be kept in the HSE Accident Book (BI 510). This book replaces previous books published by the Department of Work and Pensions and the DSS. All businesses must be using the new book from 31 December 2003 in order to comply with the provisions of the Data Protection Act which require personal details to be detached from the main book and stored separately in order to maintain confidentiality. The new book can be ordered from HSE. A sample accident record form is reproduced opposite.

B4.8   Accident studies

In paragraph B4.1, reference was made to studies of the relationships between the numbers of near-miss and damage incidents, minor accidents and serious accidents. The first of these was carried out in 1950 by H.R. Heinrich in the USA, who concluded that for every major or 'lost-time' injury there were 29 minor injuries and 300 non-injury incidents. This relationship is often presented as 'Heinrich's triangle' (see Figure B4.5).

Figure B4.3	Exceptions to reporting responsibilities under RIDDOR (Source: HSE)
All reportable events	**Responsible person**
In mines	The mine manager
In quarries or closed mines or quarry tips	The owner
At offshore installations (except reportable diseases)	The owner – mobile installations / The operator – fixed installations
At diving operations (except reportable diseases)	The diving contractor
Injuries and disease	
Of an employee at work at offshore installations or diving operations	That person's employer
Of a self-employed person at work	
• in premises under someone else's control	The person in control of the premises
• in premises under their own control	The self-employed person (or someone acting on their behalf)
Death or injury requiring removing to a hospital for treatment (or major injury occurring at a hospital) of a person who is not at work, *arising out of or in connection with work*	The person in control of the premises where, or in connection with the work at which, the accident causing the injury occurred
Dangerous occurrences	The concession owner
At wells (see para. 13, schedule 2)	The pipeline owner
At a pipeline (see para. 14(a) to (f), schedule 2)	
Involving a dangerous substance being conveyed by road (see paras 16 and 17, schedule 2)	The operator of the vehicle

Figure B4.4   Accident record BB1510

Report Number

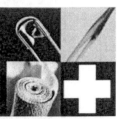

Report Number

ACCIDENT RECORD

1 About the person who had the accident

Name

Address

Postcode

Occupation

2 About you, the person filling in this record

▼ If you did not have the accident write your address and occupation.

Name

Address

Occupation

3 About the accident   Continue on the back of this form if you need to

▼ Say when it happened. Date   /   /   Time

▼ Say where it happened. State which room or place.

▼ Say how the accident happened. Give the cause if you can.

▼ If the person who had the accident suffered an injury, say what it was

▼ Please sign the record and date it

Signature   Date   /   /

4 For the employer only

▼ Complete this box if the accident is reportable under the Reporting of Injuries, Diseases and Dangerous Occurrences Regulations 1995 (RIDDOR).

How was it reported?

Date reported   /   /   Signature

© HMSO

A similar analysis was published by F.E. Bird in 1969, based on accidents at 297 organisations within the USA with the following ratios resulting:

- 1 serious or disabling injury
- 10 minor injuries
- 30 property damage accidents
- 600 incidents with no visible injury or damage.

A further study by Tye and Pearson in the UK came to similar conclusions and then, more recently, the HSE's Accident Prevention Advisory Unit (APAU) carried out a detailed analysis of 'The Costs of Accidents at Work' based upon detailed studies of five different organisations (in the oil, food, construction, health and transport sectors). This produced the following ratios (excluding the construction site):

- 1 major or 'over three day' lost-time injury
- 7 minor injuries
- 189 non-injury incidents.

The results of these studies are presented in Figure B4.5 together with an overall representation of the position, based upon all of the above studies.

B4.9   The costs of accidents

The primary purpose of the APAU study referred to above was to identify accurately the full costs of accidents (both injuries and damage incidents), occurring within the organisations studied. The study demonstrated that the vast majority of the costs incurred were ones against which the businesses would not be insured. Taking the insured costs as the insurance premiums paid over the study periods, the insured:uninsured costs ratios were 1:8 (transport), 1:11 (construction site and oil platform) and 1:36 (creamery). The accident costs represented 8.5 per cent of the tender price for the construction site and 37 per cent of the annualised profits of the transport company. The potential contents of both insured and uninsured costs are as follows:

Insured costs
- employers liability claims
- public liability claims
- major damage
- external legal costs
- major business interruption
- product liability.

Uninsured costs
- fines (from prosecutions)
- sick pay (to accident victims)
- minor loss of staff time
- minor damage to equipment or buildings
- loss of or damage to products
- clear-up costs
- minor business interruption
- interruption resulting from a prohibition notice
- loss of or dissatisfaction of customers (due to product defects or delay)
- provision of first aid and medical attention

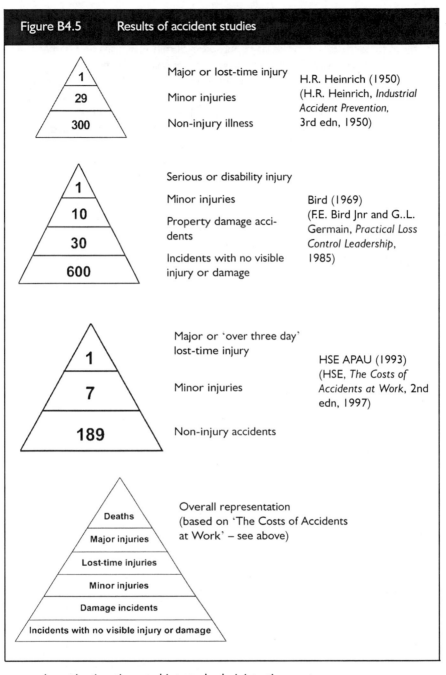

Figure B4.5   Results of accident studies

- investigation time and internal administration costs
- disruption due to loss of key staff
- hiring and training replacement staff
- time wasting by staff 'rubbernecking'
- hiring replacement equipment
- lowered employee morale
- damage to external image (to potential customers and the community).

While most businesses are aware of their insurance costs (in terms of the pre-miums paid), they are much less aware of the uninsured costs of accidents. Many of these result from the large numbers of minor injuries and damage incidents that occur – usually at a level at which the lost staff time is not recorded or the cost of damage to equipment or buildings is far below that required to trigger an insurance claim. (Such costs are often hidden in general maintenance and repair budgets.) Most businesses would find it quite difficult to quantify these uninsured costs, apart from fines and sick pay to accident victims.

Both the financial costs and the human costs of accidents can be reduced by effective accident and incident investigation systems which determine both their immediate and underlying causes and then implement appropriate reme-dial actions in order to eliminate these causes. Such systems should include the investigation of minor accidents, property damage incidents and near-miss acci-dents so that their causes can be rectified before they can cause a more serious (and more costly) accident. Accidents and incidents should be treated as an opportunity to monitor the effectiveness of health and safety management arrangements and learn from the findings.

B4.10 Components of investigation procedures

Effective investigation arrangements should include the following compo-nents. These may be within the context of a formal written procedure within a large organisation, while smaller employers should follow the same princi-ples. (Figure B3.3 on page 59 demonstrates how the management cycle can be applied to accident and incident investigation arrangements.)

Events to be investigated

The procedure should specify the types of events that should be investigated although in the case of 'near misses' some discretion must be given to those responsible for initiating and carrying out investigations. These may be defined as follows:

- all injuries requiring first aid treatment (to employees and other relevant persons, e.g. contractors, visitors, clients, volunteers)
- damage incidents (probably limited to those causing damage estimated as greater than a specified monetary value)
- other incidents with significant injury or damage potential.

While some employers might be reluctant to investigate relatively minor acci-dents, they should reflect that simple accidents can often result in straightfor-ward investigations and simple solutions, while others might reveal causes that could have potentially calamitous results.

The purpose of investigation

- To discover what happened (referring to RIDDOR and insurance needs).
- To determine why it happened (the contributory causes – see paragraph B4.11 below).
- To identify and implement appropriate remedial actions.

Responsibility for conducting investigations

Initial investigations should normally be carried out by supervisors or team lead-ers – those responsible for supervising the people, equipment, materials or environment involved. They are also most likely to be present at the time of the incident in order to carry out a prompt investigation (see below).

Larger organisations sometimes have additional arrangements for investigating more serious incidents in greater detail. These investigations are likely to involve relevant senior managers, health and safety specialists, engineers and others able to make specialist contributions. Arrangements should also be in place to involve employee representatives in the investigation process. This may be during the initial investigation (they have a right to carry out their own investigations – see paragraph B6.2), as part of serious incident investigation panels or during subsequent investigation reviews (see below).

Investigation time scale

Investigations should be initiated as soon as possible after the incident in order to gather information before memories fade or evidence is lost or disturbed. A minimum time scale (e.g. twenty-four hours) should be stipulated for the submission of the investigation report. In some cases this may need to be an interim report, pending the receipt of further information, for example where a seriously injured victim cannot be interviewed or a technical report is awaited.

Investigation report form

Procedures should be linked with a standard report form (on paper or in electronic format) which should be readily available to those responsible for investigation. Report forms should contain at least the following:

- details of who or what was involved
- date and time of the event
- the extent of injury or damage
- a description of what happened
- the investigator's conclusions as to the contributory causes (see paragraph B4.11)
- remedial actions already taken and recommendations as to further actions
- the name and signature of the investigator.

A sample report form is provided in Figure B4.6 on pages 88–89.

Routeing of reports

Procedures should specify who should receive copies of investigation reports (whether hard copies or via e-mail). This will vary depending on the management structure of the organisation involved and responsibilities for submitting external reports (see below).

Some of these persons are likely to be involved in reviewing the quality of the investigation and implementing remedial action, while others may just receive reports for information purposes. Report routeing might include the following:

- department managers
- senior management
- health and safety specialists
- human resources specialists
- employee representatives.

Figure B4.6   Accident and incident investigation report

Department _____   Section _____

Person involved _____ Surname _____ First names _____

Status of person   Employee ☐   Visitor ☐   Other (state) ☐

(Tick as appropiate)   Temp. ☐   Contractor ☐   e.g. neighbour, member of public

Date	Time	Location

Details of incident – what happened

Description of incident – what happened

A sketch or photo may be helpful. Continue on a separate sheet if necessary

Contributary causes – why you think it happened. Include both immediate and underlying causes.

What action has been taken (or will be taken) to prevent similar incidents

Name of employee representative (if involved in investigation)

Investigation carried out by	Name	Signature	Date

FOR COMPLETION BY THE DEPARTMENT MANAGER

Further action required/Progress of remedial action/Other comments

Signature

Copies to be retained by: INVESTIGATOR and DEPARTMENT MANAGER

Original to be sent to: HEALTH AND SAFETY MANAGER within 24 hours

FOR COMPLETION BY HEALTH AND SAFETY MANAGER

Actions confirmed complete on

Signature

External reporting

Arrangements should be established for submitting reports externally:

- making reports under RIDDOR (see earlier in the chapter)
- informing insurance companies
- communicating to others in the organisation, e.g. group health and safety specialists.

Implementation of remedial actions

Procedures should establish who will allocate the responsibilities for implementing remedial actions. This is likely to be the relevant manager, probably in conjunction with the health and safety specialist and others, e.g. training staff. Progress of remedial actions also needs to be monitored – this may be done by the manager, the health and safety manager or by regular reviews at health and safety committee meetings.

B4.11 Contributory causes of accidents

(For brevity, the term 'accidents' is also used here to describe non-injury incidents.)

Most accidents have multiple causes, with several factors coming together to produce an unwanted consequence (often resulting in injury or damage). F.E. Bird divided causes into four types:

PEOPLE   EQUIPMENT   MATERIALS   ENVIRONMENT

In Appendix 5 of the HSE booklet 'Successful Health and Safety Management' immediate causes were divided into:

PREMISES   PLANT & SUBSTANCES   PROCEDURES   PEOPLE

Both of these categorisations provide useful *aides-mémoire* for persons investigating accidents. However, once the IMMEDIATE CAUSES of the accident have been identified, the BASIC or UNDERLYING CAUSES must be determined so that appropriate remedial action can be taken. The same immediate cause may be a symptom of a variety of basic causes: for example, if an employee follows an incorrect working method, this may be because:

- the employee has never known the correct method (lack of training, instruction, supervision)
- the employee has forgotten the correct method (lack of supervision, refresher training)
- the employee has deliberately chosen the wrong method (lack of supervision, poor attitude).

Examples of IMMEDIATE CAUSES and BASIC/UNDERLYING CAUSES are provided in Figure B4.7. (Some basic causes relate to more than one immediate cause, but they are generally placed close to the most appropriate ones.)

B4.12 Training in investigation technique

Staff expected to be involved in investigating accidents and incidents should be provided with suitable training. This should include familiarisation with the investigation procedure (including the reasons for carrying out investigations), the investigation report form and important elements of investigation and reporting technique, including the following:

	Immediate causes	**Basic/underlying causes**
People-related	Poor technique	Lack of training
	Incorrect method	Correct method not specified
	Lack of care or attention	Inadequate supervision
	Fatigue	
	Deliberate act	Poor attitude
	Distraction	Inadequate control of external influences, e.g. visitors
	Physically unsuitable	Inadequate selection/health surveillance
	Affected by drink or drugs	Controls/PPE not specified/ available
	Control measures/ PPE not used	
Relating to equipment, materials, workplace, environment	Defective equipment or workplace • access • lighting • noise • heating	Lack of maintenance Lack of inspection
	Unsuitable equipment	Poor design Correct equipment not specified Correct equipment not available Lack of planning Insufficient resources
	Unsuitable materials	Inadequate assessments, e.g. COSHH
	Weather	Inadequate supervision

Figure B4.7 — Examples of immediate causes and basic underlying causes

Note: See Appendix 5 of HSE booklet 'Successful Health and Safety Management' for more comprehensive illustration and examples.

1. Observation of the scene and surrounding area:
   (a) checking equipment, materials and the working environment.
2. Interviewing of witnesses and others:
   (a) the injured person and any other persons involved (preferably searately)
   (b) noting any key questions down beforehand
   (c) asking open-ended questions in a friendly manner
   (d) involving technical specialists (engineers, chemists, health and safety staff), if appropriate.

3. Identifying all the contributory causes:
   (a) not jumping to conclusions too early (keeping an open mind)
   (b) considering basic, underlying causes as well as immediate causes (not just blaming the victim).
4. Report writing:
   (a) accuracy with dimensions, equipment identification, etc.
   (b) qualifying witness evidence they are not sure about (using phrases such as 'he states that' or 'he reported that')
   (c) avoiding terminology that may not be understood by outsiders
   (d) using sketches or photographs where appropriate
   (e) completing all of the report form, using 'none' or 'not applicable' where relevant.
5. Making and implementing relevant recommendations:
   (a) keeping them realistic and not going 'over the top'
   (b) implementing those actions that are within their control as soon as possible
   (c) discussing recommendations with relevant colleagues
   (d) pressing those responsible to take action.
6. Training could include relevant elements such as:
   (a) preparation of lists of key questions for hypothetical accidents
   (b) role play on interview technique
   (c) examples of good and bad investigation reports
   (d) use of appropriate videos (several are available commercially).

B4.13 Encouraging reporting

Staff can be encouraged to report accidents using some of the following techniques.

Communicating the procedure

If employees are expected to report damage and near-miss incidents as well as accidents, it is essential that the procedure is communicated effectively to them and that the importance of investigation as accident prevention is stressed.

First aid prompts

The investigation and reporting of all accidents requiring first aid treatment can be assisted by arranging for first aiders or nursing staff treating injuries to instigate investigation. This may be done informally, but the use of a simple form, which the accident victim must take back to their supervisor (possibly with a copy going to their manager or the health and safety specialist), has been known to produce excellent results. Some organisations have an investigation report form in which the first section is completed by the person providing first aid treatment, and then the form is passed on to the victim's supervisor for investigation to take place.

User-friendly report forms

The form illustrated in Figure B4.6 on pages 88–9 is quite deliberately a simple affair intended to encourage reporting. A long, complicated form will be a significant deterrent to a hard-pressed supervisor who is in two minds whether to report a near-miss incident. Any statistical analysis of reports (part of body injured, causation, etc.) should be done centrally rather than expecting the investigator to tick a variety of boxes. This approach will also aid consistency of analysis.

Positive use of statistics

Organisations that make comparisons or set targets based on the number of minor accidents occurring are likely to find that every opportunity is taken to avoid reporting such accidents, if at all possible. Presenting statistics differently (e.g. by comparing between departments the ratios of lost-time accidents to non-injury incidents reported) can produce useful results. Departments and sections reporting diligently on damage and near-miss incidents in effect gain credit, while those having serious accidents, the reporting of which cannot be avoided, show up poorly.

Removing some of the barriers

Employees will be reluctant to report incidents if they are likely to be subject to disciplinary action or open to ridicule as a result. While employers can never guarantee that disciplinary action will not be taken, they should take care to avoid having a blame culture. A more tolerant approach should be taken with those who are honest enough to submit a report than with those who are subsequently found out. The teamwork aspect of reporting incidents should be emphasised. By reporting incidents where 'no one was hurt' or 'it was only a graze', employees might be preventing accidents to their workmates in the future.

Providing incentives

The author experienced considerable success from the introduction of a monthly award for the best non-injury incident report. The award was based on the potential seriousness of the incident, the quality of the investigation, the clarity of the report and the extent of remedial action already taken. Not only did the employee making the initial report and the supervisor carrying out the investigation receive a small prize, but their win and the lessons learned from the investigation were publicised on noticeboards and in newsletters. Both the quantity and quality of incident investigations improved as a result, with the importance of reporting becoming more deeply engrained into all levels of the workforce.

Producing a system that works

Supervisors and employees will be more likely to take the trouble to carry out an investigation and submit a report if something worthwhile happens as a result. Ensuring that recommended actions are implemented (or reasons provided if they cannot be) will build up the credibility of the accident and incident reporting arrangements. People will use something that they know produces positive results.

Source materials

1.   HSE (1999) 'A Guide to the Reporting on Injuries, Diseases and Dangerous Occurrences Regulations 1995', L 73.
2.   HSE (1999) 'RIDDOR Explained' (free leaflet), HSE 31.
3.   F.E. Bird Jnr and G.L. Germain (1985) Practical Loss Control Leadership, Institute Publishing, USA.
4.   HSE (2000) 'Successful Health and Safety Management', HSG 65.

B5

Monitoring Health and Safety

B5.1 Introduction

Chapter B3 stressed the importance of the management cycle in the effective management of health and safety. Monitoring of the effectiveness of health and safety arrangements is a key stage in that management cycle. The previous chapter covered accident and incident investigation, which is in itself a reactive form of monitoring. This chapter deals with health and safety inspections, health and safety audits and other proactive types of monitoring activity. It also provides guidance on the establishment of performance standards against which the results of monitoring activity may be compared.

B5.2 Health and safety inspections

The term 'health and safety inspection' (shortened to 'inspection' in this part of the chapter) is usually used to describe a visual check of equipment, premises, working conditions, working practices, etc. Inspections are usually of relatively short duration and will take place periodically throughout the year. Some types of inspection will be carried out more frequently, e.g. pre-use checks of equipment – this kind of inspection is covered in paragraph B5.5. Union-appointed safety representatives have the right to carry out inspections as a result of the Safety Representatives and Safety Committees Regulations 1977 (SRSC). These regulations are covered in greater detail in Chapter B6.

B5.3 Inspection arrangements

In larger workplaces it will often be beneficial to describe the arrangements for inspections in a formal procedure, while in smaller organisations they might be referred to in the 'arrangements' section of the health and safety policy or just established informally. In all cases the arrangements will need to deal with the following aspects.

Where is to be inspected?

Smaller workplaces can often be easily inspected in their entirety but larger locations will often need to be divided up into separate inspection units. The aim should be to complete each inspection within a reasonable period of time (between thirty and ninety minutes is recommended). A medium-sized manufacturing plant might be divided up into the following units:

Production area A	Warehouse and plant rooms
Production area B	Offices
Maintenance workshop	Gatehouse, roadways and other external areas

How frequently should inspections be carried out?
The SRSC Regulations allow inspections by safety representatives at least every three months and this is an appropriate minimum to adopt. However, higher-risk areas or activities may need to be inspected more frequently – monthly or even weekly if the risks justify. In the manufacturing plant referred to above, the production areas and the maintenance workshop might be inspected monthly, with the other units visited quarterly.

Who should carry out the inspections?
Those responsible for carrying out inspections should be clearly identified, with arrangements for substitutes if someone is unavailable due to holiday, sickness or other commitments. Managers and supervisors should be actively involved in the inspection programme and it is usually beneficial to make inspections jointly with safety representatives rather than leaving them to carry out separate inspections. The author's preference is to pair a manager or supervisor with an employee representative (who may or may not be a union safety representative). Large groups are generally not very effective in carrying out inspections.

It may be possible to rotate the participants so that they are not always inspecting the same areas with the same partners. Preferably the participants should have received some training in hazard spotting.

What should be looked for?
Those carrying out inspections should look out for both unsafe conditions and unsafe acts.

- *Unsafe conditions* might consist of unguarded machines, defective equipment, blocked access routes, poor housekeeping and other tripping hazards, defective premises.
- *Unsafe acts* might involve failure to wear PPE, other incorrect work methods, over-exertion, speeding, misuse of equipment, horseplay or other reckless behaviour.

(It should be noted that most unsafe conditions are preceded by unsafe acts – someone removed the guard or blocked the access route.)

Inspections can be aided considerably by the use of checklists specific to the area being inspected. Such checklists can be incorporated into the report form. The contents of checklists are referred to in paragraph B5.4 below and some sample report forms are provided in Figures B5.1 and B5.2 together with a completed form in Figure B5.3 (on page 99).

Reporting and implementing remedial actions
Completion of the type of report form shown in Figures B5.1 to B5.3 not only provides confirmatory evidence that the inspection has been carried out but assists in identifying necessary remedial actions. Someone (usually a member of management) must be made responsible for identifying the individuals who will be responsible for implementing the remedial actions and ensuring that they have been completed. The type of form depicted in Figure B5.4 on page 100 (with a completed version in Figure B5.5) will aid this process.

Monitoring the programme
While the inspection programme is a monitoring activity, it in turn needs to be monitored. For example:

- Are inspections being carried out as scheduled?
- Are the intended participants taking part?
- What is the quality of the inspection reports?
- Are remedial actions being implemented?

Monitoring might be carried out by a suitable member of management, through the health and safety committee or via a health and safety audit (see paragraphs B5.6 to B5.10).

B5.4   Inspection checklists

The types of unsafe acts and unsafe conditions likely to be found will vary according to the area being inspected and the inspection checklist should reflect this. A good starting point in preparing a checklist is to look at the risk assessment for the area in question, and then identify those aspects where conditions or the application of control measures might vary or deteriorate. Common items for inclusion are the following:

Fire exit routes	Access routes
Fire protection equipment	Housekeeping standards
Guarding standards	Compliance with PPE requirements
Condition of electrical equipment	Manual handling practices

Checklist topics can be incorporated into an inspection report form (as in Figures B5.1 to B5.3) or provided on a separate card or sheet.

FIND THIS ON CD	Figure B5.1	Sample inspection report form (production activities)

ARMAGEDDON ASSOCIATES
HEALTH AND SAFETY INSPECTION – PRODUCTION AREA

Date _____ Inspection by_____

Risk topic	Specific items	Comments
A) FIRE	Evacuation routes available SC doors not wedged open Extinguishers accessible 'No smoking' rules observed Flammable materials controlled	
B) ACCESS & HOUSEKEEPING	Roadways, walkways Access to racking, cupboards, etc. Storage of materials Stacking of pallets Housekeeping standards	
C) ELECTRICAL EQUIPMENT	Fixed electrical installation Access to isolators, switches, etc. Condition of cables, plugs, sockets	
D) EQUIPMENT	Guards – position/adjustment (production machines & conveyors) Forklift trucks (condition/driving standards) Use of cranes Lifting equipment	
E) MANUAL HANDLING	Handling aids available/used (pallet trucks, trolleys) Working practices	
F) PPE COMPLIANCE	Safety footwear (all areas) Hearing protection areas/machines Use of gloves (cleaning stations)	
G) OTHER ITEMS		

DETAIL RECOMMENDED ACTIONS ON SEPARATE SHEET

Figure B5.2	Sample inspection report form (office areas)

OLDVILLE DISTRICT COUNCIL
OFFICE HEALTH AND SAFETY INSPECTION

Office location_____ Date _____ Inspection by_____

Risk topic	Specific items	Comments
A) FIRE	Evacuation routes available SC doors not wedged open Extinguishers accessible 'No smoking' rules effective	
B) ACCIDENTS etc	First aid equipment 'First aider' notices	
C) VISITORS & CONTRACTORS	Receptionist's alarm button Signing in/out records Contractors' activities	
D) ACCESS ROUTES (INTERNAL & EXTERNAL)	Corridors, walkways clear Trailing cables controlled Access to shelves, cupboards, etc. External footpaths, roadways	
E) OFFICE & KITCHEN EQUIPMENT	Condition of equipment (cables, plugs, sockets, etc.)	
F) SERVICES	Fixed electrical supply Access to isolators, switches, etc	
G) HAZARDOUS SUBSTANCES	Control of cleaning materials Gloves available/used	
H) MANUAL HANDLING	Storage of records, stationery Access to racking	
I) DISPLAY SCREEN EQUIPMENT	DSE workstations (seats, lighting, blinds, ergonomic layout)	
J) OTHER ITEMS		

DETAIL RECOMMENDED ACTIONS ON SEPARATE SHEET

Figure B5.3	Sample inspection report form (production activities): completed example	

ARMAGEDDON ASSOCIATES
HEALTH AND SAFETY INSPECTION – PRODUCTION AREA

Risk topic	Specific items	Comments
A) FIRE	Evacuation routes available	*All well maintained.*
	SC doors not wedged open	*Several propped open.*
	Extinguishers accessible	*All clear.*
	'No smoking' rules observed	*Evidence of smoking in lunchroom.*
	Flammable materials controlled	*Good control.*
B) ACCESS & HOUSEKEEPING	Roadways, walkways	*Surfaces clean & in good condition.*
	Access to racking, cupboards, etc.	*Access to tool cupboard blocked.*
	Storage of materials	*Generally good.*
	Stacking of pallets	*Stack near N door looks unstable.*
	Housekeeping standards	*Swarf build-up near some machines.*
C) ELECTRICAL EQUIPMENT	Fixed electrical installation	*Satisfactory condition.*
	Access to isolators, switches, etc.	*All kept clear.*
	Condition of cables, plugs, sockets	*Cable to bench drill damaged*
D) EQUIPMENT	Guards – position/adjustment	*Loose guard at head of B conveyor*
	(production machines & conveyors)	*Milling m/c guard incorrectly adjusted.*
	Forklift trucks (condition/driving standards)	*Satisfactory (pre-use checks up to date).*
	Use of cranes	*Satisfactory.*
	Lifting equipment	*Shackle found with wrong colour code.*
E) MANUAL HANDLING	Handling aids available/used (pallet trucks, trolleys)	*Blue pallet truck broken.*
	Working practices	*Storekeeper carrying excessively large box.*
F) PPE COMPLIANCE	Safety footwear (all areas)	*Worn by all staff.*
	Hearing protection areas/machines	*Not worn by press operator.*
	Use of gloves (cleaning stations)	*Suitable gloves in use.*
G) OTHER ITEMS	*Building cladding*	*Evidence of water ingress on east side (three places).*
	Lighting	*Several strip lights require replacement.*

DETAIL RECOMMENDED ACTIONS ON SEPARATE SHEET

Figure B5.4   Form for reporting and checking the implementation of remedial actions

HEALTH AND SAFETY INSPECTION – RECOMMENDED ACTIONS

Date of Inspection _____
Area Inspected _____

*Priority Code: A – Urgent; B – Important; C – Routine

No.	Recommended Action	Priority Code*	For Action By	Completed/Progress Details	Date	Initials

Figure B5.5   Form for reporting and checking the implementation of remedial actions: completed example

HEALTH AND SAFETY INSPECTION – RECOMMENDED ACTIONS

*Priority Code: A – Urgent; B – Important; C – Routine

Date of Inspection ___02 August 03___

Area Inspected ___Production___

No.	Recommended Action	Priority Code*	For Action By	Completed/Progress Details	Date	Initials
1.	Ensure self-closing fire doors are kept closed.	C	Supervisor	All doors closed today	9/8/03	KB
2.	Remind staff about 'no smoking' rules.	C	Manager (notice)	Notice posted	5/8/03	KB
			Supervisor – monitoring	No cigarette ends seen	9/8/03	KB
3.	Clear access to tool cupboard.	C	Storekeeper	Access now clear	9/8/03	KB
4.	Re-stack palletised components near north door. A		Supervisor	Re-stacked immediately	6/8/03	AS
5.	Remind staff re regular removal of swarf.	C	Supervisor	Done at team meeting	9/8/03	RT
6.	Repair cable to bench drill.	A	Mtnce Supervisor	Now repaired	9/8/03	KB
7.	Secure the guard underneath the head of B conveyor.	B	Mtnce Supervisor	Now secured	9/8/03	KB
8	Improve adjustment mechanism on milling machine guard.	B	Mtnce Supervisor	Modified version in place	9/8/03	KB
9.	Remove shackle with wrong colour code from use.	A		Done during inspection	2/8/03	AS
10.	Remind all staff only to use correctly. colour-coded lifting equipment.	C	Supervisor	Done at team meeting	9/8/03	RT
11.	Repair the blue pallet truck.	C	Mtnce Supervisor	Completed	9/8/03	KB
12.	Remind storekeeper to use trucks, trolleys. where appropriate.	C	Supervisor/Storekeeper	New truck ordered	7/8/03	RT
13.	Remind press operator of hearing protection rules.	C	Manager/Operator	Given informal warning	6/8/03	KB
14.	Check and repair cladding on east side of building.	C	Mtnce Supervisor	Repairs scheduled for Sept. 1	9/8/03	KB
15.	Replace all defective striplights.	C	Mtnce Supervisor	All now replaced	19/8/03	KB

B5.5   More frequent inspections

Some types of equipment or control measures need to be inspected frequently in order to ensure that they are up to the required standard. In some cases these inspections are required by specific requirements contained within regulations.

PUWER Regulation 6

This regulation requires the inspection of work equipment where there could be significant risk resulting from the following:

- incorrect installation or re-installation
- deterioration
- exceptional circumstances.

HSE guidance (see page 114) suggests that regular inspection or testing may be necessary to check interlocks, protection devices, controls, overload warnings, limit switches, etc. The need should be identified during the risk assessment process. Within industry, pre-use checks have been particularly common in respect of mobile work equipment such as:

- forklift trucks (a sample checklist is provided in Figure B5.6 opposite)
- cranes
- other specialist vehicles.

In many cases such checks will also take into account the maintenance and efficiency of the equipment as well as health and safety aspects.

PUWER Regulation 33

The regulation requires the inspection and testing of guards and protection devices on power presses before the fourth hour of any working period. This must be carried out by an appointed person who must be competent and have been adequately trained. (See page 114 for details of further information.)

Figure B5.6   Sample lift truck inspection report form

ARMAGEDDON ASSOCIATES
LIFT TRUCK DAILY INSPECTION REPORT

Each truck must be checked over before it is used each day.
Any defects affecting safe operation must be reported immediately to a Supervisor.
At the end of each week this completed form must be passed on to the Production Administrator.

LIFT TRUCK Type Plant No. WEEK COMMENCING	MONDAY ✓ or ✗	TUESDAY ✓ or ✗	WEDNESDAY ✓ or ✗	THURSDAY ✓ or ✗	FRIDAY ✓ or ✗	✓ or ✗
TYRES condition						
BRAKES working efficiently						
HORN clearly audible						
LIGHTS/REVERSE LIGHT/WARNING LIGHTS working efficiently						
REVERSE ALARM SOUNDER						
FUEL, WATER, ENGINE & HYDRAULIC OIL levels checked						
NO APPARENT LEAKS						
CONTROLS/FORKS/ATTACHMENTS working correctly						
SIGNATURE						
OTHER COMMENTS/ACTION TAKEN						

LOLER Regulation 9(2)
Where the safety of lifting equipment depends on installation conditions, it must be thoroughly examined after assembly and before being put into service at a new site or new location. This would apply to the erection of temporary lifting rigs rather than the use of mobile lifting equipment. (See page 114 for further reading.)

Construction (Health, Safety and Welfare) Regulations
Regulations 29 and 30 together with Schedules 7 and 8 contain requirements for regular inspections of the following to be carried out by a competent person:

- working platforms from where a person could fall two metres or more (see also Reg. 6)
- personal suspension equipment (see Reg. 6)
- supported or battered back excavations (see Reg. 12)
- cofferdams or caissons (see Reg. 13).

Figures B5.7 and B5.8 contain a sample inspection report form and notes on the carrying out of inspections. (See page 114 for further reading.)

It should be noted that the requirements referred to in this section of the chapter are in addition to the requirements for less frequent statutory examinations contained in a variety of regulations – these are summarised in Chapter B9.

B5.6 Health and safety audits

A health and safety audit should be a much more detailed and comprehensive process than an inspection. Ideally an audit should evaluate how well the organisation is implementing each stage of the management cycle and should include a review of the following:

- policies and procedures (and their detailed contents)
- arrangements for implementing those policies and procedures
- mechanisms in place to monitor their effectiveness
- arrangements for reviewing and acting on monitoring findings.

The audit process is likely to involve examining policies and procedures, checking records and other documents, interviewing a variety of employees (persons at different levels in the organisation as well as those with specialist health and safety roles) and conducting some sample inspections within the workplace – to discover what is actually being achieved in practice. During the audit the management systems should be compared with relevant standards such as legal requirements, HSE guidance, industry norms or generally accepted good practice.

There are a variety of health and safety audit systems available on the market, most of which are now computer-based. However, these will not always be appropriate for the needs of the organisation concerned and there may be benefits in developing a customised audit system. The sample series of audit questions contained in Figure B5.9 on page 109 has been adapted from an audit system developed by the author for use by a group of companies in the metal processing industry.

Some commercial audit systems leave too much to the judgement of the person carrying out the audit – consequently an audit conducted by an in-house specialist may present a much rosier picture than one carried out by an external

Figure B5.7   Sample inspection report form

INSPECTION REPORT

Report of results of every inspection made in pursuance of Regulation 29(1)

1.   Name and address of person for whom inspection is carried out.

2.   Site address   3.   Date and time of inspection

4.   Location and description of workplace (including any plant, equipment or materials inspected).

5.   Matters which give rise to health and safety risks.

6.   Can work be carried out safely?   Y/N

7.   If not, name of person informed.

8.   Details of any other action taken as a result of matters identified in 5 above.

9.   Details of any other action considered necessary.

10.   Name and position of person making the report.

11.   Date report handed over.

Figure B5.8   Notes on the carrying out of inspections

Construction (Health, Safety and Welfare) Regulations 1996

INSPECTION REPORT: NOTES

Place of work requiring inspection	Timing and frequency of inspections					
	Before being used for the first time	After substantial addition, dismantling or alteration	After any event likely to have affected its strength or stability	At regular intervals not exceeding 7 days	Before work at the start of every shift	After accidental fall of rock, earth or any material
Any working platform or part thereof or any personal suspension equipment	✓	✓	✓			
Excavations which are supported in pursuit of paragraphs (1), (2) or(3) of Regulation 12			✓		✓	✓
Cofferdams and caissons			✓		✓	

NOTES
General

1. The inspection report should be completed before the end of the relevant working period.
2. The person who prepares the report should, within 24 hours, provide either the report or a copy to the person on whose behalf the inspection was carried out.
3. The report should be kept on site until work is complete. It should then be retained for three months at an office of the person for whom the inspection was carried out.

Working platforms only

1. An inspection is only required where a person is liable to fall more than 2 metres from a place of work.

2. Any employer or other person who controls the activities of persons using a scaffold shall ensure that it is stable and of sound construction and that the relevant safe guards are in place before his employees or persons under his control first use the scaffold.

3. No report is required following the inspection of any mobile tower scaffold which remains in the same place for less than 7 days.

4. Where an inspection of a working platform or part thereof or any personal suspension equipment is carried out
   i. before it is taken into use for the first time; or
   ii. after any substantial addition, dismantling or other alteration;
   not more than one report is required in any 24-hour period.

Excavations only

1. The duties to inspect and prepare a report apply only to any excavation which needs to be supported to prevent any person being trapped or buried by an accidental collapse, fall or dislodgement of materials from its sides, roof or area adjacent to it. Although an excavation must be inspected at the start of every shift, only one report of such inspections is required every 7 days. Reports must be completed for all inspections carried out during this period for other purposes, e.g. after accidental fall of material.

Checklist of typical scaffolding faults

Footings	Standards	Ledgers	Bracing	Putlogs and transoms	Couplings	Bridles	Ties	Boarding	Guard-rails and toe-boards	Ladders
Soft and uneven	Not plumb	Not level	Some missing	Wrongly spaced	Wrong fittings	Wrong spacing	Some missing	Bad boards	Wrong height	Damaged
No base plates	Jointed at same height	Joints in same bay	Loose	Loose	Loose	Wrong couplings	Loose	Trap boards	Loose	Insufficient length
No sole plates	Wrong spacing	Loose	Wrong fittings	Wrongly supported	Damaged	No check couplers	Not enough	Incomplete	Some Missing	Not tied
Undermined	Damaged	Damaged			No check couplers			Insufficient supports		

auditor (a consultant or a specialist from a sister company). Audit questions containing judgemental words such as 'adequate', 'suitable', 'satisfactory' are likely to create problems – preferably the questions should be devised in such a way as to demonstrate what is adequate, suitable or satisfactory. (See paragraph B5.8.)

B5.7   Audit scope
The scope of the audit will vary to a certain extent, depending on the activities of the organisation being audited. However, there are many topics that are common to the effective management of health and safety, whatever the size or business of the organisation concerned. The system developed for the group of metal processing companies (referred to in paragraph B5.6 above) contained the following sections:

- policy, responsibilities, awareness, health and safety advice
- risk assessments
- inspections and audits
- accident and incident investigation
- operating procedures
- training
- occupational health
- emergency arrangements
- management of contractors
- engineering and purchasing
- communication with employees
- workplace inspection (to verify other audit findings).

Clearly some of these topics could be subdivided and other specific topics included where appropriate.

B5.8   Audit questions
The questions contained in the sample in Figure B5.9 (relating to accident and incident investigation) require 'yes' or 'no' answers together with a limited amount of detail to support the answer given. For most questions the decision as to 'yes' or 'no' is straightforward – the arrangements either meet the stated requirement or they do not. However, inevitably a degree of judgement on the part of the auditor has to be introduced, e.g. on the quality of investigation reports and suitability of remedial actions.

Figure B5.9   Sample health and safety audit questions

D   ACCIDENT & INCIDENT INVESTIGATION

	YES	NO
1. RIDDOR & GROUP REQUIREMENTS		

1.1 Has responsibility been clearly allocated for reporting
 under RIDDOR and to the Group Safety Manager?   ❑   ❑
Details...

1.2 Do all such persons have sufficient awareness of RIDDOR
 and Company requirements?   ❑   ❑

1.3 Is an adequate supply of F. 2508s (or an electronic
 reporting system) readily available for use?   ❑   ❑
Evidence...

1.4 Is a file of completed F. 2508s available for inspection?   ❑   ❑
Location..

1.5 Does this appear to demonstrate compliance with
 RIDDOR requirements?   ❑   ❑
Comments ....

2 INVESTIGATION PROCEDURE

2.1 Is there a documented investigation procedure?   ❑   ❑
Details...
(If NO, go to section 3)

2.2 Does the procedure contain requirements for the
 following to be investigated
- all injuries resulting in absence from work?   ❑   ❑
- specified categories of minor injuries?   ❑   ❑
- specified types of damage incidents?   ❑   ❑
- non-injury incidents with significant injury/damage potential?   ❑   ❑

2.3 Does the procedure specify a time scale for investigation
 and reporting?   ❑   ❑
Details ....

2.4 Does the procedure specify who is responsible for initial
 investigation and reporting?   ❑   ❑
Details...

2.5 Does the procedure specify to whom reports should
 be submitted?   ❑   ❑
Details...

	YES	NO

2.6 Does the procedure specify who should
- allocate responsibility for remedial action?   ❏   ❏
- monitor remedial action?   ❏   ❏

Details..

3 INVESTIGATION REPORT FORM

3.1 Is an in-house investigation report form in use?   ❏   ❏

Details..

(If NO, go to section 4)

3.2 Does the form contain adequate provision for
- details of the injury/damage/incident?   ❏   ❏
- description of the circumstances (what happened)?   ❏   ❏
- analysis of the contributory causes (why it happened)?   ❏   ❏
- recommendations for preventative action?   ❏   ❏
- signature of the person(s) investigating?   ❏   ❏
- comments and signature of the person reviewing the report?   ❏   ❏

4 TRAINING IN INVESTIGATION

4.1 Have a majority of staff involved in the investigation process
(including review) received formal training (of at least 2 hours'
duration) in investigation techniques?   ❏   ❏

Approx. % trained...

Details of training...

5 COMPLETED REPORT FORMS

5.1 Are completed investigation report forms kept on file for
at least 3 years?   ❏   ❏

Location(s)..

5.2 Do completed report forms demonstrate
- investigation within a suitable time scale?   ❏   ❏

(that specified in the procedure if one exists - see 2.1)
- satisfactory quality of investigation?   ❏   ❏
- suitable recommendations for remedial action?   ❏   ❏
- systematic monitoring/review of remedial action?   ❏   ❏

Comments ....

5.3 Do completed report forms demonstrate that adequate
numbers of investigations are being carried out of
- minor injuries?   ❏   ❏
- damage incidents?   ❏   ❏
- non-injury incidents?   ❏   ❏

('Adequate' should relate to the requirements of the location's
procedure and the numbers of such incidents expected)

Comments..

This type of audit can be made quantitative by assigning points values for each 'yes' answer that can be properly verified. The number of points allocated would reflect what the audit designer felt was important. Alternatively the designer could assign a maximum number of points that could be awarded for the question or topic and leave it to the judgement of the auditor as to how many points should be given, depending on the degree of compliance. Some audit systems use a full compliance/partial compliance/non-compliance rating for each question.

B5.9   Quantitative versus qualitative audits

Arguments can be advanced both for and against quantitative audit systems, as opposed to purely qualitative ones.

Arguments for quantitative audit systems are as follows:

- progress (or deterioration) can be measured
- target scores can be set
- comparisons between locations can be made
- poor scores clearly identify weaknesses
- scores often capture the attention of senior management.

These, however, are the arguments against quantitative audit systems:

- some managers focus on easy point scoring (rather than making more relevant improvements)
- can encourage short-term quick-fix solutions
- some systems are over-dependent on auditors' judgements
- time can be wasted quibbling about point scores
- some senior managers may put undue pressure on auditors.

B5.10   Audit reports

Whether quantitative or qualitative systems are used, the auditor should concentrate in the audit report on those areas for improvement of greatest importance to the audited organisation. A report that identifies every weakness at great length is not only demoralising but will not give much assistance in the subsequent development of an action plan. The report should also give due credit to those parts of the health and safety management system that are working well.

A well-designed series of audit questions should make it quite clear what should be contained in a good policy or procedure or what types of arrangements are likely to prove successful. However, it can often be of considerable assistance if the auditor can also provide examples of good practice to illustrate the audit report and aid subsequent action.

B5.11   Other forms of monitoring

There are many other ways in which standards of health and safety management can be monitored. Some of these concentrate on the management arrangements themselves while others monitor what is actually being achieved within the workplace. The former category includes the following:

- *Procedure compliance audits* Audits of health and safety-related procedures, carried out in a manner similar to the more comprehensive audits described above. These may be conducted as part of overall ISO 9000 procedure control systems.

- *Periodic document examinations* Checking of key health and safety documents (e.g. incident investigation reports, health and safety inspection reports, permit-to-work forms) to ensure that the relevant systems are being complied with.
- *Annual plan reviews* Periodic reviews of the progress of annual plans relating to health and safety. Such reviews should involve senior management – preferably in the form of a meeting.

Workplace-based monitoring techniques include the following:

- *Managers' health and safety tours* Tours of workplaces by senior managers in which they concentrate on health and safety matters, rather than production or product quality issues, can be very beneficial. These need not be detailed inspections but should demonstrate management commitment to and interest in health and safety. Examples of good and bad practice should be discussed with relevant employees at the time.
- *Safety surveys* Surveys can concentrate on particular pieces of work equipment (e.g. forklift trucks, portable powered tools) and their condition and methods of use.
- *Task observations* These involve a detailed observation of the work methods and equipment involved in carrying out a particular task. Consideration is given to whether the specified method is being followed, whether there are any better alternatives and whether there are any further risks that have not been considered previously.
- *Compliance surveys* Compliance with easily observable aspects of employee behaviour is checked. Most commonly this would relate to PPE requirements but compliance with traffic routes, speed limits and use of warning devices (e.g. horns) could be checked.
- *Occupational hygiene surveys* Surveys of levels of dust, fume, noise, etc., in the workplace are dealt with more fully in Chapters C2, C3 and D4.

Another alternative method of monitoring is the following::

- *Survey of employees' opinions* Opinions can be sought on a variety of topics – PPE suitability, the effectiveness of specific arrangements, reasons for non-compliance with requirements. The HSE has produced a general tool to aid in the monitoring of employees' views on the health and safety climate (see page 114 for details).

B5.12 Performance standards

There is an old adage of 'what gets measured, gets done' and this is as applicable to the management of health and safety as to other aspects of any organisation's activities. Plans, procedures and other arrangements should specify:

- What is to be achieved?
- Who is responsible?
- When should work be completed?
- What is the intended outcome?

What is to be achieved?

The work required may be a one-off or occasional activity or it may be a routine procedure or a regular occurrence. For example:

- a review of the implications of a new set of regulations – one-off
- an audit of health and safety arrangements – occasional
- workplace health and safety inspections – routine
- investigation of accidents and incidents – regular.

Who is responsible?

This may involve a designated individual or a designated position. The persons responsible should be competent to carry out the work required. Using the examples above, these might be:

- the health and safety manager
- the health and safety manager and another designated manager
- the department supervisor together with the relevant employee representative (or their designated deputies)
- the supervisor of the area where the accident or incident occurs.

When should the work be completed?

The work should be specified for completion within a definite time frame or to a regular defined frequency. Again using the examples above, this might be:

- by the end of the month
- for completion over a three-month period
- monthly (preferably during the first half of the month)
- within twenty-four hours of the accident or incident occurring.

What is the intended outcome?

The intended outcome of the work required should also be specified. Utilising the previous examples this might be:

- submission of a draft action plan for consideration by a management meeting
- submission of a detailed report containing recommended actions for review by senior management and subsequent discussion at a management meeting (on a defined date)
- submission of a completed report form to the department manager
- submission of a completed investigation report to the department manager (and/or health and safety manager if the procedure requires).

Continuing the process

In each of the examples used here, further actions will be required to continue the management process and these too will need to be subject to performance standards. These actions could involve:

- finalisation and implementation of an action plan in respect of the new regulations
- agreement and implementation of an action plan based on the results of the audit

- implementation of remedial actions identified during the inspection
- implementation of remedial actions identified as a result of the investigation.

In assigning responsibilities to persons in this way, account must be taken of their capabilities to deliver. This includes not only their own levels of competence but their needs in respect of time and resources, and for cooperation from others likely to be involved in the activity. People will not be motivated to achieve a programme of work that they perceive from the outset to be unrealistic within the time scale specified or with the resources provided.

Source materials

1. HSE (1998) 'Safe Use of Work Equipment: Provision and Use of Work Equipment Regulations 1998', ACOP and guidance, L 22.
2. HSE (1998) 'Safe Use of Power Presses: Provision and Use of Work Equipment Regulations 1998 as Applied to Power Presses', ACOP and guidance, L 112.
3. HSE (1998) 'Safe Use of Lifting Equipment: Lifting Operations and Lifting Equipment Regulations 1998', ACOP and guidance, L 113.
4. HSE (2001) 'Health and Safety in Construction', HSG 150.
5. HSE (1997) 'Health and Safety Climate Survey Tool', ISBN 071761462X.

Communication and Consultation

B6.1 Introduction

In Chapter B3, reference was made to the four Cs of effective organisation of health and safety: control, cooperation, communication and competence. This chapter deals with communication and another 'C'– consultation – something that many would argue was an essential prerequisite of achieving cooperation. It summarises the requirements of the Safety Representatives and Safety Committees Regulations 1977 and the Health and Safety (Consultation with Employees) Regulations 1996, and provides some detailed guidance on the role and organisation of health and safety committees.

The chapter also refers to other means of communicating with employees such as team meetings, 'toolbox talks' and newsletters, and also the importance of maintaining less formal methods of communication and consultation by 'walking the job' regularly.

B6.2 Safety Representatives and Safety Committees Regulations

These regulations, which were introduced during a period of Labour government, when the trade union movement was extremely strong, were restricted in their application to those workplaces where independent trade unions were formally recognised by employers for collective bargaining purposes. Strictly speaking, the appointment of safety representatives from among the employees was a matter for the trade union itself, although in practice most representatives have actually been appointed by the group of workers they represent. (Regulation 8 specifies some situations in the entertainment industry where safety representatives need not be employees.)

Appointment of safety representatives (Reg. 3)
The regulations and their accompanying ACOP leave aspects such as the numbers of safety representatives and the detailed arrangements for their appointment to be established by local negotiation.

Time off and training (Reg. 4(2))
Employers must permit safety representatives to take such time off with pay during their working hours as is necessary for performing their functions (see below) or undergoing reasonable training. What constitutes reasonable training is defined within the ACOP accompanying the regulations. This provides for initial basic training after appointment, including reference to the following:

- the role of safety representatives and safety committees
- trade unions' policies and practices
- general legal requirements

- specific requirements relating to those they represent
- workplace hazards and their elimination or control
- health and safety policies and their implementation.

Many trade unions have established their own courses for providing such basic training and there are also many TUC-approved courses available at local colleges. Some employers also provide training for safety representatives, often concentrating on risks associated with their own work activities together with their own health and safety procedures and arrangements. The ACOP also provides for additional training because of changing circumstances or legislation or because a safety representative has special responsibilities. While the ACOP provides a framework within which issues relating to time off and training can be determined, much of the detail has to be left to local negotiation and agreement.

Functions (Reg. 4(1))

Safety representatives have functions they are intended to perform, not duties that they must carry out. These include the following:

- investigating potential hazards, dangerous occurrences and employee complaints
- examining the causes of accidents
- making relevant representations to the employer
- carrying out inspections (see below)
- consulting HSE or local authority inspectors, etc.
- attending Safety Committee meetings
- being consulted by the employer on health and safety matters (planned changes, proposed training, etc.).

Inspections

Safety representatives are entitled to carry out inspections:

- of the workplace at least every three months (Reg. 5) – see paragraph B5.2 onwards
- in the event of substantial changes being made or new information received (Reg. 5)
- following notifiable accidents, diseases or dangerous occurrences (Reg. 6) see paragraph B4.10
- of documents the employer is required to keep under relevant legislation, e.g. statutory examination records, risk assessments, records of other specific assessments (COSHH, Manual Handling, etc.).

Provision of information (Reg. 7(2))

Employers must make available to safety representatives information within their knowledge, necessary for the representatives to fulfil their functions. However, there are exceptions relating to information:

- whose disclosure would be against national security interests
- whose disclosure would contravene a legal prohibition
- relating to an individual (unless consent had been given), e.g. personal health information

- where disclosure might have a serious adverse effect on the employer's (or someone else's business), e.g. product information that is a trade secret
- obtained in relation to legal proceedings.

Safety committees (Reg. 9)
Where at least two safety representatives make a request in writing, the employer must establish a safety committee – this subject is dealt with in some detail in paragraphs B6.4 to B6.7 below.

B6.3 Health and Safety (Consultation with Employees) Regulations 1996
The 1980s and 1990s saw the number of workplaces with 'recognised' trade unions greatly reduced and as a result only a minority of workers had a legal right to have safety representatives. In practice many employers offered similar rights to their employees on a voluntary basis. However, Article 11 of the EC Framework Directive required consultation and participation of workers, stating that 'Employers shall consult workers and/or their representatives and allow them to take part in discussions on all questions relating to safety and health at work.' Consequently the 1996 Regulations were introduced, ironically under a Conservative government, giving rights to employees who were not already represented by union-appointed representatives. The employer is given a choice, under Regulation 4, of either consulting all employees directly (either informally or through staff meetings) or consulting with employee representatives.

The main requirements are summarised below:

1. Duty of the employer to consult (Reg. 3). The employer must consult in good time on matters including:
   (a) changes substantially affecting health and safety
   (b) his sources of competent health and safety advice
   (c) proposed health and safety training.
2. Information (Reg. 5). Employers must provide information to employees or their representatives:
   (a) sufficient for full and effective consultation
   (b) contained in RIDDOR records.

Employee representatives are entitled (under Reg. 6) to make relevant representations to the employer on health and safety matters and to consult with HSE inspectors. Regulation 7 requires the employer to ensure that they are provided with reasonable training and to allow them time off with pay during their working hours to perform their functions and undergo training. The HSE's 'A Guide to the Health and Safety (Consultation with Employees) Regulations 1996' refers to the need to provide employee representatives with reasonable facilities and assistance. It also gives further guidance on the election of representatives and on their training. However, employee representatives have no legal right to carry out inspections of their workplace or to investigate accidents, nor can they request the establishment of a safety committee.

Most health and safety professionals would be glad to see the consolidation of the two sets of regulations, thus ending the artificial distinction between workplaces where there are recognised trade unions and those where there are not, but there are no signs of this happening in the immediate future.

B6.4 Health and safety committees and their role

Many employers have established health and safety committees, whether or not they have been formally requested to do so by union-appointed safety representatives. An effective committee can assist greatly in the management of health and safety at all stages of the management cycle (see Chapter B3) but particularly in the 'monitor' and 'review' steps in the cycle. A joint committee, in which representatives of the workforce play a significant part, can ensure:

- wider consultation on what risks are associated with work activities
- different perspectives on those risks (and on the likely effectiveness of proposed control measures)
- greater awareness of what is actually happening in the workplace (the real world!)
- the availability of more knowledge and experience
- increased ownership of and commitment to agreed solutions
- a means of routinely monitoring key health and safety systems and arrangements (investigations, inspections, training activities).

Figure B6.1 overleaf includes a short section on the purpose of the Health and Safety Committee within a development agency.

B6.5 Membership of the committee

Membership of the health and safety committee should be determined through local agreement. The aim should be to achieve a suitable balance of participants, without the numbers becoming so large as to be unmanageable or to restrict worthwhile discussion. The membership is likely to include:

- a representative of senior management (essential to demonstrate a serious commitment to health and safety)
- supervisory representation
- a health and safety specialist
- any other relevant specialists, e.g. maintenance engineer
- employee representatives (union safety representatives and/or other elected representatives).

Figure B6.1   Sample constitution of a health and safety committee

**MIDSHIRES DEVELOPMENT AGENCY
HEALTH AND SAFETY COMMITTEE**

1. THE COMMITTEE PURPOSE
The purpose of the Committee is to provide a consultative forum within the Agency on health and safety matters. While the Agency's management accepts its duties to ensure that health and safety is managed effectively, it can only do this with the active cooperation of its workforce and it recognises the many advantages in consulting with representative members of the workforce to achieve this cooperation.

2. COMPOSITION OF THE COMMITTEE
    2.1   The Committee will comprise
        - A minimum of one and a maximum of three representatives appointed by employees working at or from each Agency office. (Details of how many representatives each office is entitled to will be recorded in the minutes.)

- A minimum of two and a maximum of four representatives appointed by the Agency's Chief Executive.

2.2   Where an office has more than one representative at least one should be an employee spending a significant amount of time working outside the office.

2.3   Appointment of employee representatives will preferably be by consensus of staff working within the office, although, if necessary, a ballot will be held.

2.4   Neither employee representatives nor those appointing them need necessarily be members of any trade union.

2.5   Members unable to attend meetings or participate in other activities (e.g. inspections) may appoint deputies to replace them.

2.6   Other persons may attend Committee meetings by invitation.

2.7   Changes to the composition of the Committee or the number of representatives for an office may be agreed following a consensus of members.

3 . FUNCTIONS

The principal functions of the Committee will be to:

3.1   Discuss health and safety issues and concerns and consider how best to deal with such matters.

3.2   Provide a means of exchanging and communicating information on health and safety.

3.3   Review proposed health and safety-related procedures, instructions or documents.

3.4   Monitor the effectiveness of existing health and safety precautions.

3.5   Implement and monitor the effectiveness of the health and safety inspection programme.

3.6   Review the findings of any audits of health and safety arrangements.

3.7   Identify health and safety-related training needs of Agency staff.

3.8   Review significant accidents and incidents involving the Agency's staff or premises under its control.

3.9   Discuss any contact with enforcing agencies (HSE and local authorities).

3.10   Examine the implications of changes in legislation, published standards, etc.

3.11   Consider and promote health and safety initiatives.

4. MEETING ARRANGEMENTS

4.1   Four meetings should take place each year, normally at approximately three-month intervals.

4.2   Meeting venues may rotate between the Agency's offices.

4.3   Dates and venues of meetings for each year should normally be agreed at the final meeting of the previous year.

4.4   Changes to dates and venues will normally be notified to members at least two weeks in advance.

4.5   Meetings will normally be chaired by a senior manager nominated by the Chief Executive.

4.6   Minutes will be taken and other secretarial duties performed by a person appointed by the Agency's management (who need not be a formal member of the Committee).

4.7   A structured agenda (based on the Committee's functions) will normally be circulated to members at least one week prior to each meeting.

4.8   Minutes will normally be circulated to members within three weeks of each meeting. They will then be formally approved (or amended) at the next meeting.

4.9   Minutes of meetings should be circulated widely within the Agency. They should be distributed to appropriate members of management and arrangements made for their display on notice boards.

4.10   Given the purpose of the Committee and its functions, a provision for the taking of formal votes has been deliberately excluded from these arrangements.

Provision should be made for other persons to attend committee meetings, where appropriate, and for the use of substitutes for members unable to attend. (Figure B6.1 above provides an illustration of the membership of a committee within a development agency, whose employees are primarily carrying out administration work. An industrial plant or a workplace with shift activities would need different arrangements in order to achieve adequate representation.)

B6.6   Functions of the committee

The detailed functions of the committee will vary according to the work activities taking place within the organisation in question. Typical agenda items should include the following:

- review of accident and incident reports and accident statistics
- organisation and review of the health and safety inspection programme
- checking progress of remedial actions resulting from inspections and investigations
- review of the results of health and safety audits
- reviewing results of occupational hygiene surveys
- consideration of planned health and safety initiatives
- review of existing procedures and arrangements
- development of any new procedures, etc.
- monitoring the effectiveness of health and safety training
- review of any contact with inspectors from enforcing agencies
- provision of feedback from users on the suitability of PPE.

(Section 3 of Figure B6.1 on pages 118/19 refers to the 'Functions' of the development agency committee, several of which would be regular agenda items. Figure B6.2 below contains a typical agenda for a committee within a more industrial undertaking.)

B6.7   Meeting arrangements

The organisation of health and safety committee meetings should follow similar principles to those governing other types of meetings.

- Meetings held at a pre-determined frequency: this may be every one, two or three months depending on the size and activities of the organisation in question. There is a danger in holding meetings too frequently as well as not often enough.
- Dates and times should be planned well in advance: cancellations and postponements should be avoided as far as possible. They can be very frustrating for those who have given safety a high priority in their diary and can call into question the true level of management commitment.
- Agendas should be circulated to members well in advance: this can assist members in preparing for meetings and may prompt some into dealing with action points from the previous meeting that they had overlooked.
- A chairperson must be designated: where a senior manager chairs the meeting this can raise the profile of the committee, aid management credibility and assist in the implementation of agreed actions. However, some committees have been chaired very effectively by other members of management or by employee representatives and others have successfully rotated the role of chairperson.

Figure B6.2   Sample health and safety committee agenda

ARMAGEDDON ASSOCIATES
HEALTH AND SAFETY COMMITTEE

The next meeting will be held in Training Room B on Tuesday 6 August at 2pm.

AGENDA

1. Apologies for absence.
2. Acceptance of minutes of meeting held on 2 July.
3. Matters arising from the minutes, including
   - Item 3.4 Progress report on repairs to access road   AW
   - Item 9   Revised design of permit to work form   NT
   - Item 7.2 Report on trials of new welding gloves.   MG/GP
4. Review of July's accident and incident reports.   NT
5. Progress of outstanding remedial actions from previous investigations.   KB
6. Review of July's safety inspection reports.   NT
7. Progress of outstanding remedial actions from previous inspections.   KB
8. Review of the results of the fabrication bay noise survey (17 and 18 July).   NT
9. Review of the collective results of the lung function survey of
   welding bay employees.   DC
10. Likely changes to the Control of Asbestos Regulations and their implications. NT
11. HSE inspector's visit to the warehouse extension project.   NT
12. Proposed employee safety quiz competition.   GS

Circulation

All committee members – note responses/reports required as above.
D. Carter (Medical Centre) – item 9.
Managing Director – for information.

- Recording and circulation of minutes: this role could be carried out by a non-committee member, e.g. a secretary, although the draft minutes may need review by the chairperson prior to circulation. The aim should be to circulate minutes soon after the meeting, clearly identifying what actions are required and who is responsible for them. The committee is not a secret society and its minutes should be circulated as widely as possible – e-mail can assist greatly in this respect as well as use of more traditional methods such as noticeboards.

Since the committee is primarily a means for communication and consultation, there should not be the need for votes to be taken, other than on minor matters such as the wording of procedures, the layout of forms or the winner of the safety poster design competition. Responsibility for managing safety effectively still remains with the employer. (Section 4 of Figure B6.1 above sets out the meeting arrangements agreed for the development agency committee.)

B6.8   Communicating with all employees

Communicating with employee representatives through the medium of the health and safety committee is dealt with in paragraphs B6.4 to B6.7. Good employee representatives can provide a very effective means of communicating with their work colleagues but additional means of passing health and safety information to all employees (and means for them to communicate back) need to be available. Such communication methods include team meetings, toolbox talks and newsletters.

Team meetings

Many workplaces hold regular short meetings of employees within specific work teams, normally chaired by their supervisor or team leader. While these will often also be concerned with operational issues, quality, conditions of employment, etc., health and safety should be a regular part of such meetings. This gives management the opportunity to raise matters such as new procedures, failure to adhere to existing procedures or standards, inspection report findings, recent accidents or incidents, etc. However, employees should also have the opportunity to raise their concerns, e.g. risks not covered by existing control measures, defective equipment not rectified, unavailability of equipment, PPE, materials.

Toolbox talks

A toolbox talk is a term used to describe a short presentation made to a group of employees on a relevant health and safety topic. It would often be delivered by the supervisor or team leader but could also involve someone else such as a health and safety officer, occupational hygienist, occupational health nurse or other appropriate specialist. It could be based upon any of the following:

- a video and subsequent discussion of its contents
- a set of overhead projector slides
- briefing notes on a specific topic
- detailed consideration of an accident or incident report
- detailed review of a health and safety or operational procedure (in whole or part).

Health and safety specialists are likely to have an important part in the preparation of support materials for toolbox talks – busy supervisors and team leaders are unlikely to have time to prepare their own material on a regular basis. It is essential that the topic chosen is of relevance to the work group involved: an irrelevant topic will be a waste of everyone's time.

Newsletters

Some organisations have achieved success with newsletters dealing solely with health and safety matters but the author's preference has always been to include health and safety sections within more general newsletters. Topics to include in newsletters are those that are likely to be of interest to the workforce as a whole, rather than to small work groups. These might include:

- accident statistics and incident reporting statistics (including comparisons between departments – see paragraph B4.13)
- major health and safety initiatives
- changes to existing procedures or arrangements
- reviews of occupational health issues
- lessons learned from accident or incident investigations
- major successes within small work groups, e.g. in cracking long-standing problems or dealing with unexpected situations
- human interest stories, e.g. unusual incidents.

B6.9   Allowing employees to communicate upwards

The means of communication described so far are not always effective in allowing employees to communicate upwards. While some employees will communicate their concerns directly to supervisors, team leaders and health and safety staff or through employee representatives, this is not always the case. Some employees may wish to retain a degree of anonymity while others may not realise that there is a problem without some external prompting.

Hazard reports

Many employers achieve considerable success with anonymous hazard reporting systems where employees can 'blow the whistle' on practices or situations that they regard as hazardous without the fear of recriminations from those who have been inconvenienced or embarrassed by their reports. Such systems can utilise standard reporting forms or telephone hotlines. It is important that such reports are channelled to a person in some authority who also has the ability to differentiate between the important, the trivial and, possibly, the malicious.

Employee surveys

Reference was made in paragraph B5.11 to monitoring health and safety performance through the use of surveys. Such surveys can also be used to identify the extent of risks and problems within the organisation. For example, in relation to violence at work, employees could be asked:

While at work, have you ever been

- physically attacked?
- threatened?
- abused?

Suggestion schemes

Suggestion schemes provide employees with the opportunity to suggest the solutions to health and safety problems. In some cases the problem may already be widely recognised but in others the extent of the problem may not have been realised until the suggestion is submitted. Such schemes are generally more successful where useful suggestions are recognised in a tangible manner, usually by some financial incentive, in addition to favourable publicity within newsletters, etc.

B6.10   Other forms of communication

Walking the job

Reference was made in paragraph B5.11 to health and safety tours by senior managers which provide an opportunity for two-way communication with employees in their working environment: an opportunity for managers to demonstrate their interest and commitment and for employees to raise their concerns or voice their ideas. 'Walking the job' is also important for health and safety specialists who can sometimes gain as much from informal observations and chats with employees as from taking part in formal inspections.

Training

Training, whether taking place within a classroom environment or on the job, provides a further opportunity for communication with employees. Whatever the intended learning outcome of the training, those delivering it (supervisors, health and safety specialists, training officers, etc.) have the chance to 'win hearts and minds' in respect of health and safety, as well as learning more themselves about what actually goes on in the workplace and what concerns those working there.

Source materials

1. HSE (1996) 'Safety Representatives and Safety Committees', ACOP and guidance on the regulations, L 87.
2. HSE (1996) 'A Guide to the Health and Safety (Consultation with Employees) Regulations 1996', L 95.

Selection and Management of Contractors

B7.1   Introduction

The outsourcing of services has increased tremendously in recent years so that contractors are now used not just for major construction and engineering projects, but also to carry out much routine maintenance work, to clean offices and to maintain the security of work premises outside normal working hours.

The legislation governing the relationship between the host employer or client and the contractor has also developed considerably with the general requirements of HASAWA being followed first by the Management Regulations and then by the Construction (Design and Management) Regulations (CDM), both of which contain much more detailed requirements.

This chapter examines this legislation in some detail and then provides practical guidance on how to comply with these legal obligations in respect of the initial selection of contractors, their briefing in relation to the work they are to carry out, and their management while the work is in progress.

It must also be recognised that what might be suitable arrangements for selecting and managing contractors to carry out a major construction project will not necessarily be appropriate for dealing with the contractors who come to mend a broken window or to maintain the vending machines.

B7.2   HASAWA 1974

Section 3(1) of HASAWA states:

> It shall be the duty of every employer to conduct his undertaking in such a way as to ensure, so far as is reasonably practicable, that persons not in his employment who may be affected thereby are not exposed to risks to their health or safety.

The section also has the effect of placing obligations on employers to protect one group of contractors' employees (or visitors and other non-employees) from the activities of another group of contractors' employees. Examples of prosecutions under this section, where host employers were prosecuted as a result of accidents to employees of contractors or agencies, are provided on page 126.

The contracting employers are not without their own obligations. Section 3(1) places duties on them in respect of others who may be affected by their activities, including employees of the host organisation and any other contractors involved. This is one of the key features of HASAWA: everyone has interlocking duties to each other. It must also be remembered that section 2 of HASAWA places duties on employers in respect of their employees and this includes protecting them from the activities of contractors working around them, particularly those that their own employer has engaged.

This was an important factor in *R v. Swan Hunter Shipbuilders Ltd*, the first of the two case histories featured below. The other case (*R v. Associated Octel Co. Ltd*) is important in respect of the extent of the duties placed on employers by section 3(1) and interpretation of the term 'undertaking'.

 Control of contractors: case histories

R v. Swan Hunter Shipbuilders Ltd

In 1976 a firm of subcontractors (Telemeter Installations) was installing specialist equipment on board HMS Glasgow, a warship being built in one of Swan Hunter's shipyards on Tyneside. Their work involved the use of oxy-acetylene equipment, which they left overnight in a poorly ventilated area on board ship. (This was in contravention of regulations in force at the time as well as established good practice.) Overnight there was an oxygen leak, which created an oxygen-enriched atmosphere throughout a large part of the ship. Early the next morning a source of ignition initiated an intense fire in which eight workers were killed.

Telemeter were prosecuted by the HSE and pleaded guilty to the charges against them, being fined £15,000. The HSE also brought proceedings against Swan Hunter under section 2 of HASAWA (duties to protect their employees) and section 3(1) (duties to protect the subcontractor's employees). Evidence was given of Swan Hunter having clearly established instructions for its own staff in respect of precautions to be taken with oxy-acetylene equipment, but that this had not been provided to Telemeter or its employees.

Swan Hunter was convicted at the crown court and its appeal was subsequently rejected by the Court of Appeal, thus establishing the principle that the host employer or main contractor has a duty to coordinate activities and ensure that a safe system of work is followed, even to the extent of making sure that subcontractors' employees are properly informed and trained.

R v. Associated Octel Co. Ltd

In 1990 a specialist contractor was brought in to carry out cleaning and repair work inside a tank at Associated Octel's premises at Ellesmere Port. The work was carried out under a permit to work issued by Octel and involved an employee of the contractor entering the tank to grind internal surfaces and clean them down with a highly flammable solvent. A lamp provided by Octel was unsuitable and it broke, causing an explosion, which badly injured the contractor's employee.

Associated Octel was prosecuted under section 3(1) of HASAWA with evidence being advanced about the unsuitability of the lamp, the unsuitability of the solvent container (an old bucket), inadequate ventilation and incomplete precautions specified on the permit to work.

Octel argued that it was not the conduct of its undertaking that caused the explosion and that it had no right to control the work of a specialist contractor. However, it lost the case first at the crown court and then at the Court of Appeal. The latter ruled that Octel's undertaking was running a chemical plant and that maintenance of its premises and equipment formed part of the conduct of that undertaking. Even though the work was done by an independent contractor, Octel had to stipulate what was required to avoid risks.

B7.3 Management of Health and Safety at Work Regulations 1999

The duties of employers in relation to contractors were expanded by the Management Regulations. Regulation 3 placed a duty on employers (and the self-employed) to carry out risk assessments in respect of risks to their own employees and others not in their employment. Consequently employers must assess the risks to their own employees from contractors, as well as the risks to the contractors from the conduct of their own undertaking (their employees, activities, premises, equipment, etc.).

Regulation 5 requires the provision of 'appropriate' arrangements to ensure that the precautions and procedures identified as being necessary through the risk assessment process are properly implemented. Regulations 11 and 12 introduce more specific requirements while Regulation 13 (dealing with capabilities and training) and Regulation 15 (temporary workers) are also of relevance. The detailed requirements of the Management Regulations are contained in Chapter B2 but the key features of Regulations 11 and 12 are set out below.

Cooperation and coordination: Reg. 11
Where two or more employers share a workplace (whether on a temporary or permanent basis) they must:

- cooperate: with each other as far as necessary to comply with the law
- coordinate: take all reasonable steps to coordinate the measures they take to comply with the law
- inform: take all reasonable steps to inform the other employers of risks arising from their undertaking to employees of the other employers.

Persons working in host employers' or self-employed persons' undertakings: Reg. 12
The host must ensure that the employer of employees from an outside undertaking working in his undertaking is provided with comprehensible information on the following:

- risks to those employees arising out of or in connection with the host's undertaking
- measures taken by the host to comply with the law in relation to those employees.

The host must ensure that any person working in his undertaking (who is not one of his own employees) is provided with appropriate instructions and comprehensible information regarding risks to that person's health and safety arising out of the host's undertaking. (These requirements include information relating to emergency evacuation procedures.)

Practical means of ensuring cooperation and coordination and providing information are considered later in the chapter.

B7.4 Construction (Design and Management) Regulations 1994

These regulations apply to all large and many medium-size construction projects, as well as many activities traditionally regarded as repair or maintenance. They place duties on five key parties:

1. The client.
2. Designers.
3. The planning supervisor.
4. The principal contractor.
5. Other contractors and the self-employed.

These roles can be filled by individuals or by organisations and it is possible for an individual or an organisation to carry out more than one role, depending on how the project is organised.

Two key concepts were introduced by the Construction (Design and Management) Regulations 1994 (CDM). They are the health and safety plan and the health and safety file.

The purpose of the health and safety plan is to ensure that all of the risks expected to be associated with the project are identified in advance so that the precautions necessary to control those risks can be implemented. The outline of the plan must be prepared initially by the planning supervisor on behalf of the client and then the detail completed and implemented by the principal contractor.

The health and safety file is intended to provide an information resource for those responsible for carrying out construction and maintenance work on the structure in the future. It should contain details such as:

- 'as built' drawings and plans of the structure and its services
- specifications of materials used in the structure
- details of construction methods and design criteria
- maintenance procedures and requirements
- manuals from specialist contractors and suppliers relating to plant and equipment installed as part of the structure.

The application of the regulations, some definitions of key terms used, the duties of the five key parties and the contents of the health and safety plan are all explained in greater detail below. However, those requiring detailed information on the regulations should consult more specialised publications, particularly those available from the HSE (see page 143).

Consideration of the more general aspects of selecting and managing contractors can be found at paragraph B7.13 below.

B7.5 Definitions

Two important terms – construction work and structure – are defined in Regulation 2(1).

Construction work

This is defined as 'the carrying out of any building, civil engineering or engineering construction work'. The definition goes on to give many specific examples of what this includes, for example:

- construction, alteration, conversion, fitting out, commissioning, renovation, repair, upkeep, redecoration or other maintenance, de-commissioning, demolition or dismantling of a structure
- the preparation for an intended structure
- the assembly of pre-fabricated elements to form a structure (or its subsequent disassembly)

- the removal of a structure or part of a structure and related product or waste
- work relating to the installation, commissioning, maintenance, repair or removal of mechanical, electrical, gas, compressed air, hydraulic, telecommunications, computer or similar services, normally fixed within or to a structure.

Structure

This term is defined as:

- any building, steel or reinforced concrete structure (not being a building), railway line or siding, tramway line, dock, harbour, inland navigation, tunnel, shaft, bridge, viaduct, waterworks, reservoir, pipe or pipeline (whatever it contains or is intended to contain), aqueduct, sewer, sewage works, gasholder, road, airfield, sea defence works, river works, drainage works, earthworks, lagoon, dam, wall, caisson, mast, tower, pylon, underground tank, earth retaining structure, and any other structure similar to the foregoing, or
- any formwork, falsework, scaffold or other structure designed or used to provide support or means of access during construction work, or
- any fixed plant in respect of work which is installation, commissioning, de-commissioning or dismantling and where work involves a risk of a person falling more than two metres.

B7.6   Application of the CDM Regulations

Application of the regulations is covered by Regulation 3. All of the regulations apply to:

- construction work that is 'notifiable', i.e. the 'construction phase' from start to finish will be longer than 30 days, or will involve more than 500 person days of construction work (the planning supervisor must ensure that written notice is given to the HSE, as required by Regulation 7 – form 10 (rev.) contains the specified information)
- any (non-notifiable) work involving five or more persons carrying out construction work at any one time
- all work in relation to demolition or dismantling of a structure.

The requirements on designers (Regulation 13) apply to all construction work (whatever the size of the project).

Exemptions

Construction work in respect of which the local authority is the enforcing authority is exempt from all the CDM Regulations. (This would be minor, non-notifiable, internal work carried out by persons who normally work on the premises in occupied offices, shops, etc. – see appendix 3 to the ACOP.) Construction work carried out for domestic clients in their own residences is exempt from most regulations.

Do the regulations apply?

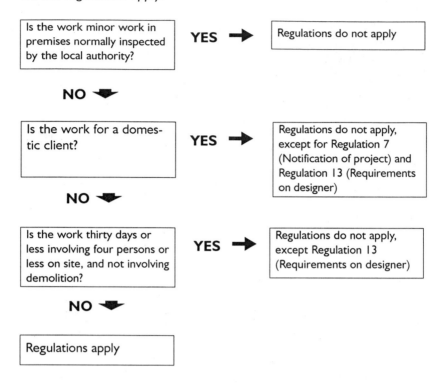

Is the work minor work in premises normally inspected by the local authority?	**YES** ➡	Regulations do not apply

NO ⬇

Is the work for a domestic client?	**YES** ➡	Regulations do not apply, except for Regulation 7 (Notification of project) and Regulation 13 (Requirements on designer)

NO ⬇

Is the work thirty days or less involving four persons or less on site, and not involving demolition?	**YES** ➡	Regulations do not apply, except Regulation 13 (Requirements on designer)

NO ⬇

Regulations apply

B7.7 Duties of the client

The client is the organisation or individual for whom the construction work is being carried out. Regulation 4 allows the client to appoint a competent and adequately resourced agent to carry out their duties under CDM. Clients may also appoint themselves as planning supervisor and/or principal contractor, providing that they have the necessary competence. They may also carry out design work. Clients' duties are contained in Regulations 6, 8, 9, 10, 11 and 12 and can be summarised as follows:

- appointing a competent and adequately resourced planning supervisor and principal contractor
- satisfying themselves that designers and other contractors they engage are competent and adequately resourced
- providing the planning supervisor with relevant information about the construction premises (existing structures and the site)
- allowing sufficient time for design and construction work
- ensuring that construction work does not start until the health and safety plan has been developed to a sufficient extent
- keeping information in the health and safety file available for future construction work (or handing on to a new owner).

B7.8 Duties of designers

A designer is anyone who prepares or amends designs for construction work including preparing drawings, design details, specifications, bills of quantities, specifying equipment or materials and carrying out related analysis or calculations. Designers' duties are contained in Regulation 13 and can be summarised as follows:

- taking reasonable steps to ensure clients are aware of their CDM duties and relevant HSC guidance
- preparing designs with adequate regard to health and safety and information supplied by the client
- providing adequate information with the designs
- cooperating with the planning supervisor and other designers
- providing information for the health and safety file.

Designers must avoid or combat risks to persons:

- carrying out construction or cleaning work
- who may be affected by such work

at any time – i.e. during initial work and future construction and cleaning work.

Concern has been mounting that many designers are not fulfilling their obligations under CDM and in 2003/4 the HSE launched a series of 'designer awareness days' together with a range of practical guides and a new website (www.safetyindesign.org). Further information for designers is also available via the construction homepage on the HSE website (www.hse.gov.uk).

However, the HSE clearly intend to take a tougher line with designers who do not comply with their duties. In 2004 an architect was successfully prosecuted for allowing the use of 36 kg blocks (which had to be handled manually on site) when he could have specified lighter blocks in his design. As well as being fined, the architect was ordered (under section 42 of HSWA) to attend a training course to increase his awareness of health and safety issues in design.

B7.9 Duties of the planning supervisor

The planning supervisor's role is mainly to coordinate health and safety aspects during the design phase of the projects. Planning supervisors' duties are contained in Regulations 7, 14 and 15 and can be summarised as follows:

- ensuring that an adequate pre-tender or pre-construction health and safety plan is prepared in sufficient time
- being able to provide adequate advice to the client and any contractors on issues relating to competence and the allocation of resources
- being able to advise the client on the initial contents of the construction phase health and safety plan (and whether it has been developed sufficiently for work to start)
- taking reasonable steps to ensure cooperation between designers
- ensuring, so far as is reasonably practicable, that designers comply with their duties in respect of avoiding or combating risks and providing information
- ensuring that the project is notified to the HSE
- ensuring that a health and safety file containing adequate information is prepared and then delivered to the client.

B7.10 Duties of the principal contractor

The principal contractor is the contractor who is assigned responsibility of the management of the project during the construction phase. In some cases clients may choose to do this themselves and this is acceptable under CDM, providing that they are competent. Duties of principal contractors are contained in Regulations 15, 16, 17 and 18 and can be summarised as follows:

- satisfying themselves that the designers and contractors that they engage are competent and adequately resourced
- ensuring that a suitable construction phase health and safety plan is prepared (before construction work starts), implemented and kept up to date
- taking reasonable steps to ensure cooperation between all contractors
- ensuring, so far as is reasonably practicable, that contractors and their employees comply with site rules
- taking reasonable steps to ensure that only authorised persons are allowed on site
- displaying the HSE notification on site
- providing relevant information about site risks to contractors and the self-employed
- ensuring that contractors' employees receive adequate health and safety information and training
- ensuring that arrangements are in place to receive advice from workers or their representatives and consult workers about health and safety matters
- providing the planning supervisor with information for the health and safety file.

B7.11 Duties of contractors and the self-employed

Duties of contractors are contained in Regulations 7, 8, 9 and particularly Regulation 19. Self-employed persons also have many similar duties. Contractors' duties can be summarised as follows:

- satisfying themselves that any contractors or designers that they engage are competent and adequately resourced
- cooperating with the principal contractor
- providing the principal contractor with relevant information about risks arising from their activities (e.g. from risk assessments or method statements)
- complying with rules in the health and safety plan and reasonable directions from the principal contractor
- informing the principal contractor about RIDDOR incidents
- providing information for the health and safety file
- notifying relevant projects for domestic clients to the HSE
- ensuring that they are aware of the names of the planning supervisor and the principal contractor and relevant contents of the health and safety plan.

B7.12 The health and safety plan

The requirements for what must be included in a health and safety plan are contained in Regulation 15 but considerable additional guidance is available in HSE publications, particularly appendix 3 of 'Managing Health and Safety in

Construction'. The regulations and guidance refer to the pre-tender plan and the construction phase plan but the author prefers to regard the plan as a continuum rather than having two distinct phases.

During the pre-tender stage the skeleton of the plan will be set out as management arrangements are outlined and significant risks and other issues identified, together with likely or intended control measures. Then, prior to work commencing and during the progress of the construction work, the plan will be steadily fleshed out. Details of management arrangements will be set down, along with details of what control measures will be taken and how those measures will be implemented. New risks and issues will continue to be identified (e.g. resulting from information received from specialist subcontractors, or from practical experience of unforeseen circumstances) and appropriate control measures specified and implemented.

Figure B7.1 opposite contains a checklist of items, which might be included in a health and safety plan. While the general project description (section A) is likely to be applicable to most projects, the extent of the other sections will depend on the size, complexity and risks of an individual project.

A health and safety plan for a major project may well refer to many of the items in the checklist but a plan for a small, simple project may only need to refer to a handful of topics. Organisations regularly carrying out projects of a similar nature and involving similar risks may well wish to use Figure B7.1 as a basis for preparing their own customised health and safety plan checklist.

FIND THIS ON CD	Figure B7.1	Health and safety plan contents

A)	PROJECT DESCRIPTION
	General description of the work involved
	Intended start date and likely duration
	Client
	Planning supervisor
*	Principal contractor
*	Other contractors and self-employed persons
*	Designers and consultants
	Existing records and plans available (including previous health and safety files)
B)	MANAGEMENT AND COMMUNICATION ARRANGEMENTS
	General project management
	Vetting of contractors (and reviewing risk assessments and method statements)
	Exchange of design information and review of design changes
†	Liaison between client/planning supervisor/principal contractor/contractors
†	Accident and incident investigation and reporting (inc. RIDDOR)
†	On-site training (inc. induction of workers)
†	Permit-to-work systems, etc.
†	Monitoring of work practices, site conditions, compliance with site rules
C)	SITE-RELATED MATTERS
†	Fire and emergency procedures (inc. maintaining emergency access routes)

†   Site rules (inc. general PPE requirements, speed limits, safety signs)
†   General site access arrangements
†   Site security (perimeter security, security staff)
       Adjacent land use
       Presence of services – electricity, gas, water, sewage, telecom (inc. above and below ground)
       Ground conditions (inc. slopes, unstable land, contaminated land)
       Existing structures (stability, fragile surfaces, falling materials, underground installations)
†   Hazardous materials (asbestos in structures, contamination, stored substances)
†   Client's activities on or close to site (risks to client's staff or from client's activities)
†   Toilet and welfare facilities
†   First aid

D)   WORK-RELATED MATTERS
†   Provision and maintenance of services (electricity, water, lighting, compressed air, gas)
†   Provision and maintenance of plant and equipment (inc. portable tools)
       General fire precautions
       Vehicle access routes (width, headroom, parking)
       Pedestrian access routes (segregation from traffic, protection of openings)
       Work at heights (scaffolding, MEWP, other access equipment)
       Work close to fragile surfaces, water, moving traffic
       Falling materials or equipment
       Lifting operations and lifting equipment
       Materials handling and manual handling
       Excavations and poor ground conditions
       Work-specific PPE requirements (and related signs)
       General material storage and waste disposal
       Storage, use and disposal of hazardous or flammable substances
       Removal or disturbance of asbestos, contaminants, contaminated land
       Other health risks (dust, noise, vibration)
       Weather-related risks
       Plant testing or commissioning issues (e.g. pressure testing, electrical commissioning)

*   Details to be inserted once known
†   Where construction work is carried out on or close to an active workplace of the client, then the client is likely to have a major input into how these aspects of the project are dealt with.

B7.13 The variety of contracted services

Whether within the context of a CDM project or more general contract work, there is a considerable variety in the types of services available and also in the sizes of the organisations providing them. The different types of contracted services include the following.

Construction and engineering project work

These types of projects may last over a considerable period of time and many will be subject to CDM. Some contractors working in this field may operate as project managers (likely to be principal contractors under the CDM Regulations), often carrying out a considerable proportion of work using their own workforce. However, others will mainly act as subcontractors in specialist fields (e.g. steelwork, flooring, heating and ventilation, scaffolding).

Routine maintenance and repair

Some of the contractors working in this area will also be involved in project work (e.g. electrical services, landscape gardeners) but many will maintain and repair specific types of equipment (e.g. forklift trucks, agricultural equipment, office equipment). An increasing number of companies are providing buildings and facilities management services to deal with maintenance and repair issues on behalf of tenants.

Plant hire

Mobile equipment (MEWPs, mobile cranes, etc.) may be hired out separately or with operators provided by the hire company.

Long-term services

Services such as cleaning, security and catering often involve contractors' employees having a permanent presence on their clients' premises, often at times when the client's own workforce is not there. Many haulage contractors have long-term contracts with their clients.

Professional and consultancy services

Architects, engineers, surveyors, accountants, medical practitioners, professional planning supervisors, health and safety consultants and training providers come into this category. Their role may be to carry out a single task (e.g. deliver a training course or carry out an inspection), to work on a specific project, or to deliver a regular service (e.g. carry out medical examinations on a weekly basis). Some such professionals are employed within larger organisations whereas others operate as self-employed 'sole traders'.

Employment agencies

Many agencies provide temporary staff for both short- and long-term periods. These staff may have specialist skills (e.g. computer analysts, project engineers), carry out administration tasks (temporary secretaries or telephonists) or perform manual work.

B7.14 The need for a flexible approach

A flexible approach is essential if the variety of contracted services described above is to be managed effectively. It might be reasonable to expect a mobile crane company carrying out a lifting operation to submit a generic risk assessment and a project-specific method statement. However, while a small company tendering to provide office-cleaning services may have a risk assessment, it would be unlikely to understand what a method statement was. A medical practitioner or a self-employed health and safety consultant would be unlikely to have either.

There will also be considerable variation in the nature of the client organisation. The owner of a small business may deal with a small number of regular contractors and be able to exercise close personal control over their work. However, a local authority or a large industrial plant will have to deal with a multiplicity of contractors working in many different locations. Clearly much more formalised procedures will be necessary in the latter case.

In some situations the client may have little choice as to whether to use a particular contractor. The contractor may have such a dominant position in a particular service area or may be the only one available to deal with an urgent situation. That is not to say that a contractor without the necessary competence (or appropriate insurance) should be used in such circumstances but some flexibility should be built into procedures. The price to pay for using a contractor who is not on the 'approved' list may be exercising close supervision over that contractor in order to ensure satisfactory standards are maintained.

B7.15 Selecting and approving contractors

Many large client organisations have lists of contractors who have been approved for health and safety purposes – one of the benefits of this can be that those contractors who do not maintain satisfactory standards can have their approval withdrawn. Smaller employers are recommended to take a similar approach. The extra administrative work involved can be justified by the reassurance that the contractors being used do appear to be competent and possess the necessary insurances – and also being able to demonstrate to others that these aspects have been checked.

Figure B7.2 opposite contains an example of a contractor's 'health and safety approval' questionnaire.

This has quite deliberately been kept brief so that it would not be unduly burdensome for any of the variety of contractors described in paragraph B7.13 to complete. It does require the submission of some support documentation, such as a health and safety policy and risk assessments. These documents would be relatively brief for small contractors carrying out low-risk activities or 'not applicable' for those with less than five employees.

Completed questionnaires and support documentation must then be reviewed by someone competent to do so, rather than just being filed away. The employer's competent source of health and safety assistance (see paragraph B2.6) is likely to have an important role here. Unsatisfactory or incomplete responses must be followed up and those contractors failing to eventually provide a satisfactory response should not be approved.

For some types of contracting work additional details may be required such as a project-specific risk assessment or method statement. This may be within the context of a CDM health and safety plan (see paragraph B7.12), indeed a potential principal contractor may have to demonstrate a capability to deliver the management systems and precautions set out in the pre-tender health and safety plan.

The sample form in Figure B7.2 contains a section for the contractor to refer to previous clients to whom reference can be made. Such references can be useful in providing practical experience of their standards as opposed to mere documentary evidence. Some client organisations seek annual confirmation from contractors that their insurance has been renewed – computer-generated letters can make this task relatively easy. (Further guidance on insurance is contained in paragraph A3.12.)

B7.16 Communicating with contractors

A wide range of types of information may need to be communicated to contractors and their employees, depending on the work they are to perform and the location of the work.

FIND THIS ON CD	Figure B7.2	Example of contractor's health and safety questionnaire

Name of company ....   Tel. no. ....
Address ....
....

1. Type of services provided	
2. Approx. number of employees	
3. Relevant accreditations e.g. CORGI, asbestos removal	
4. Relevant qualifications/experience of the company and/or its staff	

5. INSURANCE (copies of policies may be requested)	Insurer	Maximum Insured £	Date of Expiry
– public liability			
– professional indemnity			
– employer's liability			

6. Do you have a health and safety policy? (If YES, please provide a copy)	YES / NO
7. Do you have documented risk assessments for the services you provide? (If YES, please provide copies)	YES / NO

8. In the last five years, have you
  – been prosecuted for an offence under health & safety law
  – been served with an improvement or prohibition notice?
  (If YES, please provide details)
	YES/NO
	
	YES/NO

9. Please provide details of two clients to whom reference can be made as to your health and safety standards.	Company Name of contact Tel. no.
	Company Name of contact Tel. no.

APPLICATION COMPLETED BY .... (Name) .... (Signature) ....(Date)

For administrative use
APPROVED BY .... (Name) .... (Signature) ....(Date)

Standards and work-related requirements

The work they are to perform may need to conform to designated standards or may be subject to specific controls, for example:

- relevant HSE codes of practice or guidance
- British Standards or other relevant standards
- standards previously contained in tender documents or method statements
- relevant PPE standards
- responsibilities for supplying necessary services or equipment
- use of permit-to-work systems (if relevant)
- sanctions, which may be imposed in cases of non-compliance with standards.

General information for contractors

Contractors and their employees may need to be informed about some or all of the following site-related topics:

- site layout and access, including speed limits, one-way systems, parking and storage areas
- security arrangements, including pass systems, searches
- first aid and welfare facilities – canteens, toilets, washing facilities
- availability (or otherwise) of client resources – power, other services, equipment
- emergency procedures – fires and other types of emergency
- site rules and procedures – PPE, smoking, inspections, permits to work
- sanctions for non-compliance – with site rules, etc.

Special or unusual risks

Regulation 12 of the Management Regulations requires hosts to provide information about risks 'arising out of or in connection with the conduct of the host's undertaking'. This will generally relate to risks that are not immediately apparent or are unusual.

What constitutes a 'special or unusual risk' will depend on the level of knowledge or previous experience of the contractor and his employees. Examples of the types of risks that may need to be pointed out are as follows:

- transport – presence of rail traffic, forklifts, cranes
- dangerous access – fragile roofs, etc., risks from water
- electrical – high voltage equipment, unprotected conductors, earth-free areas
- buried or overhead services – gas, electricity, water, compressed air, sewage
- hazardous substances – chemicals, pathogens, asbestos
- noise – damaging or distracting
- radiation – ionising radiation, lasers, UV
- flammable or explosive substances – gases, liquids, dusts
- molten metal risks or hot metal
- presence of people or animals – young, elderly, disabled, disturbed, dangerous.

B7.17 Methods of communicating with contractors

There are several methods available for communicating necessary information to contractors and their employees.

Purchase orders

Order forms are a useful means of communicating some types of information to contracting employers, particularly specific standards for the work to be performed and possible sanctions for non-compliance with those standards or with site rules. However, forms (or accompanying documents) containing considerable detail of health and safety requirements are seldom read and, in the author's experience, are sometimes considerably out of date.

Handbooks and briefing documents

Many larger employers routinely supply contractors' handbooks or shorter briefing documents to contracting employers and sometimes to each individual employee of the contractors. (The style and content that is suitable for employees will often be different to that for their employers.) The content of such handbooks and documents would generally be that referred to in paragraph B7.16 above, with the communication of information about emergency procedures and special or unusual risks particularly important. However, simply distributing handbooks or documents will seldom achieve effective communication; they should be supplied in the context of meetings or contractors' induction training sessions (see below) with topics of particular importance or relevance being pointed out. It is prudent to collect signatures acknowledging receipt and awareness of the contents.

Meetings

Meetings with contractors (and sometimes their employees) are a valuable means of ensuring effective communication. Such meetings would normally take place prior to the contract work commencing and periodically throughout the project, the frequency relating to the nature of the work and the health and safety performance of the contractor. (This would be an important component of liaison arrangements between the various parties involved in CDM projects.) Similarly there should be periodic meetings between clients and contractors providing long-term services, at which health and safety is reviewed.

Induction of contractors' employees

The host must ensure that visiting employees of contractors are properly briefed about relevant health and safety risks. Many hosts prefer to do this themselves rather than rely on the contracting employer. The nature of the induction will depend on factors such as the type of work to be performed, the degree of risk present and the size of the host organisation.

It must also take into account the capabilities or experience of the contractor's employees and the numbers of workers to be inducted. The content of the induction should be relevant, providing general information about the work location and its health and safety arrangements as well as highlighting special or unusual risks.

Delivery through an informal briefing will often be appropriate at smaller workplaces (possibly using a checklist as an aide-mémoire). More structured

| FIND THIS ON CD | Figure B7.3 | Example of a checklist and record form for inducting contractors and visitors |

HKD PHARMACEUTICALS
CONTRACTOR/VISITOR INDUCTION RECORD

Name ....   Valid until ....

Company/organisation...

Signature ....   Mark ✓ for items covered
N/A if not applicable

GENERAL TOPICS   (Relevant to most contractors or visitors)

1.   GENERAL SITE LAYOUT
2.   LOCATION OF EATING & TOILET FACILITIES
3.   FIRST AID FACILITIES
4.   REPORTING ACCIDENTS & INCIDENTS
5.   FIRE & EVACUATION PROCEDURE
6.   NO SMOKING RULES
7.   RESTRICTIONS ON OTHER IGNITION SOURCES / EQUIPMENT
8.   GENERAL PERMIT TO WORK REQUIREMENTS
9.   SITE PERSONAL PROTECTIVE EQUIPMENT (PPE) RULES
10. RESTRICTIONS ON USE OF VEHICLES ON SITE
11. RESTRICTIONS ON USE OF HKD EQUIPMENT
12. LIMITS ON ACCESS
13. RESTRICTIONS ON EATING
14. HOUSEKEEPING & WASTE DISPOSAL
15. SECURITY REQUIREMENTS / LIABILITY TO SEARCH

TASK / SERVICE SPECIFIC TOPICS   Note
key points where these are relevant

16. ACCESS & PARKING ARRANGEMENTS
17. STORAGE OF MATERIALS
18. USE OF HAZARDOUS SUBSTANCES
19. CONTROL OF HOT WORK
20. OTHER TOPICS

I have given an induction on the topics above to the person named who has passed the standard HKD Induction Test.

Signature .... Name .... Date ....

Figure B7.4   Example of a multiple-choice questionnaire to test understanding of the content of the induction

HKD PHARMACEUTICALS
INDUCTION TEST

a) 111	b) 333	c) 999

a) operate the nearest break-glass alarm	b) telephone the emergency number
c) try to put out the fire	d) tell your supervisor

a) prepare to evacuate	b) evacuate now

a) the contractors' mess hut	b) the canteen	c) the gatehouse

a) the contractors' mess hut	b) in their own cabins	c) in their vehicles

a) portable phones	b) cameras	c) radios	d) cigarette lighters	e) none of these

a) eye protection	b) tyvek overalls	c) safety helmet	d) safety footwear

a) use of HKD equipment	b) use of HKD workshop facilities
c) removal of contractors' own waste	d) vehicle access onto site

a) production areas	b) the contractors' mess hut	c) plant rooms

a) with HKD permission	b) never	c) if the waste is hazardous

a) verbal permission from security	b) verbal permission from their supervisor
c) a signed pass-out	

a) if something has gone missing	b) at any time	c) if he is under suspicion

| | | Figure B7.5 | Example of a contractor's work completion form |

Contractor's work completion form

Contract company		
Project	Order no.	
Report submitted by	Date	
	Comments	
1.	SYSTEMS OF WORK	
1.1	Suitability of work methods	
1.2	Conformance with method statement (where applicable)	
1.3	Compliance with permit-to-work system	
1.4	Conformance with PPE standards	
1.5	Housekeeping standards	
2.	CONTRACTOR'S MANAGEMENT OF HEALTH & SAFETY	
2.1	Availability of employees for induction	
2.2	Quality and attitude of supervision	
2.3	Response when health and safety action required	
2.4	Capabilities of employees	
3.	SIGNIFICANT INJURIES, INCIDENTS etc.	
3.1	Injuries to contractor's employees or others	
3.2	Damage or near-miss incidents	
3.3	Adverse comments by HSE inspectors	
3.4	Environmental incidents or problems	
3.5	Security incidents or problems	
4.	COMPLETION INSPECTION OF WORK AREAS	
4.1	Workplace left clean and tidy	
4.2	Guards replaced where appropriate	
4.3	Waste materials removed	
4.4	Any other residual risks?	
4.5	Borrowed equipment returned in good condition	
4.6	Contractor's cabin key returned	
5.	OTHER COMMENTS (positive or negative)	
6.	HEALTH AND SAFETY RATING (award 0–10 marks) *(0 – should never be used again / 10 – excellent performance)*	

approaches may be necessary at larger locations with greater risks. The issue of contractor handbooks or information sheets and the highlighting of relevant sections can be carried out during the induction. Some organisations utilise videos or computer-based training packages.

At high-risk work locations the host may wish to test understanding of relevant points through oral or written tests or by the use of interactive computer systems. The host may also take the opportunity to check that the contractor's employees do have the qualifications, capabilities or experience claimed for them. In some situations medical screening of contractors' staff may also be appropriate. Records should be kept of inductions given to contractors' employees. Attendance at induction may be directly linked to contractors' pass systems. (The host may wish to put expiry dates on passes in order to ensure attendance for refresher training where this is considered necessary.)

Figure B7.3 provides an example of a checklist and record form for inducting contractors and visitors at a pharmaceutical plant; Figure B7.4 on page 141 contains a simple multiple-choice questionnaire intended to test understanding of the content of the induction.

B7.18 Managing and monitoring contractors

As with other aspects of the selection and management process, the degree of management control required will be dependent on the circumstances and particularly on the nature of their work. Control elements that may be appropriate include:

- identification by both the host (or principal contractor) and the contractor of the person responsible for day-to-day liaison and supervision
- control of contractors' access using passes or a signing-in system
- use of permits to work, either to control all work done by contractors or limited to high-risk activities (see Chapter B8)
- regular inspections of contractors' work areas both formally (see Chapter B5) and informally
- reporting to the host (or principal contractor) of significant accidents or incidents involving contractors
- periodic meetings to review contract activities
- inspections of contractors' work areas on completion of their work.

Formal work completion reports can help build up an overall picture of contractors' health and safety management standards and assist in deciding whether they are invited to tender again or retain 'approved' status. (Figure B7.5 opposite contains an example of such a report form.)

Source materials

1.   HSE (2001) 'Managing Health and Safety in Construction: Construction (Design and Management) Regulations 1994 (CDM)', ACOP and guidance, HSG 224.
2.   HSE (1995) 'Designing for Health and Safety in Construction'.
3.   HSE (1995) 'A Guide to Managing Health and Safety in Construction'.
4.   HSE (1999) 'Having Construction Work Done? Duties of Clients under the CDM Regulations' (free leaflet), MISC 193.

Safe Systems of Work

B8.1   Introduction

The common law duty of employers to provide their employees with a 'safe system of work' was incorporated into HASAWA in section 2(2)(a) which requires 'the provision and maintenance of plant and systems of work that are, so far as is reasonably practicable, safe and without risks to health'.

Similar requirements for safe systems of work have been introduced either by regulations relating to asbestos, carcinogens (COSHH), confined spaces, electricity and lifting operations or through their associated Approved Codes of Practice. References to the need for safe systems of work are also contained in a number of HSE guidance documents.

This chapter explains how safe systems of work are determined through the process of risk assessment and describes a number of alternative means of specifying safe systems of work and ensuring their implementation.

B8.2   Determining safe systems of work

A useful definition of a safe system of work is 'the work method resulting from an assessment of the risks associated with a task and the identification of the precautions necessary to carry out the task in a safe and healthy way'. For routine work activities this risk assessment should have been carried out some time previously and may have resulted in the safe system of work being specified formally, e.g. through a task procedure, a list of instructions or in the form of a simple notice or sign.

However, many work activities are highly variable, in both the nature of the work itself and the types and extent of the risks involved. Such activities need to be subjected to a process that has become known as 'dynamic risk assessment', in which an assessment of the risks has to be made by someone on the spot who is competent both to recognise the risks that are present and to identify the precautions necessary to control those risks. This dynamic assessment may be carried out by those who are carrying out the work or by those responsible for supervising or managing the work. The concept of dynamic risk assessment is described more fully in paragraph C1.21.

Those making dynamic risk assessments will often be utilising only their own knowledge and experience, but the process can be aided by the provision of checklists or forms that identify many of the important factors to be considered during the risk assessment, including some of the control measures that might be appropriate. An even more formal way of carrying out a dynamic risk assessment and ensuring that the resultant safe system of work is implemented is to use a permit-to-work system. The use of checklists and permits to work is described more fully later in the chapter.

B8.3   Means of specifying safe systems of work

A safe system of work can be specified in many different ways. Some of the more common methods include:

- oral instructions
- signs and notices
- written instructions and rules
- task procedures
- activity checklists
- permits to work.

The most appropriate method of specifying a safe system of work for a given task or activity will depend upon a number of factors, including the following:

- the scale and variety of the risks involved
- how frequently the task or activity is carried out
- the complexity and variability of the task or activity
- how easily the risks can be identified
- the complexity and variability of the control measures required
- the capabilities of those carrying out the task or activity.

B8.4   Oral instructions

Where staff do not already have the capability of recognising risks and identifying appropriate control measures for themselves, they can be aided by the provision of simple but relevant safety instructions. For example:

That drum in the photocopier may be hot, so keep clear of it when you're clearing blockages.

The edges of some of the door panels are sharp, so make sure you use the leather fingered gloves.

More than one box is too heavy to carry on your own. Use the trolley if there are several that need moving.

B8.5   Signs and notices

Safety signs and notices can also be used to communicate simple but necessary information in order to ensure the implementation of safe systems of work. Examples include:

Danger – confined space. Oxygen and gas tests required before entering.

Danger – overhead cables. No vehicles more than 2 metres high beyond this point.

Caution – work in progress overhead. Safety helmets must be worn in this area.

The Health and Safety (Safety Signs & Signals) Regulations 1996 specify standard colours and designs for signs:

- PROHIBITIONS – things that must not be done (round shape, black pictogram on white background, red edging and diagonal line)
- WARNINGS – about specific types of dangers (triangular shape, black pictogram on yellow background, black edging)

- MANDATORY – things that must be done (round shape, white pictogram on blue background)
- EMERGENCY ESCAPE or FIRST AID – signs (rectangular or square shape, white pictogram on green background)
- FIRE FIGHTING – signs (rectangular or square shape, white pictogram on red background).

B8.6 Written instructions and rules

In some situations simple, sequential sets of instructions may be fixed to pieces of equipment or posted in work areas. Alternatively key health and safety points may be summarised in written instructions or within safety rule books or handbooks. For example:

DUST COLLECTOR COMPARTMENT INSPECTION

Inspection may only be carried out by trained, authorised staff.
Isolate the compartment by key at the control panel.
The compartment door must be tied back securely.
A second person must remain outside the compartment door.
The person entering must wear

- a dust mask (type ABC 123)
- full body clothing
- general purpose gloves.

(Note: A permit to work must be obtained for re-bagging work or entry into ductwork.)

B8.7 Task procedures

The various elements of a safe system of work can be incorporated into a detailed procedure describing how the task should be carried out in a step-by-step sequence. As well as covering important health and safety points the procedure should also include other relevant matters such as:

- prevention of damage to equipment
- quality considerations (raw materials, product specifications)
- environmental and waste matters.

Such procedures should only be prepared after carrying out detailed observations of the task being carried out and consultation, with workers actually performing the task as well as with technical and other specialists. Figure B8.1 on page 148 provides an example of a procedure for one stage in the process of manufacturing reinforced concrete. The use of this type of detailed task procedure is more relevant for manufacturing processes where the same operations should be performed day after day in the same way.

B8.8 Activity checklists

In some situations a safe system of work can best be ensured by the use of a checklist which must be completed prior to the activity in question starting. The sample checklist provided in Figure B8.2 on page 149 is for use by persons leading cycling tours off-road (on tracks and trails). While this might seem a little unusual in relation to most work activities, the elements contained in the checklist have much in common with mainstream work situations.

The first two parts of the checklist are concerned with ensuring that minimum standards are maintained in respect of the availability of trained staff to lead the activity and of the necessary minimum equipment that must be available.

The remainder of the checklist deals with more variable aspects of the activity:

- the other participants (their capabilities and equipment)
- the route – the activity to be performed
- the prevailing weather conditions
- and the interrelationships between each of the above.

In completing the checklist the group leader is carrying out a dynamic risk assessment of these factors, taking into account his or her own knowledge and experience and also any guidelines and standards that have been provided by the employer.

B8.9 Permits to work

A permit to work (PTW) provides a very formalised method of ensuring that a safe system of work is followed for a particular task. PTW systems are more commonly used within industrial plants or for major construction projects, although they may also be relevant for controlling minor work such as entry into sewer systems or other confined spaces. They are particularly appropriate for tasks where:

- the risks involved are high or not readily apparent, e.g. high voltage electrical work, exposure to sources of radiation
- extensive precautions are required, e.g. complex isolations, atmospheric testing, specialist types of protective equipment
- those carrying out the task may not recognise all the risks (or be capable of implementing all of the precautions).

Workers need to be made aware of those tasks for which a PTW is required. This can be done through use of the following:

- signs and notices, e.g. 'Confined space – no entry without a permit to work'
- written instructions and rules, e.g. 'The following tasks require a permit to work ...'
- task procedures, e.g. 'A permit to work must be obtained before continuing on to the next stage of the task.'

B8.10 The PTW sequence

The issue and subsequent cancellation of a PTW must follow a strict sequence in order to ensure that all necessary precautions are in place before work starts and that all workers are removed from risk situations before the precautions are removed. This would normally involve the following stages:

1. A PTW is requested:
   (a) the nature of the task and its location are identified
   (b) clearance from other parties may be necessary (e.g. production agree that maintenance or cleaning work can be carried out).

FIND THIS ON CD

BARRINGTON CONCRETE PRODUCTS
MANUFACTURING AND HEALTH AND SAFETY PROCEDURES

ACTIVITY		KEY HEALTH AND SAFETY POINTS
6	POURING OF CONCRETE INTO MOULDS	
6.1	Ensure that the spacer plugs are in positiotn (large sections - Mould 2 only).	
6.2	Remove the safety chains and place them under the moulds.	Do not leave them in gangways creating tripping hazards.
6.3	Lift the vibrating beam into position at the anchor end of the mould to be cast, using the forklift truck and the appropiate lifting beam.	The forklift truck must be driven only by licensed drivers.
6.4	Clamp the vibrating beam into position.	
6.5	Wheel up the power pack trolley and connect it to the vibrating beam.	
6.6	Direct the concrete mixer to reverse alongside the mould. (The Technical Manager must have previously accepted that the load meets quality specifications.)	Keep clear of the vehicle while it is reversing. The vehicle should have an audible and visual reversing alarm.
6.7	Switch on the vibrating beam	Hearing protection must be worn by those in the building when the beam is vibrating.
6.8	Direct the mixer driver to start pouring concrete from the anchor end of the mould involved in the pouring operation to protect against cement burns.	Gloves must be worn by all operatives.
6.9	Eyewash units are available for use if necessary. Rake the concrete to distribute it evenly between the moulds.	As above.
6.10	Clean and smooth the moulds using a trowel. (A smooth finish is essential for product quality.)	As above.

Figure B8.2   Sample activity checklist

GUIDED OFF-ROAD CYCLING TOUR

Date......................
Start time.......................   Tour title.......................

STAFF (Minimum 2 trained)	DETAILS/COMMENTS
Leader Sweeper Other(s) Person trained in 'emergency aid'	
EQUIPMENT	
First aid kit Maps of area Tool kit/spare tubes Mobile phone	
PARTICIPANTS (Maximum 15)	
Apparent fitness/capabilities *Any special medical conditions? Need for shorter alternative route? Suitable head protection and footwear? * Waterproof clothing? (if appropriate) * Adequate water and food? Condition of cycles	
ROUTE & WEATHER	
* Condition of surfaces * Effects of weather (mud, snow, wind) * Severe gradients * Sections needing special precautions * Intended lunch location * What to do if lost	
ROUTE DETAILS & MOBILE PHONE NO.LEFT WITH	

* THESE ITEMS MUST BE COVERED DURING THE BRIEFING OF THE
PARTICIPANTS
Signed .... (Leader), ... (Sweeper)

Figure B8.3 Example of a PTW form

PERMIT TO WORK   ARMAGEDDON ASSOCIATES   Serial no. A7890

DETAILS OF PROPOSED WORK ACTIVITY

Planned start date:	Planned start time:	Expected duration:	Clearance is given for this work to take place
			Signature Production Supervisor Date

DETAILS OF ISOLATIONS CARRIED OUT

ISOLATION CHECKLIST	✓	1) ACCEPTANCE
ELECTRICITY		I accept this permit to carry out the work described above and all persons under my control will abide by them.
STEAM		
WATER		SIGNATURE
GAS		(Permit holder)
COMPRESSED AIR		Date Time
HYDRAULICS		
OTHER SERVICES		2) PERMIT ISSUE
MATERIAL FEEDS		I confirm the isolations have been carried out and other precautions described above have been taken. I am satisfied the permit holder understands any restrictions on the work.
OTHER POSSIBLE RISKS	OTHER PRECAUTIONS TAKEN/PPE REQUIREMENTS/RESTRICTIONS ON WORK	THIS PERMIT IS ISSUED
FLAMMABLE LIQUIDS/GASES		SIGNATURE
LACK OF OXYGEN		(Permit issuer)
CONFINED SPACE ENTRY		Date Time

HAZARDOUS SUBSTANCES

ACCESS/WORK AT HEIGHTS

PRESSURE SYSTEMS

OTHER RISKS

OTHER RESTRICTIONS ON
WORK

Top copy - permit issuer   Blue copy - permit holder   Yellow copy - PTW display board

3) WORK TERMINATION

The work described above has finished. All persons, equipment and materials under my control are clear of the area, which has been left in an unsafe position

SIGNATURE
(permit holder)
Date   Time

4) PERMIT CANCELLATION
THIS PERMIT IS CANCELLED

Isolations may be removed. Other precautions may be withdrawn.
Narmal activities may then continue

SIGNATURE
(authorised holder)
Date   Time

2. The permit issuer identifies the risks involved and the precautions required:
   (a) this should be aided by the PTW form (see paragraph B8.11 and Figure B8.3 on pages 152 and 150/1 respectively)
   (b) the permit issuer must be competent for the purpose (see paragraph B8.12).
3. Isolations are carried out and other precautions implemented, e.g.
   (a) isolations of power sources, services, material feeds are made *and* proved
   (b) atmospheric tests are carried out
   (c) necessary access equipment is provided
   (d) suitable PPE/RPE is made available
   (e) barriers and warning signs are put in position.
   The permit issuer is responsible for ensuring that effective precautions are taken even if he or she does not do everything personally. Details of the precautions should be entered on the PTW form.
4. The person in charge of the work (the permit holder) signs the PTW form:
   (a) they accept any rules or restrictions (e.g. on PPE to be worn or limitations on the task).
5. The permit issuer then signs the PTW form to issue the permit:
   (a) one copy should be given to the permit holder
   (b) one copy should be retained by the permit issuer (or at a central point)
   (c) a further copy may be displayed publicly (showing others what PTWs are currently in force).
   ONCE WORK IS COMPLETE (or must cease for other reasons):
6. *All* copies of the permit must be brought together.
7. The permit holder signs to cancel all copies, stating that:
   (a) work has terminated
   (b) workers, equipment, materials, etc., have been removed
   (c) the area has been left safe.
8. The permit issuer (or another authorised person) signs to cancel all copies, permitting:
   (a) the removal of isolations
   (b) the withdrawal of other precautions.
Normal activities may then be resumed.

B8.11 The PTW form

Some organisations use different types of PTW forms to deal with different types of activities (e.g. electrical work, hot work, entry into confined spaces, work at heights). The author usually prefers to operate with a single type of form since work activities can sometimes involve interlinking risks (hot work in confined spaces or electrical work at heights).

A good PTW form should provide the following:

- a checklist of the types of risks that may need to be considered
- prompts or guidance on appropriate precautions
- a logical layout to aid correct completion (with the permit issuer and permit holder signing to acknowledge their responsibilities at each stage).

As can be seen from stage 5 in paragraph B8.10, at least two copies of the PTW form will be required. Self-carbon paper is commonly used for PTW forms, with different colours aiding in the identification of the different copies. The numbering of each set of forms can be important in avoiding possible confusion between similar work activities.

An example of such a form is provided in Figure B8.3 on pages 150/1.

B8.12 The authorisation of permit issuers

Persons should only be authorised to issue and cancel PTWs if they are competent to do so – this should not be something that comes automatically on acquiring a managerial or supervisory position. Authorised persons should have the necessary knowledge and experience to enable them to:

- appreciate the purpose of the PTW procedure
- understand the steps involved in PTW issue and cancellation
- identify risks associated with activities for which they are issuing permits
- carry out and supervise isolations
- implement or supervise other necessary precautions
- assess the capabilities and attitudes of permit holders
- recognise and deal with problems.

Many larger employers have formal arrangements in place for the training of staff who are to issue PTWs and for testing their competence prior to their authorisation.

B8.13 Communication, training and implementation

However safe systems of work are specified, it is essential that they are communicated effectively to those who are expected to follow them. While training will be necessary for those following these systems of work, it will also be vital for those who are expected to determine safe systems of work through the process of dynamic risk assessment – PTW issuers (as described in paragraph B8.12 above) or those completing activity checklists (see paragraph B8.8 and Figure B8.2). Both communication and training are dealt with in greater detail in Chapter B3. The arrangements for ensuring that safe systems are implemented (including associated communication and training) should be subjected to the 'monitoring' stage of the management cycle by the application of the methods described in Chapter B5, including inspections, task observations and auditing techniques.

Health and Safety Records and Documentation

B9.1　Introduction

This chapter describes the various records and other types of documentation that must be kept for health and safety purposes. These have been divided into three categories:

1. Statutory requirements for most workplaces.
2. Statutory requirements for specific risks.
3. Desirable records and documentation.

Further guidance is provided later in the chapter on record and document retention periods. In many cases more information on specific records or documents is provided elsewhere within the book and appropriate cross-references have been included.

B9.2　Statutory requirements for most workplaces

Employers Liability Insurance certificate

Copies of ELI certificates must be displayed prominently at each place where employees who are covered by the insurance work. Some public bodies and employers who only employ close family members are exempt. (See paragraph A3.12 for further details.) Certificates must be retained for at least forty years.

Information for employees: poster or leaflet

The Health and Safety Information for Employees Regulations 1989 require employers to display the approved poster at a place reasonably accessible to employees and in such a position that it can be easily seen and read.

(Alternatively they may provide each employee with an approved leaflet, although most employers choose the poster option.) The current version of the poster (introduced in 1999) is entitled 'Health and Safety Law: What You Should Know' and contains the following sections in which details must be inserted:

- trade union or other safety representatives (see paragraphs B6.2 and B6.3)
- responsibilities for the management of health and safety (see paragraph B1.6)
- name and address of the enforcing authority (HSE or local authority) and the local Employment Medical Advisory Service (EMAS)

Health and safety policy

All employers with five or more employees must have a health and safety policy document. Chapter B1 provides full details.

General risk assessments

Employers with five or more employees must also record the significant findings of their risk assessments and any group of employees identified by it as being especially at risk (Regulation 3(6) of the Management Regulations). Further details of risk assessment requirements and how to comply with them are contained in Chapter C1.

Fire risk assessments

The 1999 version of the Management Regulations contained a specific requirement that the general risk assessment (referred to above) should also take account of the requirements of the Fire Precautions (Workplace) Regulations 1997. There can be few workplaces where there are no significant risks from fire.

RIDDOR reports

Details of RIDDOR reports submitted to the enforcing authorities must be kept for at least three years (see paragraph B4.7).

HSE Accident Book

All businesses must have an accident book in which to record work-related accidents. Since the beginning of 2004 a new-style HSE accident Book must have been in use. This ensures compliance with Data Protection requirements by providing a tear-off page for each accident record. (See paragraph B4.7 and Figure B4.4).

CDM health and safety files

Health and safety files relating to structures must be kept available for as long as the structure exists or until such time as the files are passed on to persons acquiring the structure. (Health and safety plans are only of direct relevance for the duration of the project although they may have relevance for future damages claims and be useful in devising plans for similar projects.) Chapter B7 provides further information on the CDM Regulations.

B9.3 Statutory requirements for specific risks

There are a number of statutory requirements for records to be kept which are contained in regulations dealing with specific risks. These generally fall under one of three headings:

1. Specific risk assessments.
2. Statutory examinations, inspections and tests.
3. Occupational hygiene and medical surveillance.

B9.4 Specific risk assessments

Risk assessments are required under several sets of regulations dealing with specific risks and appropriate records must be kept. Regulations commonly encountered include those dealing with:

- hazardous substances (see Chapter C2)
- noise (see Chapter C3)
- manual handling operations (see Chapter C4)
- display screen equipment workstations (see Chapter C5)
- personal protective equipment (see Chapter C6).

The nature and extent of the records to be kept varies a little between the different regulations. Other regulations of more narrow application requiring assessments include the following:

- Control of Lead at Work Regulations 2002
- Control of Asbestos at Work Regulations 2002 (see Chapter D2)
- Supply of Machinery (Safety) Regulations 1992
- Control of Major Accident Hazard Regulations 1999 (COMAH)
- Ionising Radiations Regulations 1999.

Employers needing more details of these requirements are recommended to consult specialist HSE publications (see Chapter E1).

B9.5   Statutory examinations, inspections and tests

Paragraph B5.5 contained details of frequent types of inspections required by several regulations. A number of regulations also require examinations, inspections and/or tests to be carried out periodically.

Control of Substances Hazardous to Health Regulations 2002

Under Regulation 9, local exhaust ventilation (LEV) control measures must be subjected to thorough examinations and tests at least every fourteen months (more frequently in respect of processes specified in Schedule 4 to the regulations). Other engineering controls must be examined and tested 'at suitable intervals'. Suitable records must be kept for at least five years. Non-disposable respiratory protective equipment (RPE) is also subject to general maintenance requirements contained in Regulation 9. The COSHH ACOP states that thorough examinations and tests should be carried out every month (every three months for half mask respirators used occasionally in low-risk situations). The ACOP also contains details of the records, which should be kept. Further information on the COSHH Regulations is contained in Chapter C2.

Control of Asbestos at Work Regulations 2002

Regulation 10 requires the maintenance of control measures, with detailed requirements for inspections, examinations and tests of exhaust ventilation equipment, respiratory protective equipment and vacuum cleaners contained in the ACOP. All records of maintenance and repair (including examinations and tests) must be kept for at least five years.

Control of Lead at Work Regulations 2002

These regulations contain similar requirements to those above. Full details of the requirements are set out in the ACOP booklet (see ref. 2 on page 160).

Lifting Operations and Lifting Equipment Regulations 1998

Under Regulation 9, lifting equipment that is exposed to conditions causing deterioration liable to result in dangerous situations must be thoroughly examined:

- lifting equipment used for lifting persons or an accessory for lifting (slings, shackles, etc.) – at least every six months
- other lifting equipment – at least every twelve months

OR in accordance with an examination scheme (which may specify more or less frequent examinations)
AND each time exceptional circumstances liable to jeopardise the safety of the lifting equipment have occurred (e.g. shock loading, accidental impact).

These thorough examination reports must be kept until the next report or for two years, whichever is the later. The requirements of these regulations are covered in more detail in Chapter D1. Further information is also available in an ACOP booklet (see ref. 3 on page 160).

Pressure Systems Safety Regulations 2000

These regulations apply to systems containing steam, air and other gases or fluids under pressure. Regulations 8 and 9 require written schemes of examination to be drawn up and implemented in respect of all protective devices, pressure vessels and certain parts of the pipelines and pipe work. Guidance on appropriate frequencies is provided in the ACOP booklet (see ref. 4 on page 160). The last examination report must be retained, together with any previous reports containing material information.

Provision and Use of Work Equipment Regulations 1998 (Power Presses)

Regulation 32 continues a requirement contained in the old Power Presses Regulations for the thorough examinations of power presses and their guards and protection devices. These must be thoroughly examined:

- before use for the first time after installation or re-assembly
- at least every twelve months (for presses with fixed guards only)
- at least every six months in other cases
- after exceptional circumstances that may jeopardise safety.

Reports of thorough examination must be kept for at least two years. Full details of PUWER requirements relating to power presses are contained in the ACOP booklet.

Ionising Radiations Regulations 1999

These regulations require a considerable number of records to be kept, including those relating to the maintenance and examination of engineering controls, etc., PPE (Reg. 10) and also the following:

- appointment of radiation protection advisers (Reg. 13)
- designation of controlled or supervised areas (Reg. 16)
- local rules and radiation protection supervisors (Reg. 17)
- accounting for radioactive substances (Reg. 28).

Other regulations deal with occupational hygiene and medical surveillance issues. Full details are contained in the ACOP booklet (ref. 6).

B9.6 Occupational hygiene and medical surveillance

Several sets of regulations contain requirements for the carrying out of occupational hygiene surveys and medical surveillance of exposed staff and the maintenance of associated records.

Control of Substances Hazardous to Health Regulations 2002

Regulation 10 requires air testing in the workplace where this is determined as necessary to ensure adequate control of hazardous substances or for substances or processes specified in Schedule 5 to the Regulations. Reg. 11 requires health surveillance where appropriate for the protection of employees or for those exposed to substances or processes specified in Schedule 6. Records relating to personal exposure of employees or health surveillance must be kept for forty years, other exposure records for five years. Further details are contained in Chapter C2.

Control of Asbestos at Work Regulations 2002

Regulations 18 and 21 contain equivalent requirements in respect of air monitoring and medical surveillance of employees exposed to asbestos to those referred to above in the COSHH Regulations. See ref. 1 for further details.

Control of Lead at Work Regulations 1998

Equivalent requirements for air monitoring and medical surveillance are contained in Regulations 9 and 10 (see ref. 2 on page 160).

Ionising Radiations Regulations 1999

Many detailed requirements for monitoring of radiation levels and classification and monitoring of exposed persons are contained in these regulations, particularly Regulations 19 to 26. The ACOP booklet (see ref. 6 on page 160) contains full details.

B9.7   Desirable records and documentation

There are many other records and documents relating to health and safety that should be retained, even though there may not be a statutory obligation to keep them. There are several reasons for this:

- frequent references need to be made to them (e.g. health and safety procedures or records of 'approval' of contractors)
- to demonstrate effective systems of safety management to outsiders (e.g. HSE or local authority inspectors, insurance companies, auditors)
- in case of future legal proceedings (particularly damages claims).

The periods for which such documents are retained (see paragraph B9.8) will be influenced by the above factors, particularly the possibility of legal proceedings.

Insurance details

Reference was made in paragraph B9.2 to the requirement to keep Employers Liability Insurance certificates for forty years. Details of other previous insurances (public liability, professional indemnity, product liability, motor vehicles) should also be retained.

Training records

Details should be kept of health and safety inductions given to employees (see Figure B3.5 on pages 70/1) and also of other health and safety-related training provided (see paragraphs B3.23 and B3.24). Records should also be retained of employees' training in operational procedures or their training or authorisation to carry out specific activities or operate particular items of equipment, particularly where these present a significant risk. Such records should be retained even after employees have left the organisation, in case of future damages claims.

Management and operating procedures

Current management procedures relating to health and safety, e.g. those dealing with:

- accident and incident investigation
- health and safety inspections

- selection and management of contractors
- fires and other emergencies

should obviously be readily available, as should current operating procedures (task procedures). However, previous versions of procedures should also be retained for a while, again in case of possible damages claims.

Accident and incident report forms

These records should be retained because of the possible future need for analysis but also in case of future damages claims. While retention for a period of three years should be adequate in respect of claims relating to accidents referred to in the actual reports, a longer period may be justified to counter (or confirm) allegations in claims that similar accidents or incidents have occurred previously.

Health and safety inspection reports

Previous reports should be kept for possible future analysis and again in case of future damages claims. The author recalls one damages claim purporting to be supported by a health and safety inspection report. However, the availability of a file of previous reports clearly demonstrated this document to be a crude forgery.

Specialist inspection reports

Non-statutory inspection reports such as those relating to fixed electrical installations, portable electrical appliances or gas equipment should also be retained.

Contractors' approval records

Where contractors have been approved to provide services (as described in Chapter B7), suitable records should be kept. These may simply be copies of completed versions of the type of form depicted in Figure B7.3 (page 140) together with any updated records of contractors' insurances.

Health and safety committee minutes

While these minutes may also be relevant to possible damages claims, future reference to them may also be useful in other respects, e.g. to identify the reasons for previous decisions.

Occupational hygiene and medical surveillance

Paragraph B9.6 referred to statutory duties to retain records relating to occupational hygiene surveys and medical surveillance. However, some regulations do not contain any such duties, e.g. the Noise at Work Regulations do not require audiometry to be carried out on workers exposed to noise and only require sufficient information on noise levels to determine which noise action level (if any) has been reached. Other forms of medical surveillance (e.g. vision screening) are not subject to any specific requirements.

B9.8   Record and document retention periods

Records relating to health of individuals

Occupational hygiene surveys relating to individuals and medical surveillance records are required by various regulations to be kept for forty years, as must Employers Liability Insurance certificates. This is because of the lengthy periods taken for many occupational diseases to develop.

For the same reasons it is prudent for employers to keep other health-related records for similar periods. The records would include:

- COSHH assessments
- noise assessments
- asbestos and lead assessments
- ionising radiation assessments
- other occupational hygiene and medical surveillance records.

Other types of risk assessments may also contain information relating to conditions that take some time to develop and therefore justify retention. For example:

- general risk assessments (e.g. sections relating to vibration or stress)
- manual handling assessments (e.g. long-term back problems)
- display screen equipment workstation assessments (e.g. postural or vision problems).

A similar case could be made for retaining CDM health and safety plans where the project involved potential exposure to significant occupational health risks.

CDM health and safety files
As stated in paragraph B9.2, each file must be retained for as long as the structure concerned exists or until the file is passed on to the person(s) acquiring the structure.

Other records
Some regulations require relevant records to be retained for at least five years (see paragraphs B9.5 and B9.6). In all other cases, employers are recommended to retain records for at least three years, even if there is no statutory obligation to do so. This is because of the three-year period allowed for the initiation of civil damages claims (referred to in paragraph A3.7).

The defensively minded may prefer to keep some records longer, should the availability of storage space permit. This is less of a problem where the relevant records can be kept in electronic form.

Source materials

1. HSE (2002) 'Work with Asbestos which does not Normally Require a Licence: Control of Asbestos at Work Regulations 2002', ACOP, L 27.
2. HSE (2002) Control of lead at work, ACOP, L 132.
3. HSE (1998) 'Safe Use of Lifting Equipment: Lifting Operations and Lifting Equipment Regulations 1998', ACOP and guidance, L 113.
4. HSE (2000) 'Safety of Pressure Systems: Pressure Systems Safety Regulations 2000', ACOP, L 122.
5. HSE (1998) 'Safe Use of Power Presses', ACOP and guidance, L 112.
6. HSE (2000) 'Work with Ionising Radiation: Ionising Radiation Regulations 1999', ACOP and guidance, L 121.

Risk Assessment Requirements

General Risk Assessments

C1.1 Introduction

As was explained in paragraph A2.11, a type of risk assessment has long been necessary in order to determine what was 'reasonably practicable'. In more recent times an explicit requirement for risk assessment has been incorporated into many codes of regulations, putting risk assessment at the centre of any health and safety programme. The most comprehensive requirement for risk assessment is contained in the Management Regulations and these regulations also contain specific requirements for assessments in relation to young people (under 18s) and new or expectant mothers.

This chapter examines these legal requirements and then provides guidance on the whole risk assessment process:

- planning and preparation
- risk assessment techniques
- assessment records
- following the assessment (implementation and review).

The guidance provided on many of these aspects of practical assessment is applicable not just to the general types of risk assessment required by the Management Regulations but also to the more specific types of assessments required by other regulations – several of these are referred to in Chapters C2 to C7.

Within this chapter are several checklists and examples of completed risk assessments including a 'model' risk assessment – a useful technique to apply when dealing with a number of workplaces all carrying out the same types of activities in similar ways. The concept of dynamic risk assessment (introduced in paragraph B8.2) is also explored further.

C1.2 Requirements of the Management Regulations

Three aspects of risk assessment have requirements contained within the Management Regulations:

1. General risk assessments (Regulation 3)
2. Young persons (Regulations 3 and 19)
3. New or expectant mothers (Regulations 16, 17, 18)

While the parts of Regulation 3 relating to general risk assessments are set out below (in paragraph C1.3), their practical implications are not considered until later in the chapter. Several practical aspects of assessments relating to young persons and new or expectant mothers are considered in paragraphs C1.4 to C1.7, although many of the general approaches and techniques described later in the chapter are also applicable to those specific topics.

C1.3   Regulation 3

Regulation 3(1) of the Management Regulations states:

> every employer shall make a suitable and sufficient assessment of:
>
> (1) the risks to the health & safety of his employees to which they are exposed whilst they are at work; and
> (2) the risks to the health & safety of persons not in his employment arising out of or in connection with the conduct by him of his undertaking, for the purpose of identifying the measures he needs to take to comply with the requirements or prohibitions imposed upon him by or under the relevant statutory provisions and by Part II of the Fire Precautions (Workplace) Regulations 1997.

Regulation 3(2) imposes similar requirements on self-employed persons. Regulation 3(3) requires a risk assessment to be reviewed if:

- there is reason to suspect that it is no longer valid; or
- there has been a significant change in the matters to which it relates.

Regulation 3(6) requires employers with five or more employees to record:

- the significant findings of their risk assessments; and
- any group of employees identified as being especially at risk.

(Although Regulation 3 contains the legal requirement to carry out a risk assessment in respect of fire precautions, the practical aspects of fire risk assessments are considered separately in Chapter C7.)

C1.4   Young persons

'Young person' is defined in the regulations as 'any person who has not attained the age of eighteen'. Paragraphs (4) and (5) of Regulation 3 of the Management Regulations contain the following specific requirements relating to young persons:

> (4) An employer shall not employ a young person unless he has, in relation to risks to the health and safety of young persons, made or reviewed an assessment in accordance with paragraphs (1) and (5).
> (5) In making or reviewing the assessment, an employer who employs or is to employ a young person shall take particular account of –
> a) the inexperience, lack of awareness of risks and immaturity of young persons;
> b) the fitting-out and layout of the workplace and the workstation;
> c) the nature, degree and duration of exposure to physical, biological and chemical agents;
> d) the form, range and use of work equipment and the way in which it is handled;
> e) the organisation of processes and activities;
> f) the extent of the health and safety training provided or to be provided to young persons; and
> g) risks from agents, processes and work listed in the Annex to Council Directive 94/33/EC on the protection of young persons at work.

Regulation 19 contains further requirements on the 'Protection of young persons':

(1) Every employer shall ensure that young persons employed by him are protected at work from any risks to their health or safety which are a consequence of their lack of experience, or absence of awareness of existing or potential risks or the fact that young persons have not yet fully matured.

(2) Subject to paragraph (3), no employer shall employ a young person for work –

    a) which is beyond his physical or psychological capacity;

    b) involving harmful exposure to agents which are toxic or carcinogenic, cause heritable genetic damage or harm to the unborn child or which in any way chronically affect human health;

    c) involving harmful exposure to radiation;

    d) involving the risk of accidents which it may reasonably be assumed cannot be recognised or avoided by young persons owing to their insufficient attention to safety or lack of experience or training; or

    e) in which there is a risk to health from

        i) extreme cold or heat;

        ii) noise; or

        iii) vibration,

and in determining whether work will involve harm or risk for the purposes of this paragraph, regard shall be had to the results of the assessment.

(3) Nothing in paragraph (2) shall prevent the employment of a young person who is no longer a child for work –

    a) where it is necessary for his training;

    b) where the young person will be supervised by a competent person; and

    c) where any risk will be reduced to the lowest level that is reasonably practicable.

(4) The provisions contained in the regulations are without prejudice to –

    a) the provisions contained elsewhere in these Regulations; and

    b) any prohibition or restriction, arising otherwise than by this regulation, on the employment of any person.

The term 'child' is defined in Regulation 1(2) of the Management Regulations as a person not over compulsory school age. Regulation 10(2) of the Management Regulations contains requirements relating to the supply of information to parents and guardians about risks affecting their children and related precautions (see paragraph B2.9).

C1.5   Assessing risks to young persons

The Management Regulations now allow a more flexible approach to be taken to the work of young persons than was previously the case. Most of the old age-related prohibitions on certain work activities have now disappeared. In deciding what restrictions need to be placed on work by young persons the key phrase to remember is that contained in Regulation 3(5): 'the inexperience, lack of awareness of risks and immaturity of young persons'. All of the other specific considerations contained in Regulations 3(5) or 19 can be related back to this phrase, particularly if both physical and psychological maturity are taken into account.

There are many types of work that present little or no risk to most young persons (although the same cannot always be said of young persons with special needs) while there are also work activities that create risks not just for young persons but for anyone without relevant experience or awareness. There is also a broad spectrum in the capabilities of individual young persons – some have maturity beyond their years or experience and awareness of risks acquired from family businesses or part-time jobs, whereas others demonstrate high levels of naivety and immaturity.

Figure C1.1 on page 166 contains examples of the types of work that are likely to require restrictions for the majority of young persons but it is illustrative rather than comprehensive. Employers should use it as a basis for identifying similar situations in their own workplaces and work activities. Appropriate restrictions might involve:

- outright prohibition of the activity for young persons (at least initially)
- provision of close supervision by a competent person
- provision of specific training for certain activities
- provision of specific health surveillance for the young persons.

(The wording of Regulation 19(3) allows more flexibility in respect of young persons above school leaving age.)

These restrictions can then be progressively removed as individual young persons receive training and demonstrate increased experience, awareness and maturity. However, some restrictions may need to remain after workers pass the age of 18 if they do not possess the capabilities or have not received the training required by Regulation 13 of the Management Regulations (see paragraph B2.12).

The more flexible approach of the Management Regulations largely removes the artificial distinction previously existing between under-18s and over-18s. Further guidance is available in an HSE booklet 'Young People at Work: A Guide for Employers'.

C1.6   New or expectant mothers

Relevant definitions are provided by Regulation 1 of the Management Regulations:

'New or expectant mother' means an employee who is pregnant, who has given birth within the previous six months, or who is breastfeeding. 'Given birth' means 'delivered a living child or, after twenty-four weeks of pregnancy, a stillborn child'.

The requirement for 'risk assessment in respect of new or expectant mothers' is contained in Regulation 16 of the Management Regulations, which states:

(1) Where –
   a) the persons working in an undertaking include women of child-bearing age; and
   b) the work is of a kind which could involve risk, by reason of her condition, to the health and safety of a new or expectant mother, or to that of her baby, from any processes or working conditions, or physical, biological or chemical agents, including those specified in Annexes I and II of Council Directive – 92/85/EEC on the introduction of measures to encourage improvements in the safety and health at work of pregnant workers and workers who have recently given birth or are breastfeeding,

– the assessment required by regulation 3 (1) shall also include an assessment of such risk.

(2) Where, in the case of an individual employee, the taking of any other action the employer is required to take under the relevant statutory provisions would not avoid the risk referred to in paragraph (1) the employer shall, if it is reasonable to do so, and would avoid such risks, alter her working conditions or hours of work.

(3) If it is not reasonable to alter the working conditions or hours of work, or if it would not avoid such risk, the employer shall, subject to section 67 of the 1996 Act* suspend the employee from work for so long as is necessary to avoid such risk.

(4) In paragraphs (1) to (3) references to risk, in relation to risk from any infectious or contagious disease, are references to a level of risk at work which is in addition to the level to which a new or expectant mother may be expected to be exposed outside the workplace.

* This refers to the Employment Rights Act 1996 which would require any such suspension to be on full pay (unless suitable alternative work is refused unreasonably).

Regulation 17 of the Management Regulations allows new or expectant mothers to be suspended from work at night by virtue of a certificate from a registered medical practitioner or a registered midwife, if necessary for her health or safety.

Regulation 18 states that employers need not take action under Regulation 16 until they are notified in writing by an employee that she is a new or expectant mother, nor need they maintain action if:

- a certificate of pregnancy is not produced in a reasonable time
- the employer knows the employee is no longer a new or expectant mother, or
- the employer cannot establish that she remains a new or expectant mother.

Where risks to new or expectant mothers are identified and these cannot be avoided by normal preventive and protective measures, the choices are as follows (in descending order of preference):

- alter working conditions or hours of work (if it is reasonable to do so and would avoid the risks)
- identify and offer suitable alternative work
- suspend from work on full pay.

HSE guidance suggests that suspension from night work (as provided for by Regulation 17) should not be necessary and that suspension can be avoided in other cases if the right approach is taken.

C1.7 Assessing risks to new or expectant mothers

The HSE booklet 'New and Expectant Mothers at Work: A Guide for Employers' provides a wealth of guidance on potential risks to new or expectant mothers and means of avoiding such risks. These risks are summarised in Figure C1.2. They include risks to the unborn or breastfeeding baby as well as to the mother. A practical way of meeting legal obligations is for the employer to identify such potential risks during the general process of risk assessment. These can then be included in a questionnaire or checklist of the type provided in Figure C1.3. Such a checklist can then be used to identify which, if any, of the risks present in the workplace as a whole might affect an individual employee and what action needs to be taken. General welfare issues affecting the employee (e.g. the effects of morning sickness) can also be addressed. It should be noted that the checklist is intended to be used both on first notification of pregnancy and on returning to work after giving birth. Some aspects may need further review, particularly in the later stages of pregnancy as size increases.

C1.8 Risk assessments: planning and preparation

As in most activities, planning of and preparation for a risk assessment programme is essential to the programme's eventual success. Consideration must be given to the person who is to carry out the assessments and what is expected of them.

C1.9 Who will make the assessments?

Chapter B3 considered the employer's duty to appoint competent health and safety assistance and the attributes that such a person (or persons) should possess. One would expect the source of health and safety assistance (whether an employee or an external consultant) to play a significant part in the risk assessment programme. However, in large organisations it would be impossible for a health and safety specialist to carry out risk assessments alone and, even in smaller organisations, there will be many others who can contribute to the process such as:

- managers, supervisors, team leaders
- engineers and technical specialists
- safety representatives and individual employees.

Some of the above are likely to need training in aspects such as:

- the purpose of risk assessments
- the legislation behind risk assessment
- risk assessment techniques
- recording of risk assessments.

FIND THIS ON CD	Figure C1.1	Risks to young persons

The following activities or situations are likely to require some types of restrictions for most young people

PHYSICALLY DEMANDING WORK
Work that could injure the developing body, because of
- the weights involved or the force required; or
- the repetitive nature of the activity

PSYCHOLOGICALLY DEMANDING WORK
- dealing with death, serious illnesses or traumatic injuries
- situations with a significant possibility of violence or aggression
- dealing with difficult customers or clients
- stressful decision-making

EXPOSURE TO HAZARDOUS SUBSTANCES
- toxic or carcinogenic substances (including lead and asbestos)
- which can cause heritable genetic damage or harm the unborn child
- which can chronically affect human health
- categorised as corrosive, harmful or irritant

EXPOSURE TO HARMFUL PHYSICAL CONDITIONS
- ionising radiation and non-ionising radiation, e.g. lasers, UV, IR
- extreme cold or heat presenting risks to health
- excessive noise or vibration (from work equipment – see below)
- work in pressurised atmospheres or diving work

USE OF CERTAIN TYPES OF WORK EQUIPMENT
Where the equipment presents significant risks and high levels of skill or complex precautions are required to ensure its safe use:
- woodworking or metal processing machines (fixed and portable types)
- some types of food processing machinery (slicers, etc.)
- some hand tools (sharp knives, cleavers, axes)
- mobile equipment (forklift trucks, agricultural and construction vehicles)
- firearms

HIGH RISK WORKPLACES
- work at heights (including use of ladders or harnesses)
- work in confined spaces
- work near unstable structures (e.g. in construction, demolition)

OTHER HIGH RISK ACTIVITIES
- work with potentially aggressive animals (agriculture, horse care, veterinary work)
- exposure to live electrical equipment or high voltage equipment
- work with flammable liquids and gases, or explosive materials
- exposure to high pressure materials (e.g. in pressure testing)
- handling significant quantities of cash or valuables.

Figure C1.2   Risks to new or expectant mothers

PHYSICAL AGENTS, ETC.
- manual handling activities
- ionising radiation (radioactive sources, X-ray equipment, etc.)
- high levels of radio frequency radiation (e.g. microwaves)
- work in compressed air
- diving work
- physical shocks or low frequency vibration (risks of miscarriage)
- unsuitable movement or posture (excessive standing, cramped working positions)
- excessive physical or mental pressure (causing anxiety or raised blood pressure)
- extreme heat (causing fainting or heat stress, or dehydration during breastfeeding)

BIOLOGICAL AGENTS
- agents in hazard groups 2, 3, 4 (as categorised by the Advisory Committee on Dangerous Pathogens)

- agents such as hepatitis B, HIV, herpes, TB, syphilis, chickenpox, typhoid, rubella, cytomegalovirus, chlamydia in sheep (where risks are higher than in the community in general, e.g. in laboratories, health care work, the emergency services, work with animals or animal products)

CHEMICAL AGENTS
- substances labelled with specified risk phrases (e.g. possible risk of irreversible effect, may cause cancer, may cause heritable genetic damage, may cause harm to the unborn child, possible risk of harm to the unborn child, may cause harm to breastfed babies)
- organic mercury compounds
- antimitotic (cytotoxic) drugs (used in cancer treatment)
- chemicals absorbed through the skin (inc. some pesticides)
- carbon monoxide (e.g. from incomplete combustion of gas)
- lead and lead derivatives (women of reproductive capacity are prohibited from working in most lead processing activities)

WORKING CONDITIONS/OTHER ASPECTS OF PREGNANCY
It is sometimes difficult to differentiate between actual risks to new or expectant mothers and conditions that make work difficult or unpleasant. Aspects to be considered include:
- the effects of morning sickness
- the need to leave the workplace periodically
- increasing size (posture at workstations, access to confined areas, fitting of protective clothing)
- balance, agility, dexterity, coordination, reach, speed of movement
- tiredness at certain times (availability of suitable rest facilities)

Larger employers may arrange customised training in-house although there are many colleges, training organisations and consultancies that can provide such training. The pairing of an 'insider' with an 'outsider' to carry out assessments is often beneficial. The insider (a manager, supervisor, safety representative or employee) will be familiar with the workplace, the work equipment and the work activities that are the subject of the assessment. The outsider (the health and safety specialist, external consultant, person from another department) can question the status quo and bring in a wider experience of risks, control measures and standards.

While safety representatives and individual employees can contribute considerably to risk assessment, it must not be forgotten that the carrying out of a 'suitable and sufficient assessment' is ultimately the responsibility of the employer, who must apply sound management principles to the risk assessment programme (as referred to in paragraph B3.3).

C1.10 Risk assessment units
Reference was made in Chapter B5 to dividing up workplaces for the purpose of health and safety inspections. Similar divisions may also be necessary in order to make the risk assessment programme more manageable. Separation could be according to different areas of the workplace (as shown in paragraph B5.3) but the division may also be made according to processes or product lines, or according to the different activities carried out or services provided

Figure C1.3   Example of a risk assessment checklist for new or expectant mothers

ARMAGEDDON ASSOCIATES   NEW OR EXPECTANT MOTHERS   RISK ASSESSMENT CHECKLIST

NAME OF EMPLOYEE _____   WORK LOCATION _____

	POSSIBLE RISKS TO CONSIDER	PRECAUTIONS / CHANGES AGREED
Factory work	Manual handling (assembly, stores, maintenance) Ionising radiation (non-destructive testing, laboratory) Excessive standing or awkward sitting position Heat (heat treatment area) Hazardous substances (solvent cleaning, laboratory) Significant risk of falls Other	
Administrative work	DSE workstation layout Awkward meeting times Significant travel (especially by car or air) Other	
General matters	Ability to leave workplace temporarily Availability of rest area/toilets Major pressure or stress situations Possible effects of fatigue Other	

Signature for Armageddon Associates _____   Signature (new/expectant mother) _____

Date _____   Date for further review (if any) _____

This form should be completed when the pregnancy is first notified and when the new mother returns to work.

by the organisation. Where a workplace is divided into areas (as in paragraph B5.3), there may also be some common issues that need to be addressed across the premises as a whole (e.g. electrical equipment, maintenance work).

Where a team approach is taken to risk assessment, the various assessment units can then be allocated to the different members of the team according to their capabilities and also taking into account their overall workload.

C1.11 Identifying the topics to be addressed

The ACOP to the Management Regulations defines both 'hazard' and 'risk':

- a 'hazard' is something with the potential to cause harm
- a 'risk' is the likelihood of potential harm from that hazard being realised.

The extent of the risk will depend on the following:

- the likelihood of harm occurring
- the potential severity of that harm
- the number of people who might be exposed.

The ACOP states that risk assessment involves 'identifying the hazards present and evaluating the extent of the risks involved, taking into account existing precautions and their effectiveness' – the quality of the precautions must be evaluated as well as the risks identified. It also goes on to say that insignificant risks can usually be ignored, as can risks arising from routine activities associated with life in general. For example, working at heights can be said to be a hazard for both a librarian and a scaffolder. However, the consequences of the librarian falling from a kick stool or small stepladder make the risks insignificant in comparison with those for the scaffolder.

One way to identify what needs to be addressed during the risk assessment is to carry out a brainstorming session and list the significant risks associated with the workplace or work activity being assessed. This will utilise the knowledge and experience of those carrying out the assessment. However, it must also be borne in mind that the risk assessment process is intended to identify the measures necessary in order to comply with the law. Consideration of relevant regulations may well identify a few more issues, including the need to check that precautions are up to the standards required by the regulations.

Figure C1.4 below provides a checklist containing many possible risk topics, together with some of the more common regulations that need to be considered during the assessment. As the risk assessment progresses, this initial list of topics might well change, with some topics being subdivided while others might be merged for convenience, e.g. portable tools might be divided into portable powered tools and hand tools, while ladders and scaffolding might be combined into 'work at heights'.

The purpose of this topic-based approach is to ensure that nothing of importance is overlooked during the risk assessment. It can also assist in the recording of the assessments by providing a list of logical headings and subheadings to use in the eventual records (see paragraph C1.17 onwards).

Figure C1.4   Risk assessment topics

Topics subject to regulations containing specific assessment requirements

Hazardous substances (COSHH)	See Chapter C2
Noise	See Chapter C3
Manual handling	See Chapter C4
Display screen equipment	See Chapter C5
PPE needs	See Chapter C6
Asbestos	See Chapter D2
Dangerous Substances and Explosive Atmospheres Regulations 2002 (DSEAR)	See paragraph C7.4
Lead	
Ionising radiation	
Major accidents (COMAH)	

Fire-related risks – See Chapter C7
Flammable liquids
Flammable gases
Dust explosions
Hot work (burning and welding)

Work equipment risks (PUWER and LOLER) – See Chapter D1
Production and process machines
Maintenance machines
Portable powered tools
Hand tools
Forklift trucks
Construction vehicles
Agricultural vehicles
Other mobile equipment
Lifting equipment
Lifting operations

Electricity at Work Regulations – see Chapter D3
Fixed installation
Maintenance of equipment
Electrical work
Pressure Systems Safety Regulations
Compressed air
Steam
Other pressure systems

Workplace risks (Workplace and CHSW Regulations) – See Chapter D2
Vehicle access
Pedestrian access
Work at heights

- ladders and stepladders
- scaffolds
- mobile elevating work platforms
- falling objects

Excavations

Work environment risks
Work in confined spaces

Excessive heat or cold
Work near water
Severe weather
Buried or overhead services
Work on clients' premises
Work in domestic property
Violence or aggression
Lone working and remote locations
Home working
Work abroad
Driving

Storage risks
Racking, shelving, cupboards
Stacking of materials
Tanks and silos
Waste materials

Miscellaneous risks
Pressure testing
Hand-arm vibration and Whole body vibration
Non-ionising radiation (lasers, microwaves, UV, IR)
Molten metal
Glass
Work with animals
Fatigue
Stress
Risks to the public
Risks from crowds
Work with volunteers
Work-related driving

C1.12 Useful information

There is likely to be a lot of information available that might be useful during the risk assessment. Some of this may be contained in internal documents whereas other information might be available from external sources such as the HSE. Information likely to be available internally includes the following:

- *Previous risk assessments* – Even where these are well out of date or incomplete they may still provide a basis for the new assessments. Assessments made under specific regulations may provide useful information, e.g. a COSHH assessment might specify PPE requirements for a task.
- *Documented safe systems of work* – Chapter B8 dealt with safe systems of work, which should have been identified as a result of a risk assessment process. Signs and notices, instructions and rules, task procedures, activity checklists, permit-to-work systems all need to be evaluated. Do they address all the relevant risks and are they proving effective in controlling them?
- *Accident and incident records* – Chapter B4 emphasised the importance of accident and incident investigations as a means of identifying and controlling risks. What risks have been identified through investigations and have they been controlled? In smaller workplaces even a check of the accident book should be carried out.

- *Health and safety inspection and audit reports* – Chapter B5 referred to the importance of inspections and audits in monitoring standards. Have risks identified through this monitoring process been controlled and have weaknesses in controls and management systems been addressed?
- *Training records* – The content of health and safety-related training programmes should be evaluated and checks made as to which staff have actually received training.

External information likely to be useful during risk assessments includes the following:

- *Literature from manufacturers and suppliers* – Handbooks from manufacturers and suppliers of equipment may highlight risks that had not previously been recognised, but are more likely to provide information on correct operating arrangements and maintenance regimes. Data sheets on substances and materials may identify fire or explosive risks as well as risks to health.
- *HSE information* – Chapter E1 describes the wide range of information available from the HSE – from ACOPs and guidance on specific regulations to general guidance on common work sectors (engineering, motor vehicle repair, warehousing, residential care) or on the risks associated with individual work activities.
- *Information from others 'in the business'* – Some trade associations publish their own health and safety guidance, often in consultation with the HSE. Informal contacts with others involved in the same work activity can also often yield information that can be helpful during the risk assessment.

C1.13 Risk assessment techniques

While most of the planning and preparation for the risk assessment can be carried out in the office or conference room, the assessment itself must involve a significant amount of work in the field – observing the workplace and work activities and talking to the people carrying out the work. Even where it is impossible to see the full range of workplaces and activities, some sample visits (preferably to situations of greater risk or difficulty) are essential. Consideration must also be given to whether persons other than employees may also be at risk (see paragraph C1.16).

C1.14 Workplace observation

Time taken observing people at work will generally be time well spent. While workers are less likely to adopt unsafe practices if they know they are under observation, there may be the opportunity to catch them unawares. If a worker follows an unsafe practice despite being aware of the presence of observers, this indicates a significant problem (e.g. training, supervision, motivation). Many aspects of work can be observed including:

- condition of floors, stairways, the building fabric, fencing
- the physical environment (heating, lighting, ventilation)
- condition and suitability of equipment
- effectiveness of guarding

- access routes and housekeeping standards
- storage arrangements
- presence of hazardous substances
- compliance with standards and rules (PPE, speed limits)
- conformance with safe systems of work
- manual handling techniques
- high risk activities (e.g. confined space entry, hot work, live electrical testing).

It will often be appropriate to delve a little further by:

- entering infrequently accessed storerooms and plant rooms
- checking access to remote workplaces
- switching on lighting and ventilation to see if it works
- testing the effectiveness of trip devices and interlocked guards.

However, those carrying out the assessment should take care not to endanger themselves through such activities nor to cause disruption to processes or equipment.

C1.15 Talking to workers

Informal discussions with people at work can also be very helpful, especially in revealing previously unrecognised risks or weaknesses in controls or management systems. Aspects to be investigated include the following:

- potential variations in work activities or work methods
- standards of training (awareness of risks and related control measures)
- which activities are regarded by workers as high risk
- the effectiveness of control measures (or reasons for their non-use)
- awareness of emergency arrangements
- problem activities and suggestions for improvements.

Open-ended questions (prefaced by how, what, why, what if) are particularly useful. Discussions with safety representatives, team leaders, supervisors and managers are also likely to be relevant – some of these people may already be involved as part of the team carrying out the risk assessment.

C1.16 Persons at risk

Employees
The risk assessment should take into account not only significant risks to those working on the premises (production, maintenance and administrative staff) but also employees who work away from the premises – providing services, delivering goods and materials, selling products or services or simply attending meetings. Staff working on the premises outside normal working hours – security officers, cleaners, etc. – often get overlooked, as does the need to ensure that temporary employees are provided with appropriate training (and PPE if necessary).

Others at risk
All employers must also take account of other persons who may be put at risk by their activities. Persons at risk might include the following:

- contractors and their employees (covered in some detail in Chapter B7)
- co-occupants of premises
- volunteers (often working alongside employees in many charitable organisations)
- drivers delivering or collecting goods and materials
- visitors to the premises
- customers or service users
- residents (in health care premises, hotels, etc.)
- users of neighbouring roads and footpaths
- trespassers.

The opportunity should be taken both prior to and during the risk assessment to identify which persons may be at risk and ensure that appropriate control measures are in place. A variety of examples where this was not the case is provided in the case histories on page 126.

C1.17 Risk assessment records

During the later stages of the risk assessment serious consideration must be given to how the assessment is going to be recorded. Possible record formats are discussed further in paragraph C1.19 and examples of completed records provided in Figures C1.5 and C1.6 on pages 177 and 178/9. Larger organisations may already have their own predetermined record format which must be used. The topic-based approach referred to in paragraph C1.11 can be very useful in helping to process rough notes into a presentable record format.

C1.18 Processing notes and seeking further information

Notes should be taken throughout the observations and discussions referred to earlier. The rough notes might relate to a wide variety of matters:

- risks (those that are significant and those discounted as insignificant)
- persons thought to be at risk
- control measures (those that are effective and those that are not)
- problems identified and concerns expressed
- procedures and records of relevance
- additional control measures and precautions considered necessary.

The author generally finds that the best way to process all this information is to prepare a list of headings likely to be used in the final assessment records and to give each heading a number. The relevant heading number is then inserted alongside each note in turn, using a red pen. This helps to identify all the notes of relevance to each topic (computer-based notes could be sorted under topic headings in a similar way).

At this stage it often becomes apparent that further information is required in order to complete the assessment. This might involve the following:

- further observations or enquiries in the workplace
- assessments of activities not previously in progress
- checks on procedures, records and other documents
- checks on standards required or recommended by the HSE or in other published guidance.

C1.19 Assessment record formats

The three key elements that should be contained in any record of risk assessment are the following:

1. Risks associated with the work activity or location (and who is at risk).
2. Control measures or precautions that are (or should be) in place.
3. Improvements identified as being necessary to comply with the law.

Other details (which could be provided on a separate introductory sheet) should include:

- name of the employer
- address of the workplace or base concerned
- date(s) of the risk assessment
- name(s) and signature(s) of the person(s) making the assessment
- a date by which the assessment should be reviewed.

It should be noted that there is no legal requirement for risks to be quantified or categorised as 'high', 'medium' or 'low'. In the author's view these types of systems detract from the main purpose of the assessment – to identify the precautions necessary to comply with the law.

Two sample records are provided in Figures C1.5 and C1.6. The first of these deals with the warehouse area on a much larger site and covers the use of forklift trucks. Other topics to be covered in this area might have included racking and stacking of materials, storage of any high-risk materials, manual handling and use of any other types of material handling equipment.

The format used here has three columns – one each for risks, precautions and improvements. It should be noted that the recommended improvements were to be followed up just over two months after the assessment – the period allowed should reflect the potential seriousness of the risk and the assessor's judgement on how long is likely to be required to implement the necessary improvements.

The second example (Figure C1.6 on pages 178/9) deals with grass-cutting work carried out by a town council. The assessment programme included other activities carried out by council staff (digging and weeding, site clearance work, installation of hanging baskets and decorations) and the vehicles and other powered equipment used to carry out this work. A separate assessment of the council's administrative offices was carried out. The format used in Figure C1.6 has just two columns for risks and expected precautions. Recommendations on improvements required were communicated on a separate list.

The type of task procedure illustrated in Figure B8.1 (page 148) demonstrates another way of recording a risk assessment, providing that all of the significant risks associated with the task are taken into account. However, this type of record would need to be augmented by others relating to general matters such as fire or to specific types of work equipment (e.g. forklift trucks).

Figure C1.5 Example of a risk assessment record for work using forklift trucks

ARMAGEDDON ASSOCIATES RISK ASSESSMENT CHECKLIST

AREA: WAREHOUSE	RISK TOPIC 1. FORK LIFT TRUCKS	Sheet 1 of 1
RISKS IDENTIFIED	PRECAUTIONS IN PLACE	RECOMMENDED IMPROVEMENTS
Three counterbalanced diesel lift trucks are used within the warehouse and surrounding areas. The principal risks involved are: – collisions (with pedestrians, other vehicles or structures)	All drivers have been trained and tested in accordance with the HSE ACOP and have written authorisation to drive. Trucks are maintained and repaired in accordance with a formal contract. Trucks are thoroughly examined under a formal scheme of examination (LOLER).	
– overturning of the trucks (probably due to overloading) – loads falling during manoeuvres (particularly from high racking) – insecure stacking of pallets.	A pre-use check should be carried out on each truck every day. Areas of use are level and in good condition. There is adequate provision of lighting both inside the warehouse and in external areas.	Records show this does not happen every day. Drivers must be reminded of their responsibilities. Some light fittings are out and should be replaced. Check on lighting during regular H & S inspections.
Warehouse employees are mainly at risk but other staff and some visiting drivers may also be at risk if entering the warehouse area.	Entry to the warehouse should be restricted to warehouse employees and other authorised persons. Safety footwear and high visibility vests must be worn in the warehouse (except the lunchroom and office). All trucks have a substantial overhead guard to protect from falling objects. Palletised materials should not be stacked more than four pallets high.	An unauthorised visiting driver was found in the warehouse without the required PPE – remind all hauliers of the requirements for authorisation and PPE.
SIGNATURE(S) *N. Warr*	NAME(S) N. WARR	C. PEACE DATE 20/2/05
DATE FOR NEXT ASSESSMENT	RECOMMENDATION FOLLOW-UP	
	C. Peace 20/05/05	

Figure C1.6   Example of a risk assessment record for grass cutting work

TRUMPTON TOWN COUNCIL
RISK ASSESSMENT

1. GRASS CUTTING

RISKS	EXPECTED PRECAUTIONS
Several types of grass-cutting equipment are used by council staff. These types are as detailed below.	Grass-cutting equipment may only be used by operatives who have been trained and authorised to operate that type of machine.
The choice of equipment depends upon factors such as the length of the grass, the slope and condition of the ground, weather conditions and the availability of a trained operator.	The choice of equipment must take account of the factors listed opposite. Operatives must check equipment before use. Unsafe equipment must not be used.
(The ride-on mower was obtained prior to 1998. It has no roll-over protection but there is considered to be no trapping risk and it is only used on level sports grounds.)	
	Key points to check for are:
– Strimmers	Guards in position, strimmer head tight, fuel tank and line.
– Cylinder mowers	Guards in position, fuel tank and line.
– Rotary mowers	Guards in position, fuel tank and line.
– Gang mowers	Guards in position, PTO cover in place, tractor controls, fuel line.
– Ride-on mower	Guards in position, PTO cover in place, controls, fuel line interlock mechanism working. (The ride-on mower has an interlock which cuts off the power when the driver leaves the seat).
Equipment presents risks to operatives from	
– foot contact with blades	Safety footwear must be worn at all times.
– projectiles	Eye protection (visors, goggles or safety spectacles) must be worn when using strimmers or hand mowers.
– noise	Hearing protection (muffs or plugs) is recommended when cutting.

Hazard / Risk	Control measures
Projectiles (e.g. stones, litter, animal faeces) can create risks for operatives and members of the public (including vehicle drivers).	The area must be checked prior to cutting and any debris removed (gloves must be worn for removing debris).
Cutter blades present risks to operatives (particularly when removing blockages, etc.) and possibly also to members of the public.	Guards must be in position when equipment is being operated. Operatives must ensure that the power is off and the blades have stopped moving before attempting to clear blockages or to clean or make adjustments in the vicinity of the blades. Use of gloves may be necessary. Machines should be switched off if they are to be left unattended, particularly if members of the public are in the vicinity.
Work close to roads involves risks both to and from passing traffic. These risks will be greater if the operative needs to stand in the roadway or the mower overlaps into the roadway.	Operatives must wear high visibility clothing for work close to public roads. The use of warning signs and/or cones may be necessary and operatives should be alert to the presence of traffic. The use of warning signs and/or cones may be necessary and operatives should be alert to the presence of traffic.
Fuel and oils present risks from: – fire – skin contact – pollution due to spillage.	Petrol must only be carried in approved containers, which must be kept secure when not in use. Smoking is not allowed when handling petrol. Gloves must be worn when handling petrol or oils. Containers must be kept closed when not in use. Any spillages must be cleared up promptly.

Signed   B. Cuthbert   A. Dibble

Due for review   February 2007   Date   05/02 2005

C1.20 Model risk assessments

The ACOP to the Management Regulations states that employers who control a number of similar workplaces containing similar activities may produce a 'model' risk assessment reflecting the core hazards and risks associated with these activities. However, the ACOP goes on to say that employers or managers at each workplace must:

- satisfy themselves that the 'model' is appropriate to their type of work; and
- adapt the 'model' to the detail of their own actual work situations.

This can be done by use of a format that contains columns for the risks and expected precautions, together with a third column in which 'local arrangements' can be detailed. This approach can be adopted for chains of offices or retail outlets while the author has used it successfully with a national network of motor vehicle dealerships. Use of the model risk assessment was accompanied by an audit programme that checked on the existence and quality of the local arrangements as well as compliance with the standards contained in the 'expected precautions' column.

The example provided in Figure C1.7 on page 182 deals with the inspection and maintenance of electrical equipment used in servicing workshops. Other parts of the assessment related to the following:

- the fixed electrical installation (including its inspection and maintenance)
- the choice of equipment for workshop use (air-powered, battery-powered and 110 volt equipment all being preferable to mains voltage)
- use of hand lamps (possibly near flammable liquids or in wet conditions).

C1.21 Dynamic risk assessment

The concept of dynamic risk assessment was introduced in paragraph B8.2 in relation to the determination of safe systems of work. All of us carry out dynamic risk assessments as part of our daily lives, for example each time that we cross a road on foot, drive from a side road onto a main road or decide whether to take a coat when we walk to the pub. In doing this we combine our knowledge and experience of road traffic and the vagaries of the weather with on-the-spot observations of prevailing conditions and other relevant information, such as weather forecasts. Our capabilities in relation to crossing the road safely were closely monitored by our parents before we were allowed to cross without direct supervision. Our driving abilities should have been formally tested by an examiner.

Even when quite detailed 'generic' assessments of risk have been carried out using the processes described already in this chapter, workpeople still have to carry out dynamic assessments of the circumstances they work in or the equipment they work with:

- in the warehouse (Figure C1.5 on page 177) the forklift drivers must use their judgement in their pre-use checks and in how they operate their trucks
- in carrying out grass cutting (Figure C1.6 on pages 178/9) the operatives must check the condition of their equipment before use and check areas for debris prior to cutting

Figure C1.7   Example of a risk assessment record for work using electrical equipment

B4   WORKSHOP ELECTRICAL EQUIPMENT

DETAIL OF RISKS	EXPECTED PRECAUTIONS	LOCAL ARRANGEMENTS
Electrical equipment used in a robust working environment such as a workshop is particularly prone to damage.	**B 4.3 INSPECTION AND MAINTENANCE** Staff are expected to check the condition of all electrical equipment before using it. They should be particularly alert for: – cables that are not properly connected to the equipment or to plugs – cables that have been damaged by heat or crushing – faulty switches or controls – the presence of burns or scorch marks – loose internal components which may cause short circuits. Any defective equipment must be taken out of use and arrangements made for its repair or replacement. Checks on electrical equipment must be made periodically as part of the quarterly Health and Safety Inspections but all local management should monitor its condition on an ongoing basis. All electrical equipment* in the workshop must be inspected and tested by a competent person (e.g. an electrician) at least every 12 months. Records of such inspections and tests must be kept on file. * This requirement does not apply to: – battery operated equipment (less than 20V) – equipment less than 50V AC, e.g. telephone equipment – double insulated equipment	Inspection and testing of electrical equipment is carried out by: The inspection and test reports are kept by:

- in the garage (Figure C1.7 on page 182) workshop staff are expected to check electrical equipment prior to use.

In order to make such judgements effectively, staff must have received appropriate instruction or training in what to look for. This can be done by making them aware of the key points (as referred to in the examples in Figures C1.6 and C1.7) or providing them with a pre-use checklist (as illustrated in Figure B5.6). The activity checklist for off-road cycling tours illustrated in Figure B8.2 is a further example of where those expected to make dynamic assessments can be assisted in the process. The permit-to-work system described in paragraphs B8.9 to B8.12 is another means of ensuring that a satisfactory dynamic risk assessment is carried out – here the permit issuers must be formally authorised before they are deemed capable of making such an assessment.

C1.22 After the assessment

Once the assessment has been completed and recorded, the results will need to be discussed with other relevant parties, some of whom are likely to be outside the assessment team. In some cases the need for action may have been identified but the precise nature of that action may still need to be determined. In other situations the importance and urgency of recommended actions may need to be emphasised to other members of management.

Those carrying out risk assessments should not hold back on making appropriate recommendations because they fear they will never be implemented and similarly they should not back down on their recommendations because of pressure from other quarters. Their role is to carry out risk assessments to the best of their abilities – responsibility for complying with the legislation rests with their employers.

Some recommendations involving major expenditure or staff time may need to be costed and alternative methods of controlling some risks may need to be examined and evaluated. What should emerge is a properly prioritised action plan that includes responsibilities and time scales for each of the actions.

C1.23 Implementation and follow-up

The risk assessment team may still need to be involved in the implementation of the action plan, either in implementing some actions themselves or in providing guidance and assistance. Follow-up of action points is important, partly in ensuring that the necessary actions actually have been implemented and partly to check that no new risks have been introduced inadvertently. The risk assessment record should be revised or annotated once actions have been completed satisfactorily.

Where actions are not carried out within the required time, whether due to indolence or genuine practical difficulties, there may be a need for a further management review of the position.

C1.24 Review

Regulation 3(3) of the Management Regulations requires a risk assessment to be reviewed if:

- there is reason to suspect that it is no longer valid
- there has been a significant change in the matters to which it refers.

An assessment may be revealed to be no longer valid because:

- something that was previously unforeseen has now occurred
- it becomes apparent that risks were higher than previously thought
- control measures prove less effective than anticipated.

This may become apparent within the organisation itself (e.g. from an accident or incident investigation or an inspection) or from external sources such as the HSE, equipment or materials suppliers or others 'in the same business'. Significant changes may relate to the following:

- equipment or materials being used
- alterations to processes, products or services provided
- the employer's own premises or locations where work is carried out
- staffing levels.

A review of the risk assessment would not necessarily require the whole assessment process to be repeated but is quite likely to identify the need for new control measures or changes to existing ones. However, the conclusion may be drawn that the existing precautions are still adequate and that no additional controls are reasonably practicable.

Most businesses are subject to a gradual process of change and it is prudent to review risk assessments periodically to ensure that they are still valid. Such frequencies might vary from annually to every five years depending on the levels of risk involved and the possibility of gradual changes taking place. The sample assessment record in Figure C1.6 on pages 178/9 schedules an assessment review to take place in two years' time.

Source materials

1. HSE (2000) 'Young People at Work: A Guide for Employers', HSG 165.
2. HSE (2002) 'New and Expectant Mothers at Work: A Guide for Employers', HSG 122.
3. HSE (2000) 'Management of Health and Safety at Work', ACOP, L 21.

COSHH Assessments

C2.1 Introduction

The requirement to carry out an assessment of the health risks from hazardous substances was introduced with the first COSHH Regulations in 1988, thus pre-dating the requirement for a general risk assessment contained in the Management Regulations. The COSHH Regulations were based on earlier regulations dealing with asbestos and lead.

Many of the approaches to risk assessment described in Chapter C1 are equally applicable to carrying out COSHH assessments. This chapter outlines the harmful effects of hazardous substances and provides a summary of the requirements of the COSHH Regulations. It then concentrates on the types of precautions available to control the risks from hazardous substances and the hierarchical way in which these control measures must be applied. Some specific practical aspects of carrying out COSHH assessments are considered and examples of COSHH assessment records are provided.

C2.2 Harmful effects of hazardous substances

The many and varied ways in which hazardous substances can damage the human body is a specialist subject in its own right and many detailed works have been published on the subject. In some cases the effects may be short-term (acute) while in others there are more long-term (chronic) effects. The same substance might have an acute effect on one part of the body and a chronic effect on a different part. Some of the more common effects include the following:

Respiratory damage:
- pneumoconiosis – a general term used to describe dust-induced lung disease, of which asbestos, silicosis and byssinosis are specific forms
- lung cancer – caused by asbestos and a variety of organic materials including pitch and tar (mesothelioma is a cancer affecting the lining around the lungs and is specifically associated with asbestos)
- asthma – sensitisation of the respiratory system which can be caused by a variety of substances including flour, grain, animal hair, wood dusts and some chemicals, including isocyanates (once commonly used as hardening agents in paints and adhesives)
- irritation of the respiratory system, often caused by inhalation of acid or alkali gases or mists.

Biological effects:
- caused by micro-organisms (bacteria, fungi, viruses and microscopic parasites), often associated with animals, birds or fish

- examples include anthrax, brucellosis, hepatitis, legionnaire's disease, leptospirosis (Weil's disease), tetanus, tuberculosis
- biological agents are classified from Group 1 (unlikely to cause human disease) to Group 4 (causes severe human disease).

Poisoning and asphyxiation:
- both acute and common effects on a variety of target organs, damaging vital physiological processes
- can be caused by metals and their compounds (e.g. lead, mercury) and many organic and inorganic chemicals
- asphyxiation may be due to lack of sufficient oxygen to sustain life or the effects of carbon monoxide on the haemoglobin in the blood.

Skin effects:
- dermatitis due to primary irritants (acids or alkalis damaging tissue or solvents and detergents removing natural oils), or due to sensitising agents (some solvents, cereals, flour, animal products, isocyanates)
- skin cancers because of exposure to mineral oil, tar, pitch or arsenic.

Further examples of occupational diseases and the substances that can cause them are contained in Schedule 3 of the HSE booklet on RIDDOR (ref. 1). Information on the potential effects of individual products should be readily available in the data sheet provided by the manufacturer or supplier or printed on packaging.

C2.3 Entry routes
An important consideration in carrying out COSHH assessments is whether hazardous substances can cause harm by accessing one of the three main entry routes into the body. These are:

1. Inhalation: damage is caused by contact with or absorption through the respiratory system.
2. Ingestion: damage is caused by contact with or absorption through the digestive system.
3. Skin: damage is caused by contact with or absorption through the skin or eyes, or possibly by penetration through needle sticks or cuts.

The potential for entry by each of these routes will depend upon the form in which the hazardous substance is present in the workplace. A solid material might present a relatively low risk but once it is machined or processed its dust may constitute a significant inhalation risk and also a possible risk by ingestion through inadvertent contamination of food, drink or cigarettes. A liquid can present risks of ingestion and through skin contact but its vapour may also create a risk via inhalation.

C2.4 A summary of the COSHH Regulations
A definition of 'substance hazardous to health' is provided in Regulation 2(1) and includes the following:

- substances designated as very toxic, toxic, corrosive, harmful or irritant under the Chemicals (Hazard Information and Packaging for Supply) Regulations (CHIP) these substances should be labelled with the standard orange and black symbols and accompanied by relevant 'risk phrases'

- substances for which the Health and Safety Commission has approved a maximum exposure limit or an occupational exposure standard
- biological agents (micro-organisms, cell cultures or human endoparasites)
- dust of any kind, at concentrations in air, averaged over an eight-hour period, equal or greater than 10 mg/m^3 (total inhalable dust) or 4 mg/m^3 (respirable dust)
- any other substance creating comparable hazards.

Given this comprehensive definition, it is always advisable to assume that the Regulations apply in cases of doubt.

The Regulations place duties on employers in relation to their employees and also on employees themselves (Regulation 8). The self-employed have duties as if they were both employer and employee. Regulation 3(1) extends the employer's duties 'so far as is reasonably practicable' to 'any other person, whether at work or not, who may be affected by the work carried on'. Consideration must be given to all of the people referred to in paragraph C1.16, taking account of the fact that some (e.g. children and the elderly) will not have the same awareness of risks from hazardous substances as employees or contractors. An early prosecution under the COSHH Regulations was of a doctor's surgery following an incident where a young child gained access to an insecurely fastened cleaner's cupboard and drank the contents of a bottle of cleaning materials, thereafter suffering serious ill effects.

The requirements of the regulations are summarised below. Full details of the regulations and the associated ACOP are provided in a single HSE booklet.

- Prohibitions relating to certain substances (Reg. 4) the manufacture, use, importation and supply of certain substances is prohibited (full details are contained in Schedule 2 to the Regulations).
- Application of Regulations 6 to 12 (Reg. 5) exceptions are made from the application of the COSHH Regulations where there are other more specific regulations controlling risks. These relate to lead; asbestos; radioactive, explosive or flammable properties of substances; substances at high or low temperatures or high pressure; substances administered in the course of medical treatment; and respirable dust in coal mines.
- Assessment (Reg. 6) an assessment of health risks from hazardous substances must be carried out in order to identify steps necessary to comply with the regulations and these steps must be implemented. The assessment must be reviewed if there is reason to suspect that it is no longer valid or there are significant changes.
- Prevention or control of exposure (Reg. 7) this must be achieved through elimination or substitution or the provision of adequate controls, e.g. enclosure, local exhaust ventilation (LEV), ventilation, systems of work, personal protective equipment (PPE). The hierarchical approach required by this regulation is explained further in paragraph C2.5.
- Use of control measures, etc. (Reg. 8) employees must make 'full and proper use of control measures, PPE etc.', and employers 'take all reasonable steps' to ensure that they do. Employees must also report defects.
- Maintenance, examination and test (Reg. 9) control measures, including PPE, must be maintained in an efficient state, efficient working order,

good repair and in a clean condition. Further details of the requirements of this regulation are provided in paragraph C2.10.

- Monitoring exposure at the workplace (Reg. 10) workplace monitoring may be necessary to ensure adequate control or protect health. The need for such monitoring is explored further in paragraph C2.14.
- Health surveillance (Reg. 11) this is required where appropriate for the protection of employees. This too is explained further in paragraph C2.14.
- Information, instruction and training (Reg. 12) must be provided for persons exposed to hazardous substances, so that they know the risks to which they are exposed and the precautions that should be taken. Information must also be provided on the results of exposure monitoring and the collective results of health surveillance.
- Arrangements to deal with accidents, incidents and emergencies (Reg. 13) – emergency procedures must be established and information on emergency arrangements made available. General guidance on emergency arrangements is provided in chapter B3.

C2.5   The hierarchy of control measures

The COSHH Regulations require a very hierarchical approach to be taken to risks from hazardous substances. In strict order of preference this is as follows:

1. Prevent exposure (by elimination or substitution).
2. Provide adequate control by measures other than PPE (with enclosure and other engineering controls preferable to control by ventilation or organisational measures).
3. Provide adequate control by providing suitable PPE.

The detailed requirement is contained in Regulation 7 which states:

(1) Every employer shall ensure that the exposure of his employees to substances hazardous to health is either prevented or, where this is not reasonably practicable, adequately controlled.

(2) In complying with his duty of prevention under paragraph (1), substitution shall by preference be undertaken, whereby the employer shall avoid, so far as is reasonably practicable, the use of a substance hazardous to health at the workplace by replacing it with a substance or process which, under the conditions of its use, either eliminates or reduces the risk to the health of the employees.

(3) Where it is not reasonably practicable to prevent exposure to a substance hazardous to health, the employer shall comply with his duty of control under paragraph (1) by applying protection measures appropriate to the activity and consistent with the risk assessment, including in order of priority:
   (a) the design and use of appropriate work processes, systems and engineering controls and the provision and use of suitable work equipment and materials;
   (b) the control of exposure at source, including adequate ventilation systems and appropriate organisational measures; and

(c) where adequate control of exposure cannot be achieved by other means, the provision of suitable personal protective equipment in addition to the measures required by sub-paragraphs (a) and (b).

It should be noted that the qualifying phrase in the hierarchy is 'so far as is reasonably practicable' – see paragraph A2.11. This means that cost factors must be taken into account – in contrast to a similar hierarchical requirement relating to dangerous parts of machinery which is qualified by 'so far as is practicable' (see paragraph D1.8).

C2.6   Prevention of exposure
Exposure can be prevented by:

Elimination of the hazardous substances
- by use of alternative work methods, e.g. use of high pressure water rather than solvents for cleaning
- changing process parameters to avoid production of hazardous emissions, waste or byproducts.

Substitution of the hazardous substances by a non-hazardous or less hazardous alternative, e.g.
- using water-based paints or inks rather than solvent-based ones
- using less hazardous solvents
- using hazardous substances in a more dilute concentration (many bleaches are now supplied in a more dilute form)
- using less hazardous forms of substances, e.g. granules rather than powder.

There may sometimes be a penalty for making such changes, e.g. water-based paints may not produce the same qualities of protection or appearance. However, elimination or substitution is often 'reasonably practicable'.

C2.7   Controls other than PPE
Paragraph (4) of Regulation 7 specifies various control measures that must be adopted to comply with the requirements of paragraph (3). These include:

- total enclosure of process and handling systems (unless not reasonably practicable)
- plant, processes and systems of work that minimise the generation of, or suppress and contain, spills, leaks, dust, fumes and vapours (this might involve the use of a combination of partial enclosure and the use of local exhaust ventilation (LEV): detailed guidance on LEV is available from the HSE – see refs 3 and 4 on page 196)
- adoption of suitable maintenance procedures
- environmental controls, e.g. appropriate general ventilation
- limitation of the quantities of hazardous substances in workplaces
- minimising the numbers of persons exposed to hazardous substances.

Paragraph (5) of the Regulation identifies control measures that must be applied to prevent exposure to carcinogens. These measures, which are summarised below, may also be appropriate to the control of other types of hazardous substances:

- provision of adequate washing facilities and other hygiene measures
- prohibition of eating, drinking and smoking in areas of potential contamination
- regular cleaning of walls and surfaces
- use of suitable signs to warn of areas of potential contamination
- safe handling, storage, transport and disposal of hazardous substances
- use of closed and clearly labelled containers (clear labelling can prevent accidental exposure and inadvertent ingestion).

Paragraph (6) of Regulation 7 refers to additional measures required to control biological agents. These are summarised in paragraph C2.13.

C2.8 Control of exposure using PPE

Control using PPE should be the final option, where prevention or adequate control of exposure by other means is not reasonably practicable. This might be the case where:

- the scale of use of hazardous substances is very small
- adequate control by other means is impractical for technical reasons
- control by other means is excessively expensive or difficult in relation to the level of risk
- PPE is used temporarily pending the implementation of adequate control by other means
- emergency situations are being dealt with
- maintenance activities are only carried out infrequently.

PPE necessary to control risks from hazardous substances might include the following:

- respiratory protective equipment (RPE)
- protective clothing
- hand or arm protection (usually gloves)
- eye protection
- protective footwear.

Paragraph (9) of Regulation 7 of the COSHH Regulations stipulates that PPE must:

(a) comply with any provision in the Personal Protective Equipment Regulations 2000 which is applicable to that item of personal protective equipment

(b) in the case of respiratory protective equipment, where no provision referred to in sub-paragraph (a) applies, be of a type approved or shall conform to a standard approved, in either case, by the Executive.

The HSE has detailed guidance available on PPE generally ('Personal Protective Equipment at Work') and RPE in particular ('The Selection, Use and Maintenance of Respiratory Protective Equipment'). Further guidance on the use of PPE is provided in Chapter C6.

C2.9 'Adequate control'

Paragraphs 7, 8 and 11 of Regulation 7 provide a definition of what is meant by the term 'adequate control' used in the Regulation, in relation to inhalation risks:

- *Substances with a Maximum Exposure Limit (MEL)* – Where an MEL has been approved (usually for the more hazardous substances), exposure must be reduced so far as is reasonably practicable and kept below the MEL at all times.
- *Substances with an Occupational Exposure Standard (OES)* – For substances with an OES, control should be such that the OES is not exceeded or if it is exceeded the reasons must be identified and the situation remedied as soon as is reasonably practicable.

Tables of MELs and OESs are published regularly in an HSE booklet, 'Occupational Exposure Limits', EH 40.

In some situations it can be fairly easy to judge whether exposure is comfortably below the MEL or OES. However, it may be necessary to carry out atmospheric monitoring as part of the COSHH assessment and regular monitoring may also be required to ensure that adequate control is maintained (see paragraph C2.14).

C2.10 Maintenance, examination and test of control measures, etc.

Regulation 9 contains both general and specific requirements for the maintenance of control measures:

(1) Every employer who provides any control measure to meet the requirements of regulation 7 shall ensure that it is maintained in an efficient state, in efficient working order, in good repair and in a clean condition.

(2) Where engineering controls are provided to meet the requirements of regulation 7, the employer shall ensure that thorough examinations and tests of those engineering controls are carried out –

    (a) in the case of local exhaust ventilation plant, at least once every 14 months, or for local exhaust ventilation plant used in conjunction with a process specified in Column 1 of Schedule 4, at not more than the interval specified in the corresponding entry in Column 2 of that Schedule;

    (b) in any other case, at suitable intervals.

(3) Where respiratory protective equipment (other than disposable respiratory protective equipment) is provided to meet the requirements of regulation 7, the employer shall ensure that at suitable intervals thorough examination and, where appropriate, testing of that equipment are carried out.

(4) Every employer shall keep a suitable record of the examinations and tests carried out in accordance with paragraphs (2) and (3) and of repairs carried out as a result of those examinations and tests, and that record or a suitable summary thereof shall be kept available for at least 5 years from the date on which it was made.

The COSHH ACOP states that, where possible, all engineering control measures should receive a visual check at least once every week. Checks may simply confirm that there are no apparent leaks from vessels or pipes and that LEV or cleaning equipment appear to be in working order. It is good practice to keep records of such checks, although not a legal requirement.

Paragraph (2) requires thorough examinations and tests of engineering controls. Requirements relating to LEV are reviewed below but for other engineering controls such examinations and tests must be 'at suitable intervals', and suitable records must be kept for at least five years. The nature of examinations and tests will depend upon the type of engineering control involved and the potential consequences of its deterioration or failure. Examples might involve:

- visual inspections of vessels and pipelines
- detailed inspections and possibly non-destructive testing of critical process vessels
- regular testing of detectors and alarm systems
- planned maintenance of general ventilation equipment
- routine checks on filters in vacuum cleaning equipment.

C2.11 Maintenance of LEV

Paragraph (2) of Regulation 9 requires most LEV systems to be thoroughly examined and tested at least once every 14 months, although Schedule 4 to the Regulations requires greater frequencies for LEV used in conjunction with a handful of specified processes. Depending on the design and purpose of the LEV concerned, the examination and test might involve the following:

- visual inspection of the LEV system
- air flow measurements using an air velocity meter (comparing results with recommended capture and duct velocities)
- static pressure measurements (comparing these with design or commissioning pressures)
- visual checks of efficiency using smoke generators or dust lamps
- filter efficiency tests.

Atmospheric monitoring may also be appropriate in some circumstances (see paragraph C2.14).

Further guidance on techniques is available in an HSE booklet ('The Maintenance, Examination and Testing of Local Exhaust Ventilation') and in the COSHH ACOP ('Control of Substances Hazardous to Health'), which also specifies details of the records that should be kept in respect of thorough examinations and tests.

C2.12 Maintenance of PPE

All types of PPE are subject to the general maintenance requirements contained in paragraph (1) of Regulation 9, while paragraph (3) of the same Regulation contains specific requirements relating to non-disposable respiratory protective equipment (RPE). Most types of PPE can easily be seen to be defective by the user although in some cases (e.g. for gloves providing protection against strongly corrosive chemicals) more formalised inspection systems may be appropriate.

For non-disposable types of RPE, the COSHH ACOP states that thorough examinations and, where appropriate, tests should be made at least once every month, although it states that for half-mask respirators, used more occasionally in relatively low risk situations, intervals of up to three months are acceptable.

For simple respirators a visual examination of the condition of the facepiece, straps, filters and valves is likely to be sufficient. However, for airline-fed RPE the quality and flow of the air supply should also be tested. For RPE supplied from compressed gas cylinders, more detailed examinations and testing will be required, including a check on the pressure in the cylinders. The COSHH ACOP ('Control of Substances Hazardous to Health') specifies details of the records that should be kept. Detailed guidance is available in an HSE booklet ('The Selection, Use and Maintenance of Respiratory Protective Equipment').

C2.13 Carcinogens and biological agents

Detailed control measures required in respect of carcinogens were summarised in paragraph C2.7 above. Paragraph (7) of Regulation 7 specifies a number of measures that are required to control exposure to biological agents. As was the case for carcinogen controls, some of these may also be relevant to controlling other types of hazardous substances. Control measures include:

- use of warning signs (a biohazard sign is shown in Schedule 3 of the Regulations)
- specifying appropriate decontamination and disinfection procedures
- safe collection, storage and disposal of contaminated waste
- testing for biological agents outside confined areas
- specifying work and transportation procedures
- where appropriate, making available effective vaccines
- instituting appropriate hygiene measures
- control and containment measures where human patients or animals are known (or suspected to be) infected with a Group 3 or 4 biological agent.

Several schedules, appendices and annexes within the COSHH ACOP ('Control of Substances Hazardous to Health') contain a variety of requirements and guidance relating to both carcinogens and biological agents.

The COSHH (Amendment) Regulations 2003 clarified the status of 17 dioxins by explicitly defining them as carcinogens. Mutagens (which can cause genetic damage) must be controlled in the same way as carcinogens. In practice this only affects triglycidyl isocyanurate (TGIC), which is the only mutagen not already classified as a carcingogen. Further detail on these amendments is contained in an updated version of the free HSE leaflet – INDG 136.

C2.14 Monitoring of exposure and health surveillance

Regulation 10 requires exposure to be monitored where requisite to maintain adequate control or to protect the health of employees. This is automatically required under Schedule 5 to the Regulations for work involving vinyl chloride monomer and spray from electrolytic chrome plating. Records of monitoring must be kept for at least five years, and for forty years where they are representative of the personal exposure of identifiable employees.

Exposure monitoring may be continuous (as required for vinyl chloride monomer) or at a specified frequency, e.g. weekly, fortnightly, monthly. The techniques used can involve direct reading instruments (some of which can incorporate an alarm if a specified concentration is exceeded), chemical indicator tubes and sampling pumps fitted with filter heads. Monitoring techniques are explored further in Chapter D4 and guidance is also available from the HSE.

Health surveillance is required under Regulation 11 where it is appropriate for the protection of the health of employees. Schedule 6 to the Regulations contains a list of substances and processes where health surveillance must be carried out, unless the employee's exposure is not significant. There are further detailed requirements in Regulation 11 as to the ways in which surveillance must be conducted and the records that must be kept (these must be kept for at least forty years).

Techniques commonly used for health surveillance include the following:

- testing of blood, urine or exhaled air (for concentrations of hazardous substances or their metabolites)
- testing of physical capabilities, e.g. lung function testing
- physical examinations or measurements
- interviews (about possible physical changes or experience of ill health).

Chapter D4 contains further information on surveillance techniques and guidance on the subject is available from the HSE.

Decisions on whether exposure monitoring or health surveillance is necessary to control exposure or protect employees' health would normally be made when the COSHH assessment is carried out or during the review of an assessment. It may be deemed appropriate if it is found that an MEL has been exceeded or if an OES was being exceeded on a regular basis.

C2.15 Practical aspects of assessment

Many of the practical aspects of the planning and preparation for COSHH assessments and the carrying out of these assessments are similar to those referred to in Chapter C1 in relation to risk assessments generally.

Assessment units

Larger workplaces will often need to be divided up into manageable units for assessment purposes. Processes may need to be separated into their constituent stages.

Gathering information

Information will need to be gathered on all the various hazardous substances to be considered in the assessment. In production processes this might involve raw materials, products, by-products, emissions and other waste products. However, substances used or evolved during maintenance activities and materials present within the fabric of buildings (surface treatments, construction materials, contamination) must not be overlooked.

Technical information on substances may be available from manufacturers' and suppliers' data sheets, HSE publications and other reference books. Previous exposure monitoring surveys or COSHH assessments may also be available.

Workplace observation

Observations in the workplace are an essential part of the COSHH assessment process. High concentrations of dust, fume or vapours in the atmosphere are indicative of problems, as are lack of cleanliness, accumulations of waste materials, etc. It is also possible to observe whether specified operating procedures are being followed, whether PPE and RPE requirements are being complied with and whether LEV and general ventilation appear to be effective. The possibility of persons other than employees also being at risk must also be considered.

Figure C2.1   Example of a COSHH assessment record

SPARTAN ELECTRONICS   COSHH RISK ASSESSMENT

RISKS IDENTIFIED	PRECAUTIONS IN PLACE	RECOMMENDED IMPROVEMENTS
Several employees carry out soldering work at benches on a semi-continuous basis.	The soldering tools at these workstations are provided with local exhaust ventilation (LEV) close to their tips. These appear to be extremely effective in controlling the fume from the soldering process.	The LEV equipment should be thoroughly examined and tested at least every 14 months.
Fume is evolved from components passing through the curing oven.	The conveyor is enclosed and provided with LEV which appears to be effective.	As above.
Work in this section also involves use of solder paste and isopropanol (for cleaning purposes).	Fumes from these materials are very limited and suitable gloves are provided.	
Lacquer (containing 'harmful' xylene) and an aerosol adhesive (containing harmful' dichloromethane) are used in small quantities.	These substances are used in the main factory area which has good standards of general ventilation and they do not appear to represent a significant risk.	

A limited range of 'non-hazardous' general cleaning products are used, including bleach and disinfectant.	These constitute no significant risk, although mixing of cleaning materials should always be avoided.	Cleaning staff must be instructed never to mix cleaning products.
Storage of substances in unlabelled or incorrectly labelled containers can present major risks.	One unlabelled container was present in the cupboard next to the soldering benches and two unlabelled spray containers were stored in the cleaner's cupboard.	Provide suitable labels for these containers. Include reference to the importance of correct labelling into the staff induction programme. Remind existing staff at team briefings.

SIGNATURE(S)	T. Blyth	NAME(S)	T. BLYTH	DATE	23/2/05

DATE FOR NEXT ASSESSMENT   May 2005   RECOMMENDATION FOLLOWx-UP / ROUTINE REVIEW

Talking to workers

This too is an important part of COSHH assessment. Employees can be questioned on whether they have experienced any health problems and can be asked their opinions on the effectiveness of the control measures provided. Their awareness of the risks associated with the substances to which they are exposed and of the precautions that should be adopted can be tested.

Further investigations

The conclusions that can be drawn following a COSHH assessment of an activity will generally fall into one of three categories:

1. There are no significant risks from the substances involved.
2. There are significant risks but these are adequately controlled.
3. Control measures are inadequate and need to be improved.

However, there will also be some situations requiring further investigation. This may involve carrying out some exposure monitoring and comparing the results with the MEL or OES. Alternatively it may require investigation into such aspects as:

- hazards associated with new substances identified during assessment
- the technical specifications of PPE provided (or what might be available)
- practicability of alternative work methods or control measures.

Two HSE publications provide general guidance on the practical aspects of COSHH assessments (see page 196 for details). COSHH Essentials is also available via the HSE website (see paragraph E1.7) and provides generic assessments and advice on control measures for common substances and tasks.

C2.16 Assessment records

As for other types of risk assessment, there is no standard format for recording a COSHH assessment. The different types of format described in paragraph C1.19 (and Figures C1.5 and C1.6) are acceptable, as is the model risk assessment approach described in paragraph C1.20. The assessment may also be incorporated within the type of task procedure depicted in Figure B8.1 on page 148. A completed example of a COSHH assessment record using the three-column format is provided on pages 194/5.

Source materials

1. HSE (1999) 'A Guide to the Reporting of Injuries, Diseases and Dangerous Occurrences Regulations 1995', L 73.
2. HSE (2002) 'Control of Substances Hazardous to Health', ACOP and Guidance, L 5.
3. HSE (1993) 'An Introduction to Local Exhaust Ventilation', HSG 37.
4. HSE (1998) 'The Maintenance, Examination and Testing of Local Exhaust Ventilation', HSG 54.
5. HSE (1992) 'Personal Protective Equipment at Work', L 25.
6. HSE (1998) 'The Selection, Use and Maintenance of Respiratory Protective Equipment', HSG 53.
7. HSE (1997) 'Monitoring Strategies for Toxic Substances', HSG 173.
8. HSE (1999) 'Health Surveillance at Work', HSG 61.
9. HSE (2004) 'Step by Step Guide to COSHH Assessment', HSG 97.
10. HSE (2003) 'COSHH Essentials: Easy Steps to Control Chemicals', HSG 193.

Noise Assessments

C3.1 Introduction

The requirement to carry out an assessment of the risks associated with noise at work also pre-dated the general requirement for risk assessment contained in the Management Regulations, with the Noise at Work Regulations being passed in 1989. In fact the risks from noise had been understood for many years previously, with HM Factory Inspectorate publishing a booklet entitled 'Noise and the Worker' in 1963.

This chapter will explain how excessive noise causes injury to the hearing and also how the 'decibel' system of noise works and some of its implications. The requirements of the Noise at Work Regulations will be summarised and some of the practical aspects of noise assessments examined. An introduction to techniques for reducing noise will be provided and some of the more common types of hearing protection described.

C3.2 How noise causes injury

Sounds are heard through the action of sound waves on the eardrum. These cause the eardrum to vibrate and activate the bones of the inner ear which transmit the sound to the cochlea. The cochlea contains a liquid which is set in motion by the sound and this motion is detected by tiny hair cells which transmit the sound to the brain.

Very loud noises (e.g. explosions) can perforate the eardrum but such cases are rare and the eardrum will usually heal up. Occupational deafness is usually due to long-term damage to the hair cells contained within the cochlea. All loud noises will have a short-term effect on the hair cells which will usually recover after a period of relative quiet. However, regular exposure to loud noises will cause permanent damage to the cells, from which they will not recover.

Normally the cells that detect sounds at the higher end of the frequency range are damaged first, affecting the victim's ability to differentiate between consonants in speech. Some sufferers become quite skilled at lip reading but cannot clearly make out speech from people who are out of their vision. This loss of hearing capability can usually be demonstrated by the use of audiometry equipment (see Chapter D4). Usually both ears will be affected, although where the exposure to noise has been predominantly from one side (e.g. in the use of certain tools) the hearing will be worse on that side.

Damage to the hearing cells can also cause tinnitus, producing continuous sounds in the ear (usually ringing, buzzing or whistling) although this may be attributable to other causes. Excessive noise may also hinder communication, resulting in stress or errors due to misheard instructions. Sudden loud noises may surprise workers, again possibly causing errors or other adverse effects such as falls.

C3.3   The decibel scale and its implications

Sound is measured in decibels (dB) with a range from zero (the faintest audible sound) up to 140 decibels – the extremely dangerous level produced by a jet aircraft taking off 25 metres away. Most noises are a mixture of sound frequencies and a correction is made to take account of the human ear's varying capabilities to hear different frequencies. This is known as the A weighting, and noise levels measured in this way are shown as dB(A). Commercially available measuring instruments usually provide readings in dB(A). Typical approximate noise levels are as follows:

- quiet office: 40 dB(A)
- busy street: 80 dB(A)
- woodworking machines: 100 dB(A).

The decibel scale is logarithmic with every increase of 10 dB(A) representing a multiplication of ten in the noise level. Consequently 100 dB(A) is actually 100 times louder than 80 dB(A). An increase of just 3 dB(A) represents a doubling of the noise. Damage to the hearing is related to both the level of the noise and the duration of the exposure. The Noise at Work Regulations are based on daily personal noise exposure levels over an eight-hour working day (denoted as $L_{EP,d}$). There are two action levels triggering off the different requirements of the regulations:

- first action level: $L_{EP,d}$ of 85 dB(A)
- second action level: $L_{EP,d}$ of 90 dB(A).

(These levels are both expected to be reduced by 5dB(A) as a result of new regulations – see paragraph C3.5)

In addition there is a peak action level set at a maximum pressure of 200 pascals for the sound wave. This takes into account the potential damage from occasional loud impulse sources, such as cartridge-operated tools. (The first and second action levels are soon likely to be brought down to 80 dB(A) and 85 dB(A) respectively.)

When the logarithmic nature of the decibel scale is taken into account, each of the following exposures will give a daily personal noise exposure at the second action level of 90dB(A), providing that activities during the remainder of the day produce no significant levels of noise.

dB(A)
90 8 hours
93 4 hours
96 2 hours
100 48 minutes
103 24 minutes

In practice noise exposure will be a lot more variable than this, which must be taken into account during the noise assessment process.

C3.4   Noise measurement

There are several types of noise measurement instruments which can be used in carrying out noise assessments. The two main types are sound level meters and dosemeters.

Sound level meters

These measure the sound level at a particular location (a piece of equipment or an employee's ear). Integrating sound level meters will measure an A-weighted Leq over a period of measurement. Non-integrating meters will provide a simple indication of A-weighted sound at a particular point in time and are therefore only suitable for measuring relatively steady sound levels. The more sophisticated instruments are provided with octave-band analysis facilities, i.e. they can measure the sound levels over different sound frequency ranges.

Dosemeters

Dosemeters provide a means of measuring the noise exposure of an individual who may be highly mobile and carrying out a range of activities. They are worn by the worker with a microphone positioned close to the ear. They will provide the noise dose received over the measurement period, which can be compared to the action levels set in the regulations.

Other facilities

Some dual-purpose instruments can be used as both sound level meters and dosemeters. More sophisticated facilities such as data recorders and frequency analysers may be necessary for more detailed assessments.

Calibration and testing

All instruments should be checked both before and after use with the appropriate calibration instrument. Both the instruments and their sound calibrators should be tested at least every two years to ensure that they continue to meet the relevant standards.

The HSE booklet 'Reducing Noise at Work' contains detailed guidance on measuring instruments and the standards for their performance and periodic testing.

C3.5 A summary of the Noise at Work Regulations 1989

Full details of the regulations and related guidance is contained in the booklet 'Reducing Noise at Work' while a free HSE leaflet summarising the requirements ('Introducing the Noise at Work Regulations') is available. Details of likely changes to the regulations are summarised at the end of this paragraph.

Interpretation (Reg. 2)

The action levels referred to in C3.3 above are defined in Regulation 2 and Part I of the Schedule to the Regulations.

Assessment of exposure (Reg. 4)

Paragraph (1) requires employers to ensure that a competent person makes a noise assessment when any of their employees are likely to be exposed to the first action level or above or to the peak action level or above. The assessment must be adequate for the purposes of:

- identifying which employees are exposed

- providing information to comply with the duties under Regulations 7, 8, 9 and 11.

Noise is likely to be at hazardous levels if people have difficulty in being heard clearly two metres away. If this is not a problem and there are no high levels of noise from explosive or impact sources, then a specific noise assessment is not needed.

Assessment records (Reg. 5)
Adequate records of assessments must be kept (see paragraph C3.9).

Reduction of risk of hearing damage (Reg. 6)
Employers must reduce the risk of hearing damage to the lowest level reasonably practicable. This may mean taking steps over and above those required by the regulations, if it is reasonably practicable to do so.

Reduction of noise exposure (Reg. 7)
Where exposure is likely to be at the second action level or above, this must be reduced, so far as is reasonably practicable, by means other than the provision of personal ear protectors. (Noise reduction techniques are summarised in paragraph C3.10.)

Ear protection (Reg. 8)
Employers must ensure, so far as is practicable, that exposed employees are provided with ear protectors:

- on the request of employees exposed to the first action level or above
- to any employees exposed at or above the second action level or peak action level.

(Types of hearing protection are described in paragraph C3.11.)

Ear protection zones (Reg. 9)
Employers must ensure, so far as is reasonably practicable, that:

- compulsory hearing protection zones are identified (using the standard blue and white signs)
- their employees wear hearing protection within these zones.

Some employers also mark zones that are above the first action level to indicate that hearing protection is recommended.

Maintenance and use of equipment (Reg. 10)
Employers must ensure, so far as is practicable, that noise reduction measures and hearing protection (where compulsory) are 'fully and properly used' and 'maintained in an efficient state, in efficient working order and in good repair'. Employees also have duties to use hearing protection (in compulsory zones) and other protective measures fully and properly, e.g. keeping acoustic enclosures closed. They must also report any defects forthwith to their employer.

Provision of information to employees (Reg. 11)
Employers must provide employees exposed at or above any of the action levels with adequate information, instruction and training on the following:

- the risk of damage to hearing that such exposure may cause
- what steps employees can take to minimise the risk
- how to obtain personal hearing protection
- their own obligations under the regulations (see Reg. 10).

Duties of manufacturers, etc. (Reg. 12)

Manufacturers, designers, importers and suppliers must provide adequate information on likely noise levels if a machine is likely to cause exposure at or above the first action level or peak action level. (This extends duties under section 6 of HASAWA.)

In 2004 the HSE published a consultation document on proposed new regulations. These must be in operation by February 2006 to comply with an EC Directive, but the consultative document referred to a possible transitional period for compliance in the music and entertainment sector. The main changes from the existing regulations are expected to be:

- both the first and second action levels to be reduced by 5dB(A) (to 80 dB(A) and 85 dB(A) respectively)
- a new 'exposure limit value' of 87 dB(A), above which employees must not be exposed
- weekly exposure averages could be used where daily exposure varies
- noise measurements must be taken for exposures above the upper action level
- specific requirements for health surveillance, including testing of hearing.

C3.6   The purpose of the noise assessment

Once it has been established that a noise problem exists, the purpose of the noise assessment is set out in Regulation 4:

- to identify which employees are exposed (and to what extent in relation to the action levels)
- to provide information to enable compliance with the regulations:
  - reduction of noise exposure (Reg. 7)
  - provision of ear protection (Reg. 8)
  - identification of ear protection zones (Reg. 9)
  - provision of information to employees (Reg. 11).

This is likely to require some noise measurement to be carried out, particularly in order to identify the level of exposure in certain areas or in relation to specific equipment or activities. However, this must only be to the extent necessary in order to identify what needs to be done about the problem – sometimes assessments go into too much detail on noise levels but fail to identify what action is required to comply with the regulations.

C3.7   Assessment by a competent person

Regulation 4 specifically states that the assessment must be made by a 'competent person'. Part 3 of 'Reducing Noise at Work' (ref. 1) provides guidance on choosing a competent person. This states that the level of expertise necessary will depend on the complexity of the situation being assessed.

Necessary knowledge, experience and skills will include the following:

- understanding the purpose of the assessment
- awareness of what information needs to be obtained
- an appreciation of their own limitations (and those of the instruments they are using)

- how to use appropriate instruments to make measurements
- how to record, analyse and explain results
- how to use information provided by others (e.g. suppliers of noisy equipment or of hearing protection).

Details of suitable training courses are provided in 'Reducing Noise at Work' (ref. 1).

C3.8 Practical aspects of assessment

The practical aspects of planning, preparing for and carrying out noise assessments are similar to those involved in general risk assessments (see Chapter C1).

Gathering information

Relevant information should be gathered in advance such as:

- areas, equipment and activities thought to be noisy
- results of previous noise measurement
- any information from suppliers of noisy equipment
- relevant HSE guidance material (see 'Source materials' at the end of this chapter).

Workplace observation and discussions

Noisy equipment and activities will often be readily apparent during visits to the workplace. However, discussions with workers can sometimes reveal:

- infrequent but very noisy activities
- noisy alternatives that must be adopted when preferred equipment is unavailable
- noise problems associated with particular products
- noise caused by inadequate maintenance of equipment.

Assessing the daily personal noise exposure

In order to assess personal noise exposures it will be necessary to determine:

- noise levels for identified areas, equipment and activities
- whether those noise levels vary
- the duration of exposure to each level of noise
- the proximity of employees to the noise levels (preferably noise levels should be measured in their hearing zone).

Enough information must be obtained to determine whether exposure reaches the first action level (85 dB(A)), the second action level (90 dB(A)) or the peak action level. (HSE guidance states that if the maximum A-weighted sound level measured with a fast response is below 125 dB(A), the maximum peak level is likely to be below the peak action level.)

Detailed guidance for competent persons on carrying out a noise assessment is contained in the HSE booklet 'Reducing Noise at Work'.

C3.9 Assessment records

The format of a noise assessment is likely to be different from more general types of assessment records because of the need to include data about sound levels. However, the principles should still be the same – the nature and extent of the risk, the control measures in place and recommended improvements should all be recorded. Key aspects include:

- the sources of the noise (equipment, activities, etc.)
- details of sound level meter and dosemeter readings taken
- estimates (or measurements) of exposure periods
- assessments of whether first, second or peak action levels are exceeded
- details of controls in place (acoustic enclosures, ear protection and related signs)
- recommended actions for reducing noise exposure
- instruments used, date of the assessment and the person carrying it out.

The HSE booklet 'Reducing Noise at Work' provides further guidance on the contents of assessment records and also contains examples of assessment reports. Figure C3.1 overleaf in this chapter also contains a sample assessment report in a format that the author has found suitable for workplaces containing a limited range of noise risks.

Although Regulation 5 of the Noise at Work Regulations only requires that assessment records are kept until a further noise assessment is made, the possibility of civil litigation some time in the future would make it prudent to retain records almost indefinitely, or at least for the forty years required in relation to some COSHH-related records.

C3.10 Noise reduction techniques

Noise control engineering is a specialist subject in its own right and 'Reducing Noise at Work' provides a good summary of the techniques available. A separate HSE booklet, 'Sound Solutions', also provides sixty case studies demonstrating where a variety of noise control techniques has been used successfully in a wide range of industrial activities.

Regulation 12 requires information to be supplied on noise likely to be generated by work equipment and it is important to study this information prior to purchase as quieter equipment may be available from other suppliers and it may even be possible to utilise alternative equipment or work methods that generate much less noise. Equipment should be checked prior to final acceptance to ensure that it meets the appropriate noise specification. A free HSE leaflet is available to provide advice on purchasing machinery. Noise in the workplace can be reduced either by reducing the degree of noise generated in the first place or by preventing its transmission to areas where people might be affected by it.

It is vibrating surfaces or vibration within a fluid that generate noise. Noise can be eliminated or reduced if the vibration can be prevented or diminished. Means of reducing noise include:

- Modifying vibrating surfaces:
  - providing stiffening for structural parts or vibrating panels
  - use of anti-vibration mountings or vibration absorbing pads
  - providing absorbent surfaces (plastic or nylon) to cushion impacts.
- Reducing vibration in gases:
  - provision of silencers to reduce turbulence at exhausts
  - use of low-noise air nozzles
  - use of large-diameter and low-speed fans.

Figure C3.1　　Sample noise assessment report

Dates of assessment:	1 & 2 November 2000	Assessment by:　D B Appleby
Instruments used:	Alpha Integrating Sound Level Meter – Model C3 (Type 2)	
	Beta Personal Dosemeter Model K1	

NOISE SURVEY RESULTS

Position	Leq (2 mins) (dBA)	Average daily exposure (hours)	Assessed action level† 0	Assessed action level† 1	Assessed action level† 2	Comments
MACHINE AREA						
Cross-cut saw	92.3	8			✓	Normally same operator
Planing machine	103.7	2			✓	Any of four trained operators
Panel trimmer						
– alone	90.2	4		✓		Usually same operator
– planer operating	93.1	Max. 2			✓	Machines often work at same time
NAIL GUN BENCH						
Central position	98.4	8			✓	Dosemeter readings 97.8 dBA (N Peacock)
Outside position	97.9	8			✓	and 97.5 dBA (R Jones)

Operators mainly wearing ear protection

Operators wearing ear protection

METAL EDGING						Operator wearing ear protectio
Press	90.6	4			✓	Same operator dosemeter reading
Strip forming m/c	89.7	4			✓	89.2 dBA (D Grant)
Rolling-in m/c						Ear protection not worn
– feed position	88.5	4		✓		} Operators switch position every hour
– take-off	85.1	4	✓			}
BACKGROUND						
– Various areas	76.3 - 83.2	8	✓			Areas near nail gun bench highest

ACTION POINTS

1)   NOISE REDUCTION

1.1 Investigate means of protecting the panel trimmer from noise from the planing machine, e.g. acoustic screens.

2)   EAR PROTECTION

2.1 The following areas must be compulsory ear protection areas:

– wood working machine area*

– nail gun bench

– press and strip forming machine area

* the panel trimmer could be downgraded to 'voluntary' if action 1.1 is successful.

2.2 Recommend employees at the rolling-in machine feed position to wear ear protection.

2.3 Continue to make hearing protection readily available to all employees.

3)   EAR PROTECTION ZONES

3.1 Provide suitable signs to indicate compulsory ear protection zones (see 2.1).

3.2 Ensure all those working in these areas wear hearing protection.

3.3 Provide a 'hearing protection recommended' sign at the rolling-in feed position.

4)   INFORMATION TO EMPLOYEES

4.1 At the next staff-briefing meeting:

– show the video from the hearing protection suppliers

– summarise the results of this survey

– remind employees about the compulsory ear protection zones.

† Action levels are expected to be reduced by new regulations - see paragraph C3.5.

Efficient maintenance is also important in preventing vibration from loose, worn or badly fitting parts and in ensuring that equipment is correctly balanced and properly lubricated.

The transmission of noise through the air and within buildings into the hearing zones of workers can be reduced by measures such as:

- Acoustic enclosures or screens:
  - placing noisy equipment within an acoustic enclosure
  - providing acoustically enclosed control panels or noise refuges for workers
  - use of acoustic screens between workers and sources of noise.
- Sound-absorbent materials:
  - applied to ceilings and walls to prevent sound reflection.
- Positioning and separation:
  - extending or directing exhaust or extraction systems away from occupied areas
  - placing workers further away from noise sources.

C3.11 Ear protection

There are three main types of hearing protection in common use:

- *Ear muffs* – Muffs consist of plastic cups filled with plastic foam or other sound-absorbent material which fit over the ears. They are sealed against the head by soft pads containing either foam or a viscous liquid. Pressure is exerted on the cups in order to maintain the seal by a band normally passing over the head, but which can also be worn at the back of the head or under the chin. Muffs that can be fitted to safety helmets with a separate pressure band for each ear are also available. The condition of the seals will deteriorate due to age or misuse and the pressure exerted by the headband will also diminish with time. Muffs are not suitable to be worn over spectacles and the presence of jewellery or long hair may also affect their efficiency.
- *Semi-inserts* – These consist of a headband with a pair of pre-moulded plastic or rubber ear caps which are pressed into the entrance of the ear canal by pressure from the headband. They can easily be worn in conjunction with spectacles.
- *Ear plugs* – These are made of compressible materials such as rubber, plastic, plastic foam or mineral down (in a plastic membrane) which are fitted into the ear canal. Re-usable plugs are often supplied in a variety of sizes to ensure that a tight fit is achieved (sometimes a different size is needed for each ear). Disposable plugs usually need some moulding by the user to achieve a good fit. Re-usable plugs can be attached to cords to prevent their loss (important in food handling).

Hygiene is important with all types of hearing protection but particularly with semi-inserts and ear plugs which go into the ear canal. Semi-inserts and re-usable plugs should be cleaned regularly and clean hands are imperative when moulding disposable plugs into shape.

The choice of hearing protection will depend upon the nature of the noise exposure, the personal preference of the user and the level of protection (attenuation) required. Ear muffs are usually the most comfortable form of protection to be worn for long periods but either semi-inserts or ear plugs may be more suitable for people intermittently exposed to noise, for spectacle wearers or for use with other types of protective equipment. Plugs can be carried in the pocket and semi-inserts worn around the neck when they are not in use.

Suppliers of ear protection provide information on the attenuation properties of their products. These vary according to the frequency of the noise, most protection being more effective against frequencies in the audible range. As a rough guide most types of hearing protection will provide attenuation of at least 15 dB(A) if fitted reasonably well. In cases of doubt there is a sophisticated calculation which can be carried out using octave band analysis of the noise exposure and the attenuation figures for the hearing protection in order to determine whether it provides adequate protection. Details of this are provided in 'Reducing Noise at Work' (ref. 1).

Source materials

1.   HSE (1998) 'Reducing Noise at Work: Guidance on the Noise at Work Regulations 1989', L 108.
2   HSE (2002) 'Noise at work. Advice for employers', INDG 362.
3.   HSE (1995) 'Sound Solutions: Techniques to Reduce Noise at Work', HSG 138.

HSE guidance on specific industry sectors is also available, including the following:
HSE (2002) 'Noise (in Agriculture)', AS 8.
HSE (1995) 'Noise in Construction', INDG 127.
HSE (1998) 'Noise in Engineering', EIS 26.
HSE (1998) 'Control of Noise at Metal Cutting Saws', EIS 27.
HSE (1998) 'Control of Noise at Power Presses', EIS 29.
HSE (1995) 'Noise Control at Foundry Shakeouts'.
HSE (2000) 'Noise Assessments in Paper Mills', PBIS 1.
HSE (1993) 'Control of Noise in Quarries', HSG 109.
HSE (1990) 'Noise Reduction at Band Re-saws', WIS 4.
HSE (1990) 'Noise Enclosure at Band Re-saws', WIS 5.
HSE (1997) 'Noise at Woodworking Machines', WIS 15.
HSE (2002) 'Reducing noise from CNC punch presses', EIS 39.
HSE (2002) 'Reducing noise exposure in the food and drink industries', FIS 32.
HSE (2002) 'Noise levels and noise exposure of workers in pubs and clubs'.

Manual Handling Assessments

C4.1   Introduction

The Manual Handling Operations Regulations 1992 were one of the so-called 'six-pack' of regulations that came into operation on 1 January 1993. Like the Management Regulations and the regulations dealing with display screen equipment (DSE) and personal protective equipment (PPE), they contained a requirement for a risk assessment to be carried out. The regulations require the following:

- identification of manual handling operations involving risk of injury
- avoidance of such operations, so far as is reasonably practicable
- an assessment of operations that cannot be avoided
- reduction of risks from such operations, to the lowest level that is reasonably practicable.

This chapter provides a more detailed summary of the requirements of the regulations and examines each of the above stages in some depth.

Although the regulations themselves are quite brief, they are accompanied by considerable guidance, particularly in relation to the assessment process. Additional guidance has also been published by the HSE on manual handling in a number of important occupational sectors (see 'Source materials' at the end of the chapter).

C4.2   A summary of the Manual Handling Operations Regulations

Interpretation (Reg. 2)
Paragraph (1) defines terms used in the regulations:

> 'Injury' does not include injury caused by any toxic or corrosive substance which:
>
> (a) has leaked or spilled from a load;
> (b) is present on the surface of a load but has not leaked or spilled from it; or
> (c) is a constituent part of a load;
> and 'injured' shall be construed accordingly.

(This distinguishes manual handling risks, e.g. slippery substances on a load surface, from risks that should be addressed under the COSHH regulations.)

> 'Load' includes any person and any animal.

(A definition of considerable relevance to those in health care, agriculture, etc.)

'Manual handling operations' means any transporting or supporting of a load (including the lifting, putting down, pushing, pulling, carrying or moving thereof) by hand or by bodily force.

Paragraph (2) imposes duties on self-employed persons in respect of themselves as though they were employees.

Disapplication of regulations (Reg. 3)
The normal shipboard activities of the crew of a sea-going ship under the direction of its master are excluded from the regulations. (However, work on board ship by a shore-based contractor is covered while the ship is in territorial waters.)

Duties of employers (Reg. 4)
These requirements were summarised above in paragraph C4.1. The full text of Regulation 4 states:

(1) Each employer shall –
  (a) so far as is reasonably practicable, avoid the need for his employees to undertake any manual handling operations at work which involve a risk of their being injured; or
  (b) where it is not reasonably practicable to avoid the need for his employees to undertake any manual handling operations at work which involve a risk of their being injured –
    (i) make a suitable and sufficient assessment of all such manual handling operations to be undertaken by them, having regard to the factors which are specified in column 1 of Schedule 1 to these Regulations and considering the questions which are specified in the corresponding entry in column 2 of that Schedule,
    (ii) take appropriate steps to reduce the risk of injury to those employees arising out of their undertaking any such manual handling operations to the lowest level reasonably practicable, and
    (iii) take appropriate steps to provide any of those employees who are undertaking any such manual handling operations with general indications and, where it is reasonably practicable to do so, precise information on –
    (a) the weight of each load, and
    (b) the heaviest side of any load whose centre of gravity is not positioned centrally.
(2) Any assessment such as is referred to in paragraph (1)(b)(i) of this regulation shall be reviewed by the employer who made it if:
  (a) there is reason to suspect that it is no longer valid; or
  (b) there has been a significant change in the manual handling operations to which it relates,
  and where as a result of any such review changes to an assessment are required, the relevant employer shall make them.

Amendments to the regulation made in 2003 added sub-paragraph (3) which requires regard to be given to the following during the risk assessment:

- physical suitability of employees to carry out the operations

- their clothing, footwear or other personal effects
- their knowledge and training
- results of relevant risk assessments under the Management Regulations
- whether workers are within groups identified as especially at risk
- results of any health surveillance.

This regulation requires a similar hierarchical approach to that contained in the COSHH Regulations, i.e. avoidance of risk should be adopted, so far as is reasonably practicable, as opposed to reduction of risk. It also accepts that in some situations (e.g. emergencies) it may not be reasonably practicable to eliminate all elements of risk.

Duty of employees (Reg. 5)

Each employee while at work shall make full and proper use of any system of work provided for his use by his employer in compliance with regulation 4(1)(b)(ii) of these Regulations.

(This supplements employees' general duties under section 7 of HASAWA.)

Regulation 6 provides for exemptions to be made by the Secretary of State for Defence in respect of home and visiting forces. Regulation 7 extends the regulations to offshore installations. Regulation 8 revoked previous legislation relating to manual handling.

C4.3   Risk of injury

Manual handling operations are well established as a major cause of workplace accidents with over a quarter of all accidents reported under RIDDOR being attributable to this cause. There is already good evidence of the beneficial effects of the regulations in reducing the toll.

Appendix 1 of the HSE guidance booklet on manual handling contains guidelines that can be used to identify those manual handling operations where there is a risk of injury as opposed to those that involve little or no risk. The HSE states that these guidelines 'will provide a reasonable degree of protection to around 95% of working men and women'. However, it must be recognised that some members of the workforce are more vulnerable – young people and new or expectant mothers (see Chapter C1) and also elderly, frail or disabled workers and those with temporary injuries.

Lifting and lowering

Guideline weights for lifting and lowering are shown in Figure C4.1(a) opposite. It must be emphasised that these are not weight limits which must not be exceeded nor are they safe weights (see the previous paragraph). If the guidelines are exceeded then a detailed assessment should be made.

The guidelines assume that the load is easy to grasp with both hands and that the operation takes place in reasonable working conditions with the handler in a stable body position. If the handler's hands enter more than one zone, the smallest weight figures apply.

The weights in Figure C4.1 apply for up to thirty operations per hour. For more frequent lifting or lowering they should be reduced:

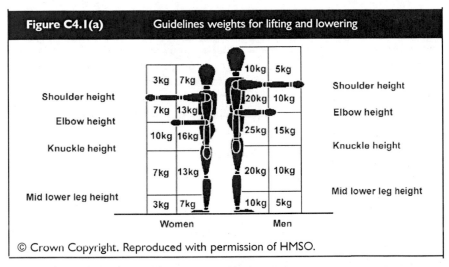

Figure C4.1(a)   Guidelines weights for lifting and lowering

© Crown Copyright. Reproduced with permission of HMSO.

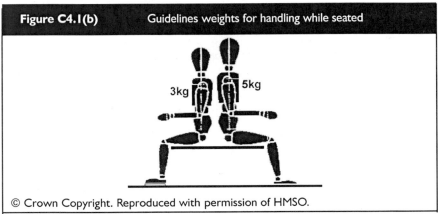

Figure C4.1(b)   Guidelines weights for handling while seated

© Crown Copyright. Reproduced with permission of HMSO.

- by 30 per cent for one or two operations per minute
- by 50 per cent for five to eight operations per minute
- by 80 per cent for more than twelve operations per minute.

More detailed assessment may also be necessary if:

- the worker does not control the pace of work (e.g. production-line work)
- there are inadequate pauses for rest
- there is no change of activity providing an opportunity to use different muscles
- the handler must support the load for any length of time.

Carrying

The HSE states that similar guideline figures can be applied where the load is held against the body and carried no further than 10 metres without resting. They may also be applied to loads carried securely on the shoulder for more than 10 metres without first having to be lifted, e.g. in unloading sacks. Longer carrying distances than 10 metres, loads held below knuckle height or loads being lifted onto the shoulder will all require assessment.

Pushing and pulling

For operations involving pushing and pulling (sliding, rolling or on wheels), guidance forces of 25 kg for stopping and starting the load and 10 kg for keeping it in motion are quoted for men, reducing to 16 kg and 7 kg respectively for women. However, judgements of what forces are actually involved are quite difficult in practice. While there is no limit for the distances involved in pushing and pulling, there should be adequate opportunity for rest or recovery.

Handling while seated

The guideline figures for handling while seated are also contained in Figure C4.1(b) on page 211.

C4.4   Identifying risk factors

Schedule 1 to the Regulations identifies various factors, which must be taken into account in assessing manual handling operations. These factors are divided into four main sections: factors associated with the tasks, the loads, the working environment or the capability of the individual performing the operation. Each of these main groupings is examined further below.

Tasks

During assessment, regard must be given to tasks that involve the following:

- holding or manipulating loads at a distance from the trunk (e.g. in reaching across obstacles)
- unsatisfactory bodily movement or posture (e.g. twisting the trunk, stooping or reaching upwards)
- excessive movement of loads (e.g. lifting or lowering loads where a change of grip is necessary, long carrying distances)
- excessive pushing or pulling of the load (e.g. large forces required, pushing or pulling below knuckle height or above shoulder height)
- risk of sudden movement of the load (e.g. due to a load suddenly becoming free or the unexpected movement of a person or animal being moved)
- frequent or prolonged physical effort (fatigue increases the risk of injury, especially in a relatively fixed posture)
- insufficient rest or recovery periods (rest breaks should be taken or job rotation introduced)
- a rate of work imposed by a process (these lessen the opportunity for brief rests or the use of different muscles).

Loads

Factors to be considered in relation to the load include whether it is:

- heavy (in relation to the guideline figures in paragraph C4.3)
- bulky or unwieldy (an awkward shape or size, lacking in rigidity, hindering vision while moving – any dimension over 75 cm may increase the risk)
- difficult to grasp (large, rounded, smooth, wet or greasy – lacking in suitable handholds)
- unstable, or with contents likely to shift (people or animals have minds of their own,* contents of partially full boxes may suddenly topple over)
- sharp, hot or otherwise potentially damaging (apart from the risk of direct injury, such loads may need to be held away from the body and present increased risk of being dropped).

*The need to avoid damaging the load will also be an important consideration when handling people or animals.

The working environment
Environmental aspects justifying particular attention are whether there are:

- space constraints preventing good posture (low work surfaces, restricted headroom, furniture and other obstructions)
- uneven, slippery or unstable floors (risks of slips, trips, falls, unexpected movement or lack of a firm footing)
- variations in level of floors or work surfaces (handling on slopes, steps or ladders increases risks, particularly on ladders or steep stairways where a firm handhold is needed)
- extremes of temperature or humidity (high temperatures increase fatigue risks, low temperatures reduce flexibility and dexterity)
- ventilation problems or gusts of wind (large loads such as panels are a major risk in high winds)
- poor lighting conditions (workers may not be able to identify sharp or rough edges or the centre of gravity of the load).

Individual capability
However, some operations may present risks to a larger proportion of the workforce. These relate to operations that require the following:

- unusual strength, height, etc. (some weights may only be capable of being handled by stronger workers or heights only reached safely by taller persons)
- special information or training (some activities may require special awareness of risks or training in the techniques to be adopted – the latter being particularly important in handling people or animals)
- the wearing of PPE (any individual may be hindered by PPE but some may be more affected than others).

C4.5 Avoiding manual handling risks
Employers must avoid the need for employees to undertake manual handling operations involving a risk of injury, so far as is reasonably practicable. There are several possible ways of doing this.

Elimination of handling
Activities such as machining or wrapping could be carried out on an item in situ rather than moving it to a different position. Tools and equipment can be transported in vehicles rather than carried by hand.

Automation or mechanisation
The regulations have encouraged much greater use of mechanical handling devices, many of which (such as lorry-mounted cranes) have become commonplace over the last decade. Options available include the following:

- forklift trucks, pallet trucks, etc.
- mobile cranes and lifting rigs
- vacuum lifting devices
- powered conveyors (both fixed and mobile)

- free-running rollers, floor-mounted trolleys, chutes
- trolleys and trucks, some with special attachments
- chain blocks, levers, lifting tools providing a mechanical advantage.

However, care must also be taken not to increase other risks, e.g. by using a large forklift truck in an area where space is severely restricted.

Reducing the weight of the load

The weights of loads can often be reduced to levels comfortably within the guidance figures referred to in paragraph C4.3, thus in effect avoiding the risk of injury. Examples of this process include the following:

- supply of paper and printed materials in smaller boxes
- other reductions in weights of packaged materials
- control of box sizes for storing archived records
- transporting large items of equipment in component parts and joining them together *in situ*
- restricting the numbers of newspapers contained in a bundle according to the number of pages in the paper (an industry code of practice has been developed on this).

C4.6   Reducing manual handling risks

Where it is not reasonably practicable to avoid manual handling risks, employers must take appropriate steps to reduce the risk of injury to the lowest level reasonably practicable.

Risk reduction measures should be considered in relation to each of the four main groupings of risk factors – the task, the load, the working environment and the capability of the individual. How far these factors can be influenced will depend on the nature of the manual handling operation and its context – for example, much of the work by the emergency services offers little scope for changing the environment and greater reliance must be placed on the design and positioning of equipment and the selection and training of the individuals carrying out the tasks.

A lot of guidance on means of reducing risks is provided in the main HSE booklet on manual handling, while the HSE booklet 'Manual Handling: Solutions You Can Handle' provides a thought-provoking range of illustrations of potential practical solutions. HSE guidance is also available on risk-reduction measures for a number of different occupational areas.

Examples of ways of reducing risks include the following:

Tasks:
- providing suitable handling aids, e.g. wheels, trolleys, levers (these may reduce risks even if some manual handling is still involved)
- storing heavier items close to waist height
- changing workplace layouts to avoid twisting, stooping, stretching
- keeping loads close to the body while being handled
- ensuring that workers can take breaks from handling activities
- utilising job rotation arrangements within work teams
- adopting team handling where appropriate (ensuring that individuals know where to get assistance when necessary).

Loads:
- reducing the weights or physical size of loads
- providing suitable handles, grips, indents, etc., to make loads easier to hold
- making non-rigid or insecure loads more stable (by using containers, support boards, carrying slings)
- removing or covering sharp edges or rough surfaces
- cleaning loads of dirt, oil or hazardous substances (enabling the load to be handled closer to the body)
- providing PPE where necessary.

The working environment:
- removing space constraints (keeping gangways and work areas clear)
- providing firm and well-maintained floors
- making arrangements for drainage or to clear up spillages
- using slip-resistant surfaces, where appropriate
- reducing risks from working at different levels (providing gradual slopes or suitable steps)
- providing a suitable thermal environment and ventilation
- reducing risks from strong winds (or powerful ventilation systems) (through relocation, alternative routes, handling aids or team handling)
- providing sufficient well-directed light.

Individual capability:
- screening workers to ensure that they are capable of carrying out higher-risk tasks
- restricting the activities of workers with insufficient physical capabilities or long-term injuries
- introducing restrictions for shorter periods for new or expectant mothers or workers with a short-term injury or illness
- encouraging physical fitness of staff engaged in manual handling
- providing workers with information and training on manual handling (see paragraph C4.9).

C4.7   Practical aspects of assessment

As for other types of assessment, those carrying out manual handling assessments must be competent for the purpose. In a small workplace with a limited range of manual handling activities, one individual may be able to conduct all of the assessments. However, the HSE guidance booklet suggests that a team approach is more likely to be necessary in larger and more complex situations. External assistance may be necessary to:

- provide training for in-house assessors
- assess novel or difficult handling risks
- devise suitable solutions to handling problems.

Larger workplaces may need to be divided up into separate units for assessment purposes and many of the approaches described in Chapter C1 in relation to general risk assessments will be equally applicable to manual handling assessments.

Gathering information
Prior to the assessment it will be necessary to identify those manual handling operations which may present risks. Many of these will be obvious to assessors   *215*

who are familiar with the workplace and the activities within it but reference to accident reports and first aid treatment records may reveal additional problems.

Consideration must also be given to employees carrying out manual handling activities away from the workplace, e.g. construction workers, peripatetic maintenance staff, delivery staff, sales personnel, staff erecting display stands at exhibitions, etc. Specific HSE guidance is available for a number of work sectors (see page 219/20).

Workplace observation

Observations in the workplace are a particularly important part of the manual handling assessment. Issues that may come to light include the following:

- non-availability of mechanical handling equipment (e.g. due to breakdown)
- failure to use mechanical handling equipment or handling aids
- individuals struggling to handle loads clearly beyond their capabilities
- poor handling techniques
- unsuitable workplace layouts
- sub-standard housekeeping.

Talking to workers

This too should be an essential part of the assessment. Workers can often provide important information such as:

- whether what is being observed is typical
- problems associated with particular products or activities
- alternative practices adopted when mechanical handling equipment breaks down
- why certain handling aids are not used
- their own level of training in handling techniques
- the unavailability of persons to provide assistance in team-handling situations.

Further investigations

Most manual handling risks will be readily apparent from observation and discussion in the workplace. Further investigation is most likely to be necessary in relation to finding practical means of reducing the risks from manual handling. This is where perusal of reference material (especially 'Manual Handling: Solutions You Can Handle' can be particularly useful. Development of solutions to some problem activities may necessitate the establishment of in-house working parties or the involvement of external specialists or equipment suppliers.

C4.8   Assessment records

HSE guidance states that, in general, the significant findings of the assessment should be recorded unless:

- it could very easily be repeated and explained at any time because it is so simple and obvious; or
- the manual handling operations are quite straightforward, of low risk, are going to last only a very short time, and the time taken to record them would be disproportionate.

As for other types of assessment, in cases of doubt it is always better to keep a record.

The HSE guidance booklet contains a sample record form in a checklist format along with a worked example. While this form may be appropriate for recording assessments of a limited range of handling tasks, the author has not found it suitable for use in larger workplaces, recommending instead that the checklist part of the form (see Figure C4.2) is kept as a separate *aide-mémoire* and then recording the findings of the assessment in a similar way as for general risk assessments, using the format illustrated in Figure C1.5 on page 177. A worked example of a manual handling assessment record using this format is provided in Figure C4.3 on page 218.

Any assessment must be reviewed if there is reason to suspect that it is no longer valid or significant changes are made to the manual handling operations.

Figure C4.2	Manual handling assessments: checklist of questions to be answered

The tasks – Do they involve:
- holding loads away from the trunk?
- twisting?
- stooping?
- reaching upwards?
- large vertical movement?
- long carrying distances?
- strenuous pushing or pulling?
- unpredictable movement of loads?
- repetitive handling?
- insufficient rest or recovery?
- a work rate imposed by a process?

The loads – are they:
- heavy?
- bulky/unwieldy?
- difficult to grasp?
- unstable/unpredictable?
- intrinsically harmful (e.g. sharp/hot)?

The working environment – are there:
- constraints on posture?
- poor floors?
- variations in levels?
- hot/cold/humid conditions?
- strong air movements?
- poor lighting conditions?

Individual capability – does the job:
- require unusual capability?
- hazard those with a health problem?
- hazard those who are pregnant?
- call for special information/training?

Other factors
Is movement or posture hindered by clothing or personal protective equipment?

Figure C4.3   Example of manual handling assessment record

AVENUE MOTORS		
AREA: PARTS DEPT.	RISK TOPIC: MANUAL HANDLING	Sheet 1 of 1

RISKS IDENTIFIED	PRECAUTIONS IN PLACE	RECOMMENDED IMPROVEMENTS
Loads are handled when being placed in storage and when being transferred to the workshops.	Handling equipment and handling aids available include:	Ensure that the lifter and pallet truck are always used for handling engines and gearboxes.
High risk loads include:	– engine/gearbox lifter (kept in workshop)	
– engines and gearboxes (heavy)	– pallet truck	
– exhaust systems (awkward)	– wheeled parts cages (most items are supplied in these)	Supply heavy-duty leather gloves for handling doors and body panels.
– doors and body panels (awkward & sharp edges)	– a small trolley	
– batteries (heavy)	– tote boxes.	
– wheels (heavy)	A wide stairway provides good access to the upper floor.	Remove stored items encroaching onto stairway.
Some larger items are supplied on pallets.	Access between the racking is generally good, well lit and clear of obstructions	Some strip lights on the ground floor require replacing.
Some items are stored on the upper floor of the Parts Dept.	Most heavy items are stored at accessible heights or on pallets on the floor.	Reposition the wheels currently stored on the top of the racking on the upper floor.
	All staff wear safety footwear.	Ensure that any new staff are trained fully in:
	Current staff are experienced, trained in handling techniques and aware of when to use handling equipment/aids or team handling.	– handling techniques
		– use of handling equipment/aids
		– when to use handling equipment/aids or team handling.

SIGNATURE(S) A. Gair	NAME(S) A. GAIR	DATE 12/2/05
DATE FOR NEXT ASSESSMENT March 2005	RECOMMENDATION FOLLOW-UP	

C4.9   Manual handling training

The HSE guidance booklet refers to a number of general points to be covered in training programmes for safe manual handling. These include the following:

- how to recognise potentially hazardous handling operations
- how to avoid hazards from manual handling
- how to deal with unfamiliar handling operations
- proper use of handling aids and PPE
- features of the working environment that contribute to safety
- the importance of good housekeeping
- factors affecting the capability of the individual
- good handling technique.

The booklet summarises some of the important elements of good handling technique. The same summary is also available in a free HSE leaflet, 'Getting to Grips with Manual Handling: A Short Guide for Employers'.

Some manual handling tasks will require more specific instruction and training, for example in:

- particular risks associated with the task (especially if these are not readily apparent)
- the use of specific manual handling aids
- the PPE required for that task
- specific techniques to be adopted.

C4.10   Information on loads

Regulation 4 requires employees to be provided with precise information on the weight of loads where this is reasonably practicable. With packaged items the weight can easily be printed on the packaging. Alternatively the weights of regularly handled loads can be provided via notices or contained within operating procedures or training material.

Where it is not reasonably practicable to provide precise information a general indication of weights must be provided. Training employees how to assess the weight of unfamiliar items provides another means of avoiding them over-exerting themselves.

A further requirement of Regulation 4 is for employees to be provided with information on 'the heaviest side of any load whose centre of gravity is not positioned centrally'. Again employees can be trained to recognise these loads for themselves but employers should draw attention to any such loads, particularly where they are regularly handled or where they may be difficult to recognise.

Source materials

1. HSE (2004) 'Manual Handling Operations Regulations 1992: Guidance on the Regulations', L 23.
2. HSE (1994) 'Manual Handling: Solutions You Can Handle', HSG 115.
3. HSE (2000) 'Handle with Care: Assessing Musculoskeletal Risk in the Chemical Industry'.
4. HSE (1996) 'Manual Handling Operations: Baggage Handling at Airports', APIS 2.
5. HSE (1999) 'Handling the News: Advice for Employers on Manual Handling of Bundles', IACL 105.

6. HSE (1999) 'Handling the News: Advice for Newsagents and Employees on Safe Handling of Bundles', IACL 106.
7. HSE (2002) 'Make a Parcel – a Save a Back', INDG 348.
8. HSE (2000) 'Manual Handling Solutions for Farms', AS 23.
9. HSE (1998) 'Handling and Stacking Bales in Agriculture', INDG 125.
10. HSE (2000) 'Backs for the Future: Safe Manual Handling in Construction', HSG 149.
11. HSE (1997) 'Preventing Injuries from the Manual Handling of Sharp Edges in the Engineering Industry', EIS 16.
12. HSE (2000) 'Moving Food and Drink: Manual Handling Solutions for the Food and Drinks Industries', HSG 196.
13. HSE (1994) 'Manual Handling in Drinks Delivery', HSG 119.
14. HSE (1996) 'Picking Up the Pieces: Prevention of Musculoskeletal Disorders in the Ceramics Industry'.
15. HSE (1994) 'Getting to Grips with Handling Problems: Worked Examples of Assessment and Reduction of Risk in the Health Services'.
16. HSE (1998) 'Manual Handling of Loads in the Health Service'.
17. HSE (2002) 'Handling Home Care: Achieving Safe, Efficient and Positive Outcomes for Care Workers and Clients', HSG 225.
18. HSE (2000) 'Manual Handling in the Railway Industry: Advice for Employers', INDG 326.
19. HSE (1999) ''Reducing Manual Handling Injuries in the Rubber Industry'.
20. HSE (1998) 'Manual Handling in the Textile Industry', IACL 103.
21. HSE (2000) 'Manual Handling Solutions in Woodworking', INDG 318.
22. HSE (2004) 'Getting to Grips with Manual Handling: A Short Guide for Employers' (free leaflet), INDG 143.

Display Screen
Equipment Assessments

C5.1   Introduction

The requirement for assessments to be made of display screen equipment (DSE) workstations was included within the Health and Safety (Display Screen Equipment) Regulations 1992. The regulations also contain requirements relating to the design of DSE workstations, breaks for DSE users, eye and eyesight tests and the provision of training and information.

This chapter summarises the requirements of the regulations and examines some of the practical issues associated with DSE workstation assessment. Much of the guidance contained in Chapter C1 on the general approaches to risk assessment can also be applied to DSE assessment.

C5.2   Risks associated with display screen equipment

The risks associated with DSE work are summarised in Annex B to the HSE booklet on the regulations. These can be summarised as follows.

Upper limb pains and discomfort

Pain and discomfort may be felt:

- in the arm, hand or shoulder (e.g. carpal tunnel syndrome, peritendinitis)
- in the back or neck.

Many of these problems are caused by poor ergonomic layout of the workstation, e.g. unsuitable positioning of the keyboard, mouse, screen or chair. However, in some cases they may result from poor keyboard technique or excessive workloads.

Eye and eyesight effects

Temporary visual fatigue may be caused leading to impaired visual performance, sore eyes or headaches. (The HSE has consistently maintained that DSE work does not damage the eyes or eyesight, nor make existing vision defects worse. However, workers may become more aware of pre-existing defects.) The causes of visual fatigue are often simply looking at the screen (at the same focal distance) for too long a period, especially without taking breaks. However, other causes can include lighting problems (glare and reflections), poor legibility or flickering of the images on the screen, or poor legibility of source documents.

Fatigue and stress

Fatigue may result in the upper-limb or vision problems referred to above, and these in turn may contribute to the potential stresses associated with DSE work. Other stress factors can include under- or over-utilisation of skills, repetitive work and social isolation. A further major factor can be the use of unsuitable or unreliable software – the frustration associated with sudden system crashes is never fully appreciated by those who have not spent the time inputting data that has now been lost!

Other possible risks

Two other possible risks have now largely disappeared as a result of changes in technology since the introduction of the regulations. Epileptic seizures (often due to flicker effects) were always rare and should have become even rarer with improvements in image stability. Dermatitis of the face and neck was thought to be associated with low humidity or static electricity – something more common in old-fashioned 'computer rooms' rather than with PCs and network terminals used today.

Risk ruled out

Much concern was expressed in the 1980s about risks of miscarriages and birth defects from electromagnetic radiation associated with DSE work. However, the HSE states very clearly that the results of scientific studies overall do not demonstrate any link. They point out that the levels of radiation from display screen equipment are well below those set out in international recommendations for protecting human health.

C5.3   A summary of the DSE Regulations

Interpretation and application (Reg. 1)
Several important terms are defined in this regulation:

- *display screen equipment (DSE):* any alphanumeric or graphic display screen, regardless of the display process involved
- *workstation:* an assembly comprising DSE, any optional accessories, any disk drive, telephone, modem, printer, document holder, work chair, work desk, work surface or other peripheral item and the immediate environment
- *user:* an employee who habitually uses DSE as a significant part of their normal work
- *operator:* a self-employed person who habitually uses DSE as a significant part of their normal work.

(Factors to be taken into account in determining whether someone is a 'user' or 'operator' are considered further in paragraph C5.5 below.)

Exclusions
The regulations do not apply to drivers' cabs or control cabs for vehicles or machinery, DSE on board a means of transport, DSE mainly intended for public operating, portable systems not in prolonged use, calculators, cash registers or other equipment with small displays and window typewriters.

Assessment of workstations and reduction of risk
Employers are required to perform a 'suitable and sufficient analysis' of those workstations

- used for their purposes by 'users' (including temps)
- provided by them and used for their purpose by 'operators'

for the purpose of assessing the health and safety risks in consequence of that use. They must then reduce the risks identified to the lowest level reasonably practicable.

The assessment must be reviewed if there is reason to suspect that it is no longer valid or there have been significant changes. Further guidance on assessment is provided later in paragraphs C5.4 to C5.6.

Requirements for workstations (Reg. 3)

All workstations must now meet the requirements laid down in the Schedule to the Regulations. However, the requirement must be relevant in relation to the health, safety and welfare of workers – for example, there is no need to provide a document holder if there is little or no inputting from documents, or an individual with a previous back problem might prefer a chair with a fixed rather than an adjustable back.

Daily work routine of users (Reg. 4)

Employers are required to plan the activities of 'users' so that their DSE work is periodically interrupted by breaks or changes of activity. Breaks should be taken before the onset of fatigue and preferably away from the screen. It is best if users are given some discretion in planning their work and are able to arrange breaks informally rather than taking formal breaks at regular intervals.

Eyes and eyesight (Reg. 5)

'Users' are given the right (at their employer's expense) to an appropriate eye and eyesight test by a competent person both before becoming a 'user' and at regular intervals thereafter. (Some employers re-test every two years for those over 50 and every five years for employees under 50.) Employees are also entitled to tests on experiencing visual difficulties that may reasonably be considered to be caused by DSE work. These tests are an entitlement for 'users' and are not compulsory.

Tests are normally carried out by opticians and involve a test of vision and an examination of the eye. (Where companies have their own vision screening facilities, 'users' may opt for a screening test to see if a full eye test is needed.) If the eye test shows a need for corrective spectacles (other than the 'user's' normal spectacles) the basic cost of these must be met by the employer. Employees would normally pay for extras, e.g. designer frames or tinted lenses.

Provision of training and information (Regs 6 and 7)

Employers are required to provide training and information for 'users' and also provide information for 'operators' at work in their undertaking. Those requirements are dealt with more fully in paragraph C5.7.

C5.4 Factors to consider during assessment

The 'suitable and sufficient analysis' or assessment of workstations required by Regulation 2 should take into account the risks of DSE work identified in paragraph C5.2 and the standards for workstations contained in the Schedule to the Regulations. The DSE assessment checklist provided in Figure C5.1 overleaf identifies some of the more important factors while Figure C5.2 provides guidance on the use of the checklist.

C5.5 Identifying 'users' and 'operators'

'Users' and 'operators' (in effect self-employed 'users') are defined by Regulation 1 as those who habitually use DSE as a significant part of their normal work. HSE guidance deliberately fights shy of specifying a set number of hours of DSE work per week for the 'user' definition to apply. Many employers use 50 per cent of the working week as a rough guide but there are other criteria to take into account.

FIND THIS ON CD	Figure C5.1	DSE workstation checklist

USER'S NAME ....
DEPARTMENT/SECTION ....

1	LIGHTING & WORK ENVIRONMENT	COMMENTS/IMPROVEMENTS REQUIRED
1.1	Is artificial lighting adequate?	
1.2	Does it cause any reflection or glare problems?	
1.3	Any reflection or glare problems from sunlight?	
1.4	Are suitable blinds available (if necessary)?	
1.5	Are temperature and ventilation satisfactory in summer and winter?	
2	SCREEN & KEYBOARD	
2.1	Is the screen set at a suitable height?	
2.2	Stable image with clear characters?	
2.3	Brightness and contrast adjustable?	
2.4	Screen can swivel and tilt easily?	
2.5	Cleaning materials available?	
2.6	Is the keyboard tiltable?	
2.7	Is there sufficient space in front of it?	
3	DESK & CHAIR	
3.1	Is the desk size adequate?	
3.2	Is there sufficient leg room under it?	
3.3	Document holder available (if required)?	
3.4	Chair comfortable and stable?	
3.5	Seat height readily adjustable?	
3.6	Height and tilt of chair back readily adjustable?	
3.7	Footrest available (if required)?	
4	HAS THE USER HAD SIGNIFICANT EXPERIENCE OF PROBLEMS WITH:	
4.1	Back, shoulders or neck?	
4.2	Hands, wrists or arms?	
4.3	Tired eyes or headaches?	
4.4	The suitability of the software used?	
4.5	Other problems?	

ADDITIONAL PROBLEMS, COMMENTS, OR IMPROVEMENTS REQUIRED?

IS A MORE DETAILED ASSESSMENT REQUIRED?   YES/NO
Signature   Name   Date

1. LIGHTING & WORK ENVIRONMENT
    1.1 Artificial lighting should be adequate to see all the documents worked with.
    1.2 Recessed lights with diffusers should not cause problems. Lights suspended from ceilings might.
    1.3 Problems can occur in the early morning or the afternoon, especially in winter when the sun is low.
    1.4 Blinds ought to be effective in eliminating glare from the sun.
    1.5 Strong sunlight may create significant thermal gain at some times.

2. SCREEN & KEYBOARD
    2.1 The top of the screen should normally be level with the user's eyes when sitting in a comfortable position.
    2.2 There should be little or no flicker on the screen. Characters should be clearly visible.
    2.3 The user should know where the brightness and contrast controls are.
    2.4 The screen should be able to swivel and tilt in order to avoid reflections.
    2.5 The user should know where to get cleaning materials for the screen and keyboard.
    2.6 Small legs at the back of the keyboard should allow its angle to be adjusted.
    2.7 Space in front of the keyboard allows hands and wrists to be rested when not keying in.

3. DESK & CHAIR
    3.1 Desks should have sufficient space to allow the screen and keyboard to be in a comfortable position and accommodate documents, document holder, phone, etc.
    3.2 There should be enough space under the desk to allow free movement of the legs.
    3.3 When inputting from documents, using a document holder helps avoid frequent neck movements.
    3.4 Chairs with castors must have at least five (four is very unstable).
    3.5 The seat height should be adjustable to achieve a comfortable position (arms approx. horizontal and eyes level with the top of the screen).
    3.6 The angle and height of the back support should be adjustable so that it provides a comfortable working position.
    3.7 DSE users who are shorter may need a footrest to help them keep comfortable when sitting at the right height for their keyboard and screen (see 3.5).

4. POSSIBLE PROBLEMS
    4.1 Problems with back, shoulders or neck might indicate that the screen is at the wrong height, an incorrectly adjusted chair or need for a document holder.
    4.2 Problems with hands, wrists or arms might indicate incorrect positioning of the keyboard or a poor keying technique. Hands should not be bent up at the wrist and a soft touch should be used on the keyboard, not overstretching the fingers.
    4.3 Tired eyes or headaches could indicate problems with lighting, glare or reflections. They may also indicate the need to take regular breaks away from the screen. Persistent problems might need an eye test.
    4.4 The software should be suitable for the work to be carried out.

The HSE guidance booklet identifies the following as factors to consider in determining whether a person is a 'user' or 'operator':

- their dependence on DSE to do the job (no alternative means readily available)
- they have no discretion as to whether or not to use DSE
- they require significant training or DSE skills to do their work
- they normally use DSE for continuous spells of an hour or more
- they use DSE more or less daily
- fast transfer of information is an important requirement
- high attention and concentration is required (the consequences of an error may be critical).

Some staff might use DSE very intensively at some times and only intermittently at others (e.g. accounts staff at 'month end' and 'year end' periods) – such employees should normally be treated as 'users'.

The HSE guidance booklet contains some useful pen portrait examples of 'definite users' (e.g. journalists, air traffic controllers), 'possible users' (e.g. check-in clerks, community care workers) and 'non-users' (e.g. senior managers) to assist employers in identification. In cases of doubt it is usually best to count people as being 'users'.

Employers have duties to assess workstations of 'users' who are homeworkers or who work on other employers' premises, and also of 'operators' using their DSE workstations (i.e. ones provided by the employer), e.g. self-employed journalists or technical specialists who work on their premises.

C5.6   Practical aspects of assessment

In workplaces where the numbers of 'users' (or 'operators') are relatively few, it will be quite easy to assess each and every workstation individually. However, this becomes much less of a practicable proposition as numbers of 'users' increase, particularly bearing in mind the frequency of moves taking place in most modern offices.

In such cases it is quite acceptable for the employer to carry out a generic assessment supported by a specific self-assessment checklist to be completed by individual 'users'. The generic assessment would take into account general standards of the environment and DSE equipment:

- lighting and blinds
- design of keyboards, availability of brightness/contrast adjustment
- design of desks and chairs.

The checklist and guidance in Figures C5.1 and C5.2 on pages 224/5 are suitable for 'users' to carry out their own self-assessments. Where this approach is used it is important for 'users' to know where to send completed checklists, and to establish mechanisms for dealing with queries and problems, including arranging for a suitable person to carry out individual assessments when required.

Selection of staff to carry out DSE assessments should follow the same principles as for any other type of risk assessment. They need not be health and safety specialists but do need to be familiar with the regulations, the risks associated with DSE work and the standards required for DSE workstations. Observations of staff at their workstations and discussions with those staff should be an integral part of the process. Action points identified during assessments need to be implemented in a systematic way and reviews of the assessment may be necessary, particularly in relation to changes (in software, hardware, furniture, work locations, the nature of the DSE work, etc.).

C5.7   Provision of training and information

Regulation 6 requires 'users' to be provided with an adequate health and safety training.

The detailed content will often need to relate to the DSE workstations in the individual's workplace and the types of DSE work carried out. It is likely to include the following:

- the causes of DSE-related problems, e.g. poor posture, screen reflections
- the user's role in recognising risks
- the importance of comfortable posture and postural change
- equipment adjustment mechanisms, e.g. chairs, contrast, brightness
- use and arrangement of workstation components (screen position, document holders)
- the need for regular cleaning of the screen
- the need to take breaks and changes of activity
- arrangements for reporting problems associated with the workstation
- information about eye and eyesight tests and breaks
- their own role in assessments (particularly using self-assessment checklists).

Specific training may be necessary when workstations are substantially modified. Handouts could be provided or posters displayed as reminders of essential points. The HSE's leaflet 'Working with VDUs' can be very useful in this respect.

Regulation 7 requires employers to ensure that all 'users' (their own employees and employees of others, e.g. agency staff) and 'operators' at work in their undertaking are provided with adequate information about

Employers must provide information on	Regulation	Own users	Other users (agency staff)	Operators (self-employed)
DSE/workstation risks		Yes	Yes	Yes
Risk assessment and reduction	2 & 3	Yes	Yes	Yes
Breaks and activity changes	4	Yes	Yes	No
Eye and eyesight tests	5	Yes	No	No
Initial training	6(1)	Yes	No	No
Training when workstation substantially modified	6(2)	Yes	Yes	No

	Figure C5.3	Example of a completed DSE workstation checklist

USER'S NAME ANNA BEAN DEPARTMENT ACCOUNTS

I	LIGHTING & WORK ENVIRONMENT	COMMENTS/IMPROVEMENTS REQUIRED
1.1	Is artificial lighting adequate?	*Good standards.*
1.2	Does it cause any reflection or glare problems?	*All fittings recessed. Diffusers fitted.*
1.3	Any reflection or glare problems from sunlight?	*Problems on some winter mornings.*
1.4	Are suitable blinds available (if necessary)?	*Vertical blinds provided – usually effective.*
1.5	Are temperature and ventilation satisfactory in summer and winter?	*Very hot on some days in summer – air conditioning unreliable.*
2	SCREEN & KEYBOARD	
2.1	Is the screen set at a suitable height?	*Screen was too low (now reset).*
2.2	Stable image with clear characters?	*Stable. Type styles and colours can be varied.*
2.3	Brightness and contrast adjustable?	*Both adjustable.*
2.4	Screen can swivel and tilt easily?	*Moves freely.*
2.5	Cleaning materials available?	*Kept in stationery cupboard.*
2.6	Is the keyboard tiltable?	*Yes – legs at back.*
2.7	Is there sufficient space in front of it?	*No – position now adjusted.*
3	DESK & CHAIR	
3.1	Is the desk size adequate?	*Yes (but equipment needed repositioning).*
3.2	Is there sufficient leg room under it?	*Adequate.*
3.3	Document holder available (if required)?	*Obtained from IT section.*
3.4	Chair comfortable and stable?	*Yes – standard model throughout offices.*
3.5	Seat height readily adjustable?	*Easily adjusted.*
3.6	Height and tilt of chair back readily adjustable?	*Adjustment mechanism stiff.*
3.7	Footrest available (if required)?	*Not required.*
4	HAS THE USER HAD SIGNIFICANT EXPERIENCE OF PROBLEMS WITH:	
4.1	Back, shoulders or neck?	*Occasionally.*
4.2	Hands, wrists or arms?	*Occasionally.*
4.3	Tired eyes or headaches?	*Regularly, especially at 'month end'.*
4.4	The suitability of the software used?	*No.*
4.5	Other problems?	*No.*

ADDITIONAL PROBLEMS, COMMENTS, OR IMPROVEMENTS REQUIRED?
1.5 Air-conditioning reliability needs to be improved.
2.1 Repositioning of equipment should overcome problems referred to in 4.1 & 4.2.& 2.7
3.3 Use of document holder should also help prevent stiff necks (see 4.1).
3.6 Handyman needs to check the back adjustment on Anna's chair.
4.3 Anna needs to make sure she takes her breaks, even at busy times.

IS A MORE DETAILED ASSESSMENT REQUIRED?   ~~YES~~/NO
Signature *A. Clarke*   Name ALAN CLARKE   Date *28/2/05*

health and safety in relation to DSE workstations and what measures the employer has taken to comply with his duties. The table below shows what information must be provided to the different 'users' and 'operators' and also provides a good guide to the responsibilities of employers in relation to each of the regulations.

C5.8   Assessment records

The checklist in Figure C5.1 on page 224 is typical of the type of record that most employers keep of DSE workstation assessments. A completed version is provided in Figure C5.3 on page 228 as an illustration. HSE guidance states that assessments need not be recorded 'in the simplest and most obvious cases' or in relation to short-term or temporary workstations. In most cases they need to be kept readily accessible for as long as they are still valid. Because of the possibility of damages claims being made relating to the long-term effects of DSE work, employers may find it beneficial to retain even obsolete assessment records for a considerable period of time. Records can be kept electronically as well as in paper form.

Source materials

1.   HSE (2002) 'Work with Display Screen Equipment: Health and Safety (Display Screen Equipment) Regulations 1992 as Amended by the Health and Safety (Miscellaneous Amendments) Regulations 2002 – Guidance on Regulations', L 26.
2.   HSE (2003) 'The Law on VDUs: An Easy Guide', HSG 90.
3.   HSE (1998) 'Working with VDUs', INDG 36.

Assessment of PPE Requirements

C6.1   Introduction

The protection of people at work through the use of personal protective equipment (PPE) is generally less effective than the elimination of risks or the control of risks at source. This is recognised in the detailed requirements of UK health and safety legislation – for example in the hierarchy of measures required to protect workers against hazardous substances which is contained in Regulation 7 of the COSHH Regulations (see paragraph C2.5). However, there will frequently be situations where the control of risks by other means is inadequate or where PPE provides the only practicable means of control.

The Personal Protective Equipment at Work Regulations 1992 require employers to carry out an assessment in order to identify such situations and also the types of PPE that will provide appropriate protection. This chapter summarises the requirements of the PPE at Work Regulations and also makes reference to other regulations containing important PPE requirements.

Factors to be considered during the risk assessment process are identified including brief descriptions of the various types of PPE available and of circumstances where the use of each might be appropriate. Different ways of recording the results of assessments of PPE requirements are also discussed.

C6.2   Regulations relating to PPE

The Personal Protective Equipment at Work Regulations 1992 replaced many outdated regulations dealing with PPE. Many of the previous regulations only applied to limited processes or activities, but the PPE at Work Regulations require employers to make an assessment of the PPE needs in respect of all types of work activity. Where the assessment shows that PPE is necessary, then the employer must provide it, maintain it, instruct and train employees in its use, and take reasonable steps to ensure that they do use it. Self-employed persons also have duties under the regulations. Several other sets of regulations also contain important PPE requirements:

- the Control of Substances Hazardous to Health Regulations 2002 (see Chapter C2)
- the Noise at Work Regulations 1989 (see Chapter C3)
- the Control of Asbestos at Work Regulations 2002
- the Construction (Head Protection) Regulations 1989
- the Control of Lead at Work Regulations 2002
- the Ionising Radiations Regulations 1999.

Requirements under most of these regulations follow a similar pattern to the PPE at Work Regulations.

C6.3   A summary of the PPE at Work Regulations

Interpretation (Reg. 2)

PPE is defined as 'all equipment (including clothing affording protection against the weather) which is intended to be worn or held by a person at work and which protects him against one or more risks to his health or safety'. This definition includes items such as safety helmets, eye protection, safety footwear, gloves, high visibility clothing, etc. Waterproof, weatherproof or insulated clothing would also be subject to the regulations where it is necessary to protect employees against health and safety risks.

Disapplication (Reg. 3)

Certain types of clothing and equipment are excluded from the application of the regulations. Among these are ordinary working clothes and uniforms (not specifically providing protection), self-defence or deterrent equipment (e.g. panic alarms), portable devices for detecting and signalling risks and nuisances (e.g. gas monitors) and equipment used during the playing of competitive sports. (PPE subject to other, more specific regulations is also excluded, e.g. hearing protection.)

Provision of PPE (Reg. 4)

Under the Management Regulations, employers are required to carry out a general risk assessment (see Chapter C1) in order to identify the most appropriate way of controlling risks. Regulation 4 of the PPE at Work Regulations requires that PPE is only considered as a last resort, i.e. after such measures as elimination at source, engineering controls or safe systems of work. Where risks cannot be 'adequately controlled by other means which are equally or more effective', the employer must provide exposed employees with suitable PPE. (Self-employed persons also have a duty to provide themselves with suitable PPE.)

Section 9 of HASAWA requires employers to provide necessary PPE free of charge. Much PPE needs to be provided as a personal issue for hygiene reasons, although in some circumstances PPE may be shared (e.g. waterproof clothing). In all cases the PPE should be readily available.

Regulation 4 also refers to factors that should be considered when selecting suitable PPE, which must:

- be appropriate for the risks involved, the conditions and the period for which it is worn
- take account of ergonomic requirements, the wearer's state of health and the characteristics of the workstation
- be capable of adjustment to fit the wearer correctly
- prevent or adequately control the risk, without increasing overall risk
- comply with appropriate standards (see paragraph C6.5 below)
- be issued personally to individuals, where necessary for hygiene or health reasons
- be compatible with other types of PPE being used.

(Regulation 5 contains this last requirement.)

Assessment of PPE (Reg. 6)

Employers must make an assessment in respect of PPE requirements which must include:

- identifying risks that have not been avoided by other means
- defining the characteristics required of PPE for it to be effective against these risks

- comparing these with the characteristics of PPE that is available
- taking into account the compatibility of PPE worn simultaneously.

Maintenance and replacement of PPE (Reg. 7)

Employers must ensure that PPE is maintained in an efficient state, efficient working order and in good repair. In some cases this might require specific arrangements to be made for its cleaning, disinfecting, inspection, examination, testing or repair. Responsibility for maintenance of PPE should be clearly identified. Simple maintenance might be carried out by the user but more complex maintenance may require specially trained personnel. In most cases it will be sufficient to follow manufacturers' instructions. Some types of PPE may need to be replaced periodically, e.g. many manufacturers recommend replacing safety helmets at least every five years. Maintenance records may be appropriate in some cases.

Accommodation for PPE (Reg. 8)

Appropriate accommodation must be provided for PPE when it is not being used. This might involve pegs (e.g. for helmets or weatherproof clothing), or more secure containers such as lockers, in order to protect PPE from contamination, damage or loss. Where PPE may become contaminated in use, it must be kept separate from other clothing and equipment.

Information, instruction and training (Reg. 9)

Employees must be given adequate and appropriate information, instruction and training in respect of the PPE provided for their use, including:

- the risks that the PPE will avoid or limit (this might include providing relevant signs)
- the purpose for which and the manner in which PPE is to be used
- arrangements for maintenance and/or replacement of the PPE.

Correct fitting or adjustment may be important for some types of PPE. Practical training as well as theoretical instruction will be required in some cases, e.g. in the use of harnesses.

Use of PPE (Reg. 10)

Employers must take all reasonable steps to ensure that PPE is properly used by employees. This will involve them taking a proactive approach to policing standards of compliance. Employees have a duty to use PPE in accordance with the training and instructions that they have received. Self-employed persons must also make full and proper use of PPE.

Reporting loss or defect (Reg. 11)

Employees must report any loss of or obvious defect in their PPE to their employer forthwith.

C6.4 Types of PPE available

The HSE guidance accompanying the PPE at Work Regulations (ref. 1) gives much detailed advice on various types of PPE which are available and the types of risk against which they provide protection. Further guidance is available in many other HSE publications, dealing with particular risks or types of work activity. The main types of PPE are outlined below.

Head protection
Safety helmets are required whenever there is significant risk from falling objects or impact with fixed objects, e.g. during construction and some types of maintenance work. Other types of head protection that may be necessary include bump caps (providing limited protection against impact in confined work environments), caps or hairnets (protecting against scalping or entanglement, e.g. from rotating machinery) and crash helmets (for use on certain types of transport, e.g. all-terrain vehicles).

Eye protection
Eye protection such as spectacles, goggles or visors will be necessary for activities and processes involving risks to the eyes and face, including the following:

- use of power-driven tools producing swarf, chippings, etc.
- exposure to molten metal and other molten substances
- activities producing intense light or other types of significant optical radiation, e.g. welding, burning, use of lasers (suitably shaded lenses are required)
- exposure to acids, alkalis and other corrosive or irritant substances
- work with liquid, gas or vapour under pressure.

Foot protection
Safety boots or shoes are likely to be required for construction, maintenance and warehouse work, heavier industrial processes and any other situation where there is risk from items falling onto the feet or from penetration through the soles or heels of footwear. Work with hot metal is likely to require the use of foundry boots with quick release fastenings and without external features such as laces which could trap molten metal. Wellington boots may also be necessary to provide protection against chemicals or in wet or muddy situations. Anti-static footwear may be necessary to prevent flammable atmospheres being ignited by static electricity.

Hand and arm protection
Gloves or gauntlets may be necessary for protection from:

- cuts and abrasions (e.g. when using sharp tools or handling items with rough or sharp edges)
- extremes of temperature (e.g. outdoor work or handling frozen or very hot items)
- skin irritation and dermatitis (e.g. from 'harmful' or 'irritant' hazardous substances)
- contact with toxic or corrosive substances
- molten metal risks (including welding and burning work).

The choice of glove material is important as it must be capable of protecting against the risk (or a combination of risks) but should also be comfortable for the wearer and suitable for the work being carried out.

Body protection
Various types of body protection may be necessary, depending on work location or work activity, including the following:

- outdoor clothing – providing protection against cold, rain, etc.
- high visibility clothing – for work on or close to roadways or to mobile work equipment

FIND THIS ON CD	Figure C6.1	Example of PPE Requirements for Various Activities Carries out by a Town Council

TYPE OF PPE	COUNCIL REQUIREMENTS
SAFETY FOOTWEAR	
Boots or shoes	To be worn at all times by outdoor staff except when in offices or amenity areas.
High leg safety boots	To be worn for chainsaw work.
PROTECTIVE CLOTHING	To be worn
Waterproof clothing	– when working in rain, snow or hail
	– when using water jets
High visibility clothing (jackets or waistcoats)	– when working close to or on public roads
Protective jacket and trousers	– for chainsaw work
Chemical suit	– for spraying activities.
EYE PROTECTION	To be worn
Safety spectacles, goggles or visors	– when using strimmers or hand mowers
	– when weeding in presence of thorns
Visors	– for chainsaw work
	– for spraying activities.
HEARING PROTECTION	Recommended for use
Ear muffs or plugs	– when cutting grass
	– in the tractor cab when the engine is running.
	Required to be worn
	– for chainsaw work.
GLOVES	To be worn
Of a suitable type – seek guidance from chargehand in cases of doubt	– when handling petrol or oils
	– when removing glass, needles or sharp objects
	– for weeding work or work close to thorns, thistles or irritant plants
	– when handling cement
	– when handling rough items
	– for spraying activities.
RESPIRATORY PROTECTION – seek guidance from chargehand on suitable type	To be worn
	– when spraying hazardous substances (if recommended on the data sheet).
	May be necessary
	– for some painting work
	– for application of wood preservatives
	– during the re-opening of graves.
HARNESS inertia protection)	*May* be necessary for some work on mobile (with elevating work platforms. (Chargehand will advise)

- heat-resistant or flame-resistant clothing – for work near furnaces or molten metal
- overalls and aprons – to protect against hazardous substances, splashes of hot or molten material
- specialist clothing – e.g. for work with chainsaws
- life jackets or buoyancy aids – for work on or close to water
- harnesses or fall arrestors – for work at heights.

Respiratory protection
Required under the COSHH Regulations (see Chapter C2) and also possibly to provide protection against asbestos or lead.

Hearing protection
Required under the Noise at Work Regulations (see Chapter C3).

C6.5   Standards for PPE
The Personal Protective Equipment (EC Directive) Regulations 1992 (amended in 1994) contain requirements for most types of PPE to satisfy certification procedures, meet type-approval requirements and carry a CE marking. The standards that must be met depend upon the simplicity or complexity of the design and the severity of the risks against which the PPE is intended to protect. Many of the requirements that must be met and the CE marking system are extremely difficult for the non-specialist to follow and there is a bewildering variety of items that have BS or EN specifications. However, most manufacturers and suppliers have well-illustrated catalogues available to assist in identifying the PPE that will be appropriate for a particular activity or situation – not just in the technical sense but also from ergonomic and other practical standpoints. However, before purchasing items it is always sensible to check with technical representatives that the PPE is suitable for the proposed application.

C6.6   Practical aspects of assessment
In practice the need for many types of PPE will be identified during general risk assessments or as part of COSHH, noise or manual handling assessments. However, it is important to ensure that no aspects of the provision and use of PPE have been overlooked:

- Have all PPE needs been identified (including general needs such as bad-weather clothing)?
- Are employees fully aware of PPE requirements (through instruction, training, signs, notices, etc.)?
- Do employees know how to obtain and use PPE?
- Are arrangements for maintenance and storage of PPE satisfactory?

C6.7   Assessment records
HSE guidance accompanying Regulation 6 ('Personal Protective Equipment at Work') states that PPE assessments need not be recorded 'in the simplest and most obvious cases which can easily be repeated and explained'. However, it is generally preferable to record the assessment:

- in order to demonstrate that the assessment has been carried out
- as an aid to informing employees what types of PPE are required for which activities and situations

- as a point of reference for the effective enforcement of PPE standards (which may ultimately require disciplinary action).

PPE assessments may be recorded:

- within task procedures (Figure B8.1 on page 148 contains some PPE requirements)
- within general risk assessments (Figures C1.5 and C1.6 on pages 177 and 178/9 respectively both refer to PPE requirements)
- within other specific risk assessments (particularly COSHH, noise and manual handling)
- as a separate document.

Figure C6.1 contains details of the PPE requirements for various activities carried out by a town council (including the grass cutting referred to in the assessment record contained in Figure C1.6 on pages 178/9).

As for other types of risk assessments, PPE assessments must be reviewed if it is suspected that they are no longer valid or where significant changes take place.

Source Materials

1. HSE (1992) 'Personal protective equipment at work. PPE at Work Regulations 1992', L 25.
2. HSE (2003) 'A short guide to the Personal Protective Equipment at Work Regulations', INDG 174.

Fire

C7.1 Introduction

The Fire Precautions (Workplace) Regulations 1997 and the Fire Precautions Act 1971 are currently the two most important pieces of legislation relating to fire safety within the workplace and this chapter outlines the requirements of both. It is expected that the Fire Precautions Act will eventually disappear, to be replaced by an approach based upon risk assessment.

The chapter provides a summary of the key factors to be taken into account in carrying out a fire risk assessment:

- fire detection
- fire warning
- fire escape routes
- evacuation procedures
- fire fighting
- fire prevention

It also contains an example of a completed record of a fire risk assessment.

C7.2 The Fire Precautions (Workplace) Regulations 1997

Only three of the twenty-two regulations contained within what we will refer to as the 'Fire Regulations' actually relate directly to fire safety – the remainder deal with administrative matters. These regulations cover:

- fire fighting and fire detection (Reg. 4)
- emergency routes and exits (Reg. 5)
- maintenance of fire equipment and devices (Reg. 6).

The regulations apply to most workplaces whatever their size. However, the following types of workplaces are excluded:

- workplaces on construction sites (covered by separate regulations – see paragraph D2.2)
- ships (as defined in the Docks Regulations), including ships under construction or repair
- mines, other than surface buildings
- offshore installations
- means of transport (vehicles, aircraft, locomotives, rolling stock, trailers, etc.)
- agricultural or forestry land away from main buildings
- domestic premises
- workplaces used only by the self-employed.

(Despite these exclusions some level of fire precautions will still be necessary in excluded workplaces in order to comply with the general requirements of HASAWA and the Management Regulations.)

There is no specific requirement for risk assessment contained within the regulations themselves but they are specifically referred to in the general requirement for risk assessments set out in Regulation 3(1) of the Management Regulations (see paragraph C1.3 on page 169). In common with most other fire safety legislation, the regulations are enforced by the fire authority (as opposed to the HSE or local authority). Some of the detailed requirements of Regulations 4, 5 and 6 are referred to later in the chapter.

C7.3   Fire Precautions Act 1971

A fire certificate is required under the Act for hotels or boarding houses providing sleeping accommodation for more than six people (employees or guests) or with sleeping accommodation for employees or guests other than on the ground or first floors. Factories, offices, shops or railway premises must apply for a certificate when more than twenty people are at work at any time, or if more than ten are at work at any time elsewhere than on the ground floor. (In multi-occupancy buildings all workers are taken into account.) Those factories storing or using explosive or highly flammable materials must also apply for certificates.

Certificates issued by the fire authority under the Act normally contain detailed requirements in respect of each of the factors of fire protection referred to in paragraph C7.1. Where fire authorities consider that premises represent a low fire risk, they may exempt them from requiring a certificate once the employers have fulfilled their obligation to submit an application.

C7.4   Other fire safety requirements

Further fire safety requirements are contained in the following pieces of legislation.

Management of Health and Safety at Work Regulations 1999
In addition to the requirement for a risk assessment relating to fire safety, the Management Regulations require procedures to be established for dealing with situations of serious and imminent danger, such as fires (Reg. 8), and necessary contacts to be arranged with external services, including the fire service (Reg. 9). Employees must also be provided with relevant information on emergency procedures (Reg. 10). In shared workplaces employers and the self-employed must cooperate and coordinate their precautionary measures, such as those relating to fire prevention and fire evacuation (Reg. 11), while Regulation 12 places requirements on host employers to ensure that visiting employees are provided with information about evacuation procedures (paragraphs B2.7 to B2.11 contain further details of these regulations).

Dangerous Substances and Explosive Atmospheres Regulations 2002
These Regulations apply only to workplaces where there are substances with the potential to create risks from fire, explosion, exothermic reactions, etc. Such substances include flammable solvents, liquefied petroleum gas (LPG) and explosive dusts (wood dust, flour, etc.). The Regulations require an assessment – to be made of activities involving dangerous substances and for measures to be taken to eliminate, reduce or control fire or explosion risks. Precautions required as a result of the assessment might include the following:

- substituting the dangerous substance by a substance or process that eliminates the risk
- reducing the quantities of dangerous substances to a minimum

- avoiding or minimising releases, or controlling them at source
- preventing the formation of an explosive atmosphere
- avoiding ignition sources
- controlling usage parameters (e.g. temperature) to avoid danger
- keeping incompatible substances apart
- preventing the spread of fire or explosions
- providing explosion reliefs within equipment
- identifying potential explosive atmosphere areas with signs
- providing emergency equipment and establishing appropriate procedures
- providing information, instruction and training to employees (and others where necessary).

The HSE has published a free leaflet specifically on DSEAR ('Fire and Explosion: How Safe is Your Workplace?') together with a series of booklets containing Approved Codes of Practice (ACOPs) and guidance. Four of these deal specifically with

- design of plant, equipment and workplaces;
- storage of dangerous substances;
- control and mitigation measures; and
- safe maintenance, repair and cleaning procedures.

A fifth and more extensive ACOP and guidance booklet provides comprehensive information on all aspects of DSEAR.

Control of Major Accident Hazards Regulations 1999 (COMAH)
The COMAH Regulations apply when specified quantities of dangerous substances are present in workplaces, including some flammable or explosive materials. The regulations require both on- and off-site emergency plans to be prepared.

Confined Spaces Regulations 1997
The 'specified risks' referred to in the regulations include those of fire or explosion. Any entry into a confined space must be in accordance with a safe system of work (see Chapter B8) and arrangements must be in place to deal with possible emergencies.

The Explosives Act 1875
Premises manufacturing or storing explosives must be licensed with the HSE or the local authority. Fire safety requirements are usually contained within the licence.

Building Regulations 1991
The Building Regulations (and their equivalents in Scotland and Northern Ireland) apply to new buildings and also to extensions or material alterations of existing buildings and a material change of use of a building. They can impose fire safety requirements including aspects such as means of escape in case of fire, fire resistance of the structure and other precautions to reduce fire spread.

Licensing requirements
Some types of workplaces, particularly those likely to attract members of the public in significant numbers, are subject to licensing control from the local authority or licensing magistrate. The most common types are those involved in

the sale of alcohol, music and dancing, theatrical performances, the showing of films, gambling, sporting activities and other types of public entertainment. Such licences can impose additional fire safety requirements.

Registration schemes

Some uses of premises are required to be registered with the local authority or other official bodies. These include the following:

- nursing and residential care homes
- children's homes
- independent schools
- childcare activities.

Such registration schemes will usually impose fire safety requirements.

C7.5   Factors in fire risk assessment

The main factors to be taken into account during a fire risk assessment are fire detection, fire warning, fire escape routes, evacuation procedures, fire fighting and fire prevention. Such fire safety measures will need to be supported by signs and arrangements made to ensure that they are satisfactorily maintained. Employees and others must also be kept informed, particularly about fire evacuation procedures and their role in fire prevention.

'Fire Safety: An Employer's Guide' was published by the HSE and the Home Office to provide detailed guidance to employers on their legal obligations in relation to fire matters and on how to carry out a fire risk assessment. Paragraphs C7.6 to C7.11 below summarise some of the key principles to be observed in the risk assessment process.

C7.6   Fire detection and warning

Regulation 4(1) of the Fire Regulations requires a workplace to be equipped with fire detectors, to the extent that it is appropriate, in order to safeguard the safety of employees in case of fire. The means of detection must discover any fire quickly enough to allow occupants to escape to a safe place. In small workplaces where exits are visible and distances to be travelled are small it is likely that any fire will be detected visually (or by smell) and no additional precautions would be necessary.

However, consideration must be given to the possibility of a fire developing in unoccupied parts of the workplace such as storage areas, kitchens, plant rooms, etc. If such a fire could develop undetected and cut off escape routes for occupants, then detection equipment should be provided. This is particularly important where only a single escape route is available to occupants. A situation to watch out for is that of a room within a room (e.g. a small windowless office in the corner of a workshop or storeroom). In some cases it may be sufficient to provide a window so that occupants would be aware of a developing fire but in others detection equipment might be required. (Insurance companies may also insist on the provision of detection equipment in order to reduce property losses, including those resulting from fires when the premises are unoccupied.)

Interlinked detection and alarm systems

Larger and higher-risk workplaces would normally be provided with smoke detectors linked into the fire alarm system. Detectors would be provided at key positions on escape routes (corridors and staircases) as well as in locations where fire is more likely to break out (plant rooms and kitchens) and to safe-guard cul-de-sac situations.

Heat detectors may be more suitable than smoke detectors where there is a possibility of spurious alarms, e.g. smoke detectors triggered by process fumes. BS 5839 Part I contains the standards for such systems.

Domestic smoke alarms

In lower-risk environments adequate protection may be provided by a single domestic smoke alarm. It is also possible to interlink domestic alarms to pro-vide a detection and alarm network, although such alarms are manufactured to less demanding standards than more sophisticated detection/alarm equipment and may be more sensitive, triggering off false alarms. BS 5446 Part I contains the standards for domestic alarms. These alarms should be tested weekly and cleaned annually with batteries replaced annually (or in accordance with the manufacturer's recommendations for extended-life batteries).

Advice should be sought prior to installing detection equipment – fire authorities are usually willing to provide such advice, as are equipment suppliers.

Fire warning

The requirement contained in Regulation 4(1) of the Fire Regulations to provide fire alarms in order to safeguard the safety of employees in case of fire, like that for fire detectors, is qualified by the phrase 'to the extent that it is appropriate'. In smaller workplaces a simple shout of 'Fire' will be quite ade-quate, providing it can be heard in all parts of the premises, including store-rooms and toilets.

In slightly larger workplaces it may be sufficient to provide a single com-bined alarm point comprising a manual call point, sounder, battery and charger. Such units are generally preferable to manually operated sounders as someone must stay in the premises in order to operate the latter.

Larger buildings are more likely to need an electrically operated alarm sys-tem with break-glass call points situated on exit routes and alongside final exits. Sufficient sounders must be provided for the alarm to be heard throughout the premises. Additional visual alarms may be necessary in very noisy workplaces or in premises occupied by persons with hearing difficulties (both workers and others, e.g. residents in care homes). Any automatic fire detection equipment would normally be linked to the alarm system – again BS 5839 Part I contains the appropriate standards.

In some workplaces a two-tier system of alarms is used. One type of sound (e.g. an intermittent ringing) indicates a potential fire problem, which must be investigated, while a different sound (e.g. a continuous ringing) requires evacua-tion. Public address systems can also be utilised as part of the fire alarm system – this is normally the case where large numbers of the public are present, e.g. at places of entertainment, sports facilities or large retail centres. Voice alarm systems should comply with BS 5839 Part 8.

Fire alarm call points that are not readily apparent should be indicated by appropriate signs, but there is no automatic requirement to provide a sign for every call point. Weekly tests should be carried out of all fire alarm equipment, including self-contained units and manual sounders. In the case of electrical alarm systems, different break-glass call points should be used each week in rotation in order to trigger off the alarm. A small plastic tool (or an Allen key for older alarms) is usually required to activate the call point. A second person should be in position at the control panel to silence the alarm sounders once the test has been carried out.

It is a good idea to carry out such tests at the same time each week so that occupants assume that an alarm at any other time is the real thing. Where alarms are directly linked to call out the fire service, suitable steps will have to be taken to avoid calling them out each time the alarm is tested.

Alarms and any associated fire detection equipment should also be inspected and tested by a competent person at least annually and preferably quarterly. Fire certificates issued under the Fire Precautions Act will contain requirements on the inspection and testing of fire alarms and fire detectors.

C7.7   Fire escape routes

Regulation 5 of the Fire Regulations contains several requirements relating to emergency routes and exits. Detailed guidance on interpretation of these requirements is provided in 'Fire Safety: An Employer's Guide'. Much of this guidance is based on categorisation of workplaces as high, normal or low fire risk.

High risk

These are defined as areas where:

- highly flammable or explosive substances are stored or used (other than small quantities)
- unsatisfactory structural features are present (e.g. lack of fire separation, long and complex escape routes)
- there are high-risk activities carried out (e.g. large kitchens, use of naked flames)
- there is significant risk to life in case of fire (e.g. sleeping accommodation, hospitals, residential care homes).

Normal risk

Areas where outbreak of fire is likely to remain confined or only spread slowly, allowing people to escape to a place of safety.

Low risk

Areas where the risk of fire occurring is low, or the potential for fire, heat and smoke spreading is negligible.

General principles for escape routes

The HSE booklet states some general principles which should be followed in relation to escape routes, including the following:

- there should normally be alternative means of escape (other than in small workplaces or from some rooms of low or normal fire risk)
- escape routes in one direction only should be avoided wherever possible
- escape routes should be independent of each other and take people away from a fire
- routes should always lead to a place of safety
- routes should be wide enough for the number of occupants
- escape routes and exits should always be available for use and kept clear of obstructions.

Within these principles the booklet provides guidance on maximum distances of the escape route from any occupied part of the workplace to a 'storey exit'. These vary according to whether more than one route is available or only a single escape route is provided (the latter distances obviously being shorter) and whether the fire risk of the area is high, medium or low.

The term 'storey exit' is defined as 'An exit people can use so that, once through it, they are no longer at immediate risk. This includes a final exit, an exit to a protected lobby or stairway (including an exit to an external stairway) and an exit provided for means of escape through a compartment wall through which a final exit can be reached.' However, even a final exit may not lead to a 'place of safety' (defined as 'a place beyond the building in which a person is no longer in danger from fire'). Care needs to be taken that final exits do not lead to walled or fenced compounds from which escape is not possible.

Doors should normally open in the direction of travel, particularly if there are large numbers of people in the premises, and special arrangements may be necessary to take account of the presence of persons with disabilities.

Fire protection

Escape routes are often very reliant on an adequate degree of fire protection, particularly in relation to corridors, lobbies and stairways. The integrity of this fire protection is highly dependent upon self-closing fire doors – the wedging open of fire doors can allow fires and products of combustion to spread rapidly through buildings. However, self-closing doors may be held open by automatic door release mechanisms linked to smoke detectors and fire alarm systems, providing that the doors can also be closed manually (and will close in the event of power failure). Such devices are now quite common in hospitals, hotels and residential care homes.

Items prohibited on escape routes

Items creating potential fire hazards should not be located in protected escape routes or in corridors or stairwells serving as the sole means of escape from any part of a workplace. Such prohibited items include the following:

- portable heaters
- heaters with naked flames or radiant bars
- gas pipes or gas-fired equipment
- cooking appliances
- temporarily stored items
- upholstered furniture
- electrical equipment such as photocopiers
- vending or gaming machines.

Emergency lighting

Escape routes should have sufficient lighting for people to get out safely. Emergency lighting may be necessary where there is no natural daylight (e.g. in internal corridors or stairways) or where there would be insufficient borrowed light from other sources (e.g. streetlights) during hours of darkness.

The emergency lighting should operate on localised or complete failure of the normal lighting and should be sufficient to illuminate the escape routes and locate fire alarm call points and fire fighting equipment. (In small workplaces and outside locations the use of torches may be sufficient.) Automatic emergency lighting should be checked by a competent person in accordance with schedules set out in BS 5266 Part 1 (ref. 4) and the manufacturer's recommendations.

Fire exit signs

Escape routes and exit doors, which are not in common use or which may not be clearly evident, should be indicated by suitable signs (complying with the Health and Safety (Safety Signs and Signals) Regulations 1996). Account should also be taken of the need to provide additional signs where buildings are regularly used by visitors or members of the public, who may not be familiar with their layout.

C7.8   Evacuation procedures

Fire evacuation procedures for small workplaces should be fairly straightforward but in other situations they will need to be more complex. This is particularly likely to be the case where persons other than employees are likely to be present (e.g. in retail premises or places of entertainment), where persons are likely to be asleep (e.g. in hotels) and where persons are likely to need assistance in evacuations (e.g. in hospitals or residential care homes). Procedures are likely to involve the following.

Action on discovering a fire

In straightforward situations this would normally mean activating the fire alarm or shouting 'Fire'. However, more complex procedures might involve alerting relevant staff by means of pagers, personal telephones, radios or coded messages on a public address system.

Reaction to the alarm

In simple situations this would normally mean evacuating the premises using the nearest safe escape routes. Some staff may be required to carry out specific tasks such as:

- telephoning the fire service (where this is not done automatically)
- taking out staff lists and/or visitor books (for roll-call purposes)
- checking that designated parts of the building have been evacuated.

In certain types of workplaces staff may also be required to take actions such as:

- carrying out emergency shutdowns of equipment
- escorting members of the public out of the premises
- providing assistance to those who are unable to escape unaided (e.g. in hospitals or residential care homes).

In larger premises the design and fire separation of the building may require only the part immediately affected by fire to be evacuated with other parts remaining on standby. Evacuation of hospitals and residential care homes often follow this approach.

Assembly points and roll calls

Normally an assembly point (or points for larger premises) should be designated where staff and others should assemble in cases of evacuation. A roll call should be carried out using staff lists and/or visitor books (if available), from verbal information provided by those present, particularly those staff who may have been responsible for checking or sweeping out the building.

Someone must be appointed to coordinate the roll call and to liaise with the fire service and other emergency services on whether any staff or other persons cannot be accounted for or are known to be trapped. Staff involved in conducting the roll call and checking or sweeping the building are often called fire wardens or fire marshals. Administrative staff and others who are expected to be present most of the time are often better choices for these posts than more senior staff who are more likely to be away. However, deputies should still be appointed to carry out these tasks in case of sickness or holiday absence of the fire warden or fire marshal.

Instruction and training

All staff must be provided with instruction on the fire evacuation procedure, as should contractors' staff and any visitors who are to be left unaccompanied. Such instruction should normally be supported by 'Fire Action' notices posted around the premises. A simple example of such a notice is provided in Figure B3.4 on page 68.

However, regular drills should also be carried out to ensure that the procedure would prove effective in case of a real fire. Fire certificates will often specify that evacuation drills should be held every three, six or twelve months and similar frequencies should be adopted for uncertificated workplaces, depending on the level of fire risk. Records should be kept of dates of fire drills, and the times taken for evacuation and completion of the roll call. Drills can be made more challenging by simulating a situation where a main escape route cannot be used.

Additional training should be provided to ensure that staff who have been allocated specific tasks in the evacuation procedure are familiar with their responsibilities and how to carry them out.

C7.9   Fire fighting

Fire certificates usually specify what fire fighting equipment should be provided in the premises but in other cases this must be determined as part of the risk assessment process. The nature, types and quantities of fire fighting equipment will relate to the size and structure of the premises, the materials stored or used there and work activities taking place.

All portable fire extinguishers are now predominantly coloured red although extinguishers within the UK are still supplied with colour coded labels, which indicate their contents. The colour codes and suitability of different types of extinguishers are shown in the table below.

The use of hose reels or sprinkler systems may be necessary in larger premises, particularly where significant quantities of flammable materials are present. Other fixed fire fighting systems containing carbon dioxide or other extinguishing gases, foam or dry powder may be necessary to extinguish particular types of fires, e.g. those involving computers or other electrical equipment.

Most fire fighting equipment is provided to protect property rather than to safeguard employees (which, strictly speaking, is what the Fire Regulations are concerned with). However, pressure from insurers and the need to preserve business assets should ensure that a minimalist approach is not taken.

Fire fighting equipment should be located in positions where it can be seen clearly and from which a safe escape can be made, if necessary. Typically

Extinguisher type	Colour code	Suitability
Water	Red	Wood, paper, textiles NOT electrical or flammable liquid fires
Foam	Cream	Flammable liquids Also can be used for wood, paper, textiles NOT electrical fires
Carbon dioxide	Black	Flammable liquids and best for electrical fires NOT paper fires
Dry powder	Blue	Flammable liquids, electrical fires Also can be used for wood, paper, textiles
Vaporising liquid	Green	Flammable liquids and electrical fires
Fire blanket	–	Flammable liquids in containers, e.g. deep-fat friers, chip pans

they should be on escape routes or alongside exit doors. Suitable signs should be provided to indicate extinguishers, etc., which may be hidden from view by other equipment or in alcoves, but signs are not required to indicate the position of every piece of fire fighting equipment.

Staff who are expected to use fire fighting equipment should be provided with instruction and training. This should include the suitability of different extinguishers for various types of fires (see the table above) and should also involve practical techniques – there is no substitute for hands-on practice.

Equipment should be inspected regularly to ensure that it is still in position and has not been discharged or damaged. A more detailed annual service of extinguishers and hose reels should be carried out by a competent person. Extended services should be conducted every five years and equipment overhauled or replaced every twenty years. Sprinkler systems and other fixed fire fighting systems should be tested and serviced in accordance with the supplier or manufacturer's instructions.

C7.10 Fire prevention

For a fire to occur there must be three things present:

1. Fuel (something to burn), i.e. flammable solids, liquids or gases.
2. Oxygen (always present in the air).
3. A source of ignition (to start the fire), e.g. naked flames, cigarettes, electrical sources, hot surfaces.

Fire prevention measures should take account of all three of these aspects.

Potential sources of fuel

All flammable materials should be stored and used safely with particular attention being paid to flammable liquids and gases. Quantities of flammable liquids and gases in workrooms should be kept to a minimum.

The Dangerous Substances and Explosive Atmospheres Regulations 2002 (DSEAR) contain many requirements of relevance to the storage and use of substances creating risks from fire or explosion – see paragraph C7.4. Good standards of housekeeping, including control of waste materials, are always important in fire prevention. Potential risks from arson must be considered in positioning of waste containers and storage facilities.

Potential sources of oxygen

While oxygen will always be present in the air, other potential sources of oxygen need to be controlled. Oxygen cylinders and piped supplies used for welding and burning (or in health care) require precautions to be taken to prevent leaks (by regular checks on connections) or to minimise the consequences of any leaks that do occur (by situating cylinders and equipment in well-ventilated areas). Oxidising materials (e.g. peroxides, chlorates) must also be used and stored safely – well separated from potential sources of fuel.

Potential sources of ignition

Sources of ignition should be identified and controlled. This might involve the following:

- strict restrictions on smoking
- controls on 'hot work' (burning, welding, etc.)
- regular maintenance (to prevent short circuits or overheating)
- careful positioning of heating appliances
- special precautions to avoid risks from static electricity
- controlled burning of waste materials
- checks on work areas after work is completed (for smouldering materials)
- precautions to prevent arson.

C7.11 Practical aspects of assessment

Much of the practical guidance on risk assessment contained in Chapter C1 can be applied to fire risk assessments. Those carrying out the assessment should be competent for the purpose (including having an awareness of their own limitations) and have appropriate reference materials available – particularly 'Fire Safety: An Employer's Guide'. Other references may be appropriate for assessments in particular work sectors (see refs 10–14). In the case of larger workplaces, up-to-date drawings of the premises will be useful (especially in preparing the assessment record), as will a copy of the fire certificate, where one is in force. During the assessment, checks will need to be made as to whether the certificate is up to date and whether the terms of the certificate are being complied with. Larger workplaces may need to be divided into separate assessment units, as for other types of risk assessments.

Fire risk assessments will involve a higher proportion of observation than other kinds of assessments, in checking that standards contained in the fire certificate or recommended by the HSE are being met. Consideration needs to be given to the possible need for workers to escape from all parts of the workplace, including overhead crane cabs and other control cabins or from pits and basement areas. Discussions with workers may be necessary in order to determine the level of fire prevention measures being taken (e.g. the control of flammable materials, including waste and the elimination of sources of ignition), the reasons for fire protection devices being misused (e.g. wedging open of fire doors) and their awareness of fire evacuation arrangements. Records will need to be checked to ensure that fire protection equipment is being maintained,

FIND THIS ON CD	Figure C7.1	Possible format for recording a fire risk assessment

ARMAGEDDON ASSOCIATES
THE RISK ASSESSMENT

Location: Head Office, Nirvana Business Park
Persons at risk: Armageddon staff, visitors to the premises and contractors' staff (including permanent cleaning staff).

	PRECAUTIONS/COMMENTS	ACTION POINTS
FIRE DETECTION	Smoke detectors in upper and lower corridors, at the top of the main staircase and in both kitchens. Maintenance contract with Acme Fire Protection (annual check).	
FIRE WARNING	Electrical alarm – break-glass points at exits from both upper and lower corridors, emergency exits (upper and lower) and main exit from building. Sounders in both corridors. Maintenance contract with Acme Fire Protection (quarterly visits). Alarm should be tested weekly by Office Manager.	1 & 2
FIRE ESCAPE ROUTES	Via main protected stairway (self-closing fire doors at ground and first floor level) and by external emergency escape stairway (first floor) and emergency escape door (ground floor). Both emergency exit doors, well indicated with exit signs and easily openable with 'panic bolts'. All doors open in direction of travel and are of adequate width. Travel distances comfortably within recommended standards for normal fire risk. Emergency lighting provided in both corridors and main stairway. (External escape stairway well illuminated by street lighting.) Lighting maintenance contract with Acme Fire Protection (annual check).	3 & 4
EVACUATION PROCEDURE	Clearly displayed on notices throughout the building. Assembly point outside the library (next door). Fire wardens and deputies appointed for each section. Receptionist expected to take visitors' book to assembly point. (Fire alarm directly linked to Fire Brigade.) Fire evacuation drills should be held at least once per year (initiated by Office Manager). Evacuation arrangements covered in induction of new staff. (Not all cleaning contractor's staff aware of procedure.)	5 & 6

FIRE FIGHTING	Adequate provision of water and carbon dioxide fire extinguishers throughout both floors of the building. (Mainly situated close to fire exit doors.) Maintenance and annual inspection of fire extinguishers carried out by Premier Fire Extinguishers. Adequate provision of water and carbon dioxide fire extinguishers throughout both floors of the building. (Mainly situated close to fire exit doors.) Maintenance and annual inspection of fire extinguishers carried out by Premier Fire Extinguishers.	
FIRE PREVENTION	Smoking is not allowed within the building (covered smoking area provided in the car park). Paper and other waste stored in wheelie bins in a covered area in the car park. The access to this area is locked outside working hours.	
		RESPONSIBILITY
ACTION POINTS	1. The fire alarm should be tested every week. (Records show several gaps in the testing programme.) 2. Alarm tests should be carried out using break-glass call points in rotation. (Some tests had only involved testing the sounder at the main panel.) 3. Self-closing fire doors must not be wedged open (the door at the head of the main staircase was wedged). 4. Office furniture partially obstructing the upper corridor must be removed. 5. Fire evacuation drills should be held at least every twelve months. (No drill has been held for nearly two years.) 6. Ensure cleaning contractor's staff (and any other contractors or temps) are made aware of evacuation procedure.	AJ AJ AJ & all staff KW AJ AJ

ACTION POINTS TO BE CHECKED: April 2005
NEXT ROUTINE REVIEW SCHEDULED FOR: February 2007

ASSESSMENT BY: A. SPARK & C. BURNS   (Names)
                   A. Spark   C. Burns   (Signatures)
DATE: 2 February

PROGRESS ON ACTION POINTS

Checked by   (Name)   (Signature)

inspected and tested to the appropriate standards and that fire evacuation drills are being held as required.

C7.12 Assessment records

There is no standard format for recording fire risk assessments and the key HSE publication (ref. 1) provides no guidance on the subject. The risk assessment formats depicted in Figures C1.5 and C1.6 on pages 177 and 178/9 can be used if required, although the general risks associated with fire are fairly self-evident and therefore much of the space in the 'Risks' columns would be wasted. However, more detail would be required on risks associated with highly flammable liquids or flammable gases used or stored in the workplace. A possible format for recording a fire risk assessment is provided in Figure C7.1 overleaf. This relates to a fairly simple office building with no special fire risks and contains sections dealing with the key aspects of fire precautions referred to in paragraphs C7.6 to C7.10.

A lot of narrative in fire risk assessment records can be avoided by including marked-up drawings showing the various elements of fire protection (smoke and heat detectors, fire alarm call points and sounders, self-closing fire doors, emergency lighting, fire fighting equipment, signs, etc.). Any narrative can then be limited to comments by the assessor(s) on whether these protection measures complied with the standards contained in the HSE booklet (ref. 1) or fire certificate. Some detail should also still be provided on maintenance, inspection and testing arrangements and on fire evacuation drills and fire prevention measures.

Source materials

1. HSE/Home Office (1999) 'Fire Safety: An Employer's Guide'.
2. BS 5839: Fire detection and alarm systems for buildings.
3. BS 5446: Components of automatic fire alarm systems for residential premises. Part 1: Specifications for self-contained smoke alarms and point-type smoke detectors.
4. BS 5266: Emergency lighting. Part 1: Code of practice for the emergency lighting of premises other than cinemas and certain other specified premises used for entertainment.
5. HSE (2002) 'Fire and Explosion: How Safe is Your Workplace?', INDG 370.
6. HSE (2003) 'Design of plant, equipment and workplaces', DSEAR 2002, ACOP and Guidance, L 134.
7. HSE (2003) 'Storage of dangerous substances', DSEAR 2002, ACOP and Guidance, L 135.
8. HSE (2003) 'Control and mitigation measures', DSEAR 2002, ACOP and Guidance, L 136.
9. HSE (2003) 'Safe maintenance, repair and cleaning procedures', DSEAR 2002, ACOP and Guidance, L 137.
10. HSE (2003) 'Dangerous Substances and Explosive Atmospheres Regulations 2002', ACOP and Guidance, L 138.
11. HSE (1996) 'Dispensing Petrol: Assessing and Controlling the Risk of Fire and Explosion at Sites Where Petrol is Stored and Dispensed as a Fuel', HSG 146.
12. HSE (2000) 'Safe Use of Petrol in Garages', INDG 331.
13. HSE (1997) 'Fire Safety in Construction Work', HSG 168.
14. HSE (1995) 'Fire Safety in the Paper and Board Industry'.
15. HSE (2003) 'Unloading petrol from road tankers', L 133.

Other Key Areas of Health and Safety

Work Equipment

DI.I   Introduction

All types of equipment used at work are subject to the Provision and Use of Work Equipment Regulations 1998 (PUWER) and this chapter summarises the contents of those regulations, explaining some of the requirements (particularly those relating to dangerous parts of equipment) in some detail.

The chapter also summarises the contents and some of the practical implications of the Lifting Operations and Lifting Equipment Regulations 1998 (LOLER). However, there are also several other sets of regulations that deal with specific types of work equipment, including the following:

- Pressure Systems Safety Regulations
  - see HSE's 'Safety of Pressure Systems: Pressure Systems Safety Regulations 2000'
  - paragraph B9.5 also provides limited information on their requirements.
- Electricity at Work Regulations
  - see Chapter D3.
- Health and Safety (Display Screen Equipment) Regulations 1992
  - see Chapter C5.
- Personal Protective Equipment at Work Regulations 1992
  - see Chapter C6.
- Supply of Machinery (Safety) Regulations 1992
  - see paragraphs D1.4 and D1.7 and HSE's guide 'Buying New Machinery: A Short Guide to the Law'.

DI.2   Provision and Use of Work Equipment Regulations 1998 (PUWER 1998)

These regulations replaced PUWER 1992, which in turn replaced parts of the Factories Act and many other pieces of legislation dealing with equipment safety. The 1998 regulations included additional requirements relating to mobile work equipment and power presses. Several sets of older regulations have now been completely revoked, including those dealing with unfenced machinery, power presses, abrasive wheels and woodworking machines.

The regulations are accompanied by an Approved Code of Practice as well as HSE guidance, all contained in a single HSE booklet (see ref. 3 on page 312).

Several terms are defined in Regulation 2. Work equipment is defined as 'any machinery, appliance, apparatus, tool or installation for use at work (whether exclusively or not)'. This definition includes the following:

- powered and unpowered tools, e.g. a portable drill, a hammer or a knife
- single machines, e.g. a photocopier or a circular saw
- apparatus, e.g. laboratory apparatus, photoelectric sensors

- lifting and access equipment, e.g. a lift truck, lifting sling, elevating work platform or ladder
- vehicles, e.g. cranes, dumper trucks, forklift trucks (road traffic legislation takes precedence over PUWER on public roads and in public places).

(Livestock, substances such as cement, structural items such as walls or stairs, and private cars are not classified as work equipment.)

'Use' in relation to work equipment is defined as 'any activity involving work equipment and includes starting, stopping, programming, setting, transporting, repairing, modifying, maintaining, servicing and cleaning'.

Regulation 3 deals with the application of the regulations and responsibilities under the regulations.

Application

PUWER applies to the following:

- all work equipment used where HASAWA applies
- work equipment used in common parts of shared buildings and work sites, e.g. industrial estates and construction sites
- all work done by the employed and self-employed, including work in private homes (except for domestic work in private households)
- homeworkers
- offshore oil and gas installations.

The application to work equipment on board ships is more complex and the HSE booklet should be consulted (see ref. 3 on page 312).

Responsibilities

Employers are responsible for equipment provided for use or used by any of their employees. This responsibility includes equipment that may be provided by others – hire companies, contractors or even by the employees themselves (common practice for specialist tradesmen).

Persons having control of work equipment, or who use or supervise or manage the use of work equipment or the way in which work equipment is used at work, all have responsibilities depending on the extent of their control. Self-employed persons have duties in respect of work equipment they use at work.

Consequently in many situations there will be overlapping responsibilities, as is the case in HASAWA itself. Hired equipment used on a construction site will be the responsibility of the hire company and the main contractor (if they hired it), but also the responsibility of the employers of any person who uses it, and of any self-employed person using it.

D1.3 General requirements under PUWER (Regulations 4 to 10)

The general requirements contained in Part II of the regulations (Regs 4 to 10) apply to all types of work equipment.

Suitability of work equipment (Regulation 4)

Work equipment must be suitable for the purpose for which it is used or provided. This must be addressed from three main aspects:

1. The equipment's initial integrity.
2. The place where it will be used.
3. The purpose for which it will be used.

Selection of suitable work equipment should be dealt with as part of the risk assessment process. The following are examples of these principles in practice:

- An open-bladed knife should not be used for a task where scissors or shears would be much more suitable.
- An overhead crane should not be used for towing a vehicle.
- In wet environments or flammable atmospheres air-powered equipment should be chosen in preference to electrical equipment.
- Equipment with a combustion engine should not be used where ventilation is poor.

Maintenance (Regulation 5)

Work equipment must be 'maintained in an efficient state, efficient working order and in good repair'. Such maintenance might range from simple visual checks on hand tools to detailed planned preventive maintenance programmes for complex plants. Particular attention should be paid to safety-critical parts. The aim is to maintain the performance of the equipment so that people are not put at risk. Although there is no requirement to keep maintenance records, such records are recommended by the HSE for high-risk equipment. Where records do exist, they must be kept up to date.

In the case of hired equipment (or equipment that is borrowed or shared), responsibility for maintenance must be clearly established. In the case of a hired forklift truck, the user organisation may carry out daily pre-use inspections (see Regulation 6) and minor maintenance work, with the hire company responsible both for carrying out periodic inspection and more complex maintenance, and for arranging statutory examinations under LOLER. All maintenance work must be done by persons who are competent for the purpose (see Regulations 8 and 9).

Inspection (Regulation 6)

Where work equipment safety depends on the installation conditions it must be inspected:

- after installation and before being put into service
- after assembly at a new site or location.

Checks may need to be made on the stability of the equipment, the correct functioning of interlock guards or of other safety devices. Fairground equipment is regularly re-installed or assembled at new locations, as are many types of construction equipment.

Work equipment exposed to potentially dangerous deterioration must be inspected:

- at suitable intervals
- after potentially dangerous exceptional circumstances.

Once again fairground equipment should be subject to regular checks but so also should other equipment where workers regularly approach danger zones and are heavily dependent upon interlocks or proximity protection devices. Potentially dangerous exceptional circumstances justifying inspection might involve impact (e.g. from a passing piece of mobile equipment), the effects of flooding or water damage, major modification or repair or even just a long period of disuse.

Inspections must be made by a competent person and be carried out to the extent appropriate for the purpose. Some testing may be required, e.g. of interlocks. Records must be made of inspections under this regulation and these must be kept until the next inspection.

The extent and content of the records kept should be appropriate to the type of work equipment and the nature of the inspection. Where work equipment moves around, it must be accompanied by physical evidence of its last inspection – a copy of the record or, in simple cases, a tag, colour code or label.

This regulation does not apply to inspections of power presses and their guards, etc., lifting equipment, winding apparatus in mines or scaffolding or excavation supports used in construction, all of which are subject to more specific inspection requirements.

Specific risks (Regulation 7)

Where use of work equipment is likely to involve specific risks, then its use and any repairs, modifications, maintenance or servicing of the work equipment may need to be restricted to designated persons.

This is particularly likely to be the case where physical safeguards (such as provision of guards) cannot adequately control the risk and reliance has to be placed on safe systems of work performed by persons who have received the necessary information, instruction and training (and demonstrated an appropriate level of skill). Examples of such activities would include use of chainsaws (see also Reg. 9), use of meat cleavers, mounting of abrasive wheels and discs, etc. The requirements of this regulation should also be considered in relation to the need to place restrictions on the work activities of young people (see paragraphs C1.4 and C1.5).

Information and instructions (Regulation 8)

Adequate health and safety information and, where appropriate, written instructions pertaining to the use of work equipment must be available to those using the equipment and also to those supervising or managing its use. Whether this information is in writing or verbal will depend on the complexity and risk of the equipment.

It should include the following:

- the conditions in which and methods by which the equipment may be used
- foreseeable abnormal situations and action to be taken in such situations
- any conclusions drawn from experience in using the equipment.

This requirement overlaps with similar general requirements contained in HASAWA and the Management Regulations. The importance of communicating effectively with employees was stressed in Chapter B6. Health and safety information must be made available to managers and supervisors so that they are aware of the risks involved in relation to work equipment under their control and the precautions necessary to minimise those risks.

Training (Regulation 9)

Adequate training (for health and safety purposes) on the methods of use, relevant risks and precautions relating to work equipment must have been received by those using the equipment and those supervising or managing its use. The ACOP includes specific requirements on training to drive self-propelled work equipment and for the use of chainsaws.

Again this requirement overlaps with requirements in HASAWA and the Management Regulations relating to training and capabilities (see Chapter B3 in relation to health and safety training). The training of managers and supervisors in respect of work equipment under their control is important – they may not be capable of operating a forklift truck or a chainsaw themselves but they should be capable of recognising when these or any other types of work equipment are being used unsafely.

Conformity with community requirements (Regulation 10)

Employers must ensure that items of work equipment conform at all times with essential design and construction requirements contained in regulations listed in Schedule 1 to PUWER. These regulations implement various European Product Directives, the most significant of which is the Machinery Directive which was enacted in the UK by the Supply of Machinery (Safety) Regulations 1992 (since amended). Other regulations deal with electrical equipment, lifts, medical devices, pressure vessels, gas appliances and construction plant.

Where an essential requirement applies to the design or construction of work equipment then the corresponding requirements contained in Regulations 11 to 19 and 22 to 29 do not apply. However, those regulations still apply to any aspects of work equipment safety not covered by the essential requirements.

Manufacturers should fix a CE marking to equipment to show that it complies with the essential health and safety requirements and also issue a 'declaration of conformity'. However, these are only claims by the manufacturer that the equipment complies and are not a guarantee of conformity. There is increasing concern about the veracity of such claims and general abuse of the CE system. Purchasers are encouraged to check the standards of equipment that they are buying and the HSE has produced an excellent free leaflet, 'Buying New Machinery: A Short Guide to the Law' to aid in this process. In cases of doubt employers should seek specialist advice.

The regulation only applies to work equipment provided for the first time after 31 December 1992. Older equipment must still comply with the specific requirements contained in Regulations 11 to 19 and 22 to 29 of PUWER.

D1.4 Specific requirements of PUWER (Regulations 11 to 24)

Although most of these regulations only apply to equipment not covered by European Product Directives (see Regulation 10 above), their requirements closely match those of the Directives (particularly the Machinery Directive). Compliance with requirements of PUWER provides a good indicator that a piece of work equipment will comply with the requirements of the relevant product directive. (For detailed interpretation of the regulations, the PUWER ACOP and guidance booklet should be consulted).

Dangerous parts of machinery (Regulation 11)

Effective measures must be taken to prevent access to any dangerous part of machinery (or rotating stock bar) or stop its movement before any part of a person enters a danger zone. This must be done through a hierarchy of measures involving (in descending order of preference): fixed guards; other guards or protection devices; jigs, holders, push-sticks or similar protection appliances. Such information, instruction, training and supervision as is necessary must be provided. These requirements are explained in greater detail in paragraph D1.8.

Protection against specified hazards (Regulation 12)

Measures must be taken to prevent or adequately control certain specified hazards:

- articles or substances falling or being ejected from work equipment
- rupture or disintegration of parts of work equipment
- work equipment catching fire or overheating
- unintended or premature discharge of articles or substances
- unintended or premature explosions of work equipment (or articles or substances within it).

So far as is reasonably practicable, these should be measures other than the provision of PPE or of information, instruction, training or supervision.

These types of hazards should be identified as part of the risk assessment process and appropriate control measures introduced. This might involve the following:

- screens to protect people from swarf from machine tools
- precautions to prevent or minimise the effects of dust explosions
- measures to prevent abrasive wheels bursting (e.g. not overspeeding wheels, ensuring that wheels are only mounted by trained persons)
- guards that will contain fragments of abrasive wheels if they should burst.

High or very low temperature (Regulation 13)

Work equipment, parts of work equipment and any article or substance produced, used or stored in work equipment, which is at a high or very low temperature must have protection where appropriate, so as to prevent injury by burn, scald or sear. Preferably this should be by the use of engineering measures such as the use of insulation and shielding. These should be applied to all accessible surfaces – sometimes insulation is applied for energy conservation purposes to large vessels or long pipe runs but not to more awkwardly shaped sections which may still be accessible.

However, in many cases (e.g. hotplates and items heated on them, soldering irons, hand-held clothes irons) it is impossible to prevent contact by engineering means, and the use of warning signs, PPE or the provision of information, instruction, training and supervision will be necessary.

Controls and control systems (Regulations 14 to 18)

The provision of controls is required 'where appropriate'. They are not generally appropriate for equipment with no moving parts or those where risk of injury is negligible. Only limited controls will be appropriate for human-powered equipment. The regulations contain some detailed requirements in respect of controls for starting and stopping equipment (including emergency stops) and for making significant changes in operating conditions. Other types of control and the design of control systems are also covered.

Isolation from sources of energy (Regulation 19)

Where appropriate, work equipment must be provided with suitable means to isolate it from all sources of energy. These means must be clearly identifiable and readily accessible. Measures must also be taken to prevent risk from re-energisation. This requires cutting off the energy supply and ensuring that inadvertent reconnection is not possible, e.g. by using locking-off devices. Use of permit-to-work systems may also be appropriate (see paragraphs B8.9 to B8.12). Isolation of electrical equipment is covered in paragraph D3.4.

Stability (Regulation 20)

Where necessary for health and safety, work equipment must be stabilised, by clamping, bolting or other measures. As well as preventing items such as storage racks or machines falling over, this might be required to prevent the movement of powered machinery or to counter the effects of the wind or the movement of persons on access equipment, e.g. scaffolds or ladders. Mobile work equipment may need to be stabilised by counterbalances, outriggers, etc.

Lighting (Regulation 21)

Suitable and sufficient lighting, which takes account of the operations to be carried out, must be provided where work equipment is used. This might require the provision of local or temporary lighting to ensure safe operation in addition to existing general lighting.

Maintenance operations (Regulation 22)

Equipment must be constructed or adapted to ensure, as far as reasonably practicable, that maintenance operations involving risk can be carried out with the equipment shut down. Otherwise appropriate measures must be taken to protect those carrying out maintenance operations, e.g. the use of temporary guards, limited movement controls, crawl speed facilities, etc.

Markings (Regulation 23)

Where appropriate for reasons of health and safety, work equipment must be marked in a clearly visible way, e.g. abrasive wheel speeds, safe working loads, gas cylinders, storage vessels, pipelines (with details of their contents). Markings must comply with the requirements of the Safety Signs and Signals Regulations or other relevant published standards.

Warnings (Regulation 24)

Where appropriate for reasons of health and safety, work equipment must incorporate warnings or warning devices. These must be unambiguous, easily perceived and easily understood. In addition to conventional signs, audible or visual devices may be required, such as reversing alarms on vehicles, start-up sounders, indicator lights, etc. Signs must also comply with the Safety Signs and Signals Regulations.

D1.5 Mobile work equipment (Regulations 25 to 30)

The requirements of PUWER that apply to all types of work equipment (see paragraphs D1.3 and D1.4) also apply to mobile work equipment – these requirements are additional. Mobile work equipment may be self-propelled, towed or remote controlled and may incorporate attachments.

Employees carried on mobile work equipment (Regulation 25)
Such mobile work equipment must be suitable for carrying persons and, so far as reasonably practicable, incorporate other safety features. The ACOP and guidance refer to several aspects to be taken into account, including the following:

- the provision of seats (for drivers and any passengers)
- fitting of cabs and other enclosures, barriers or guard-rails
- possible need for falling object protective structures (FOPSs)
- possible need for seat belts or other restraining systems
- speed adjustment for certain activities
- guards or barriers to provide protection from wheels or tracks.

Rolling over of mobile work equipment (Regulation 26)
Risks from mobile work equipment rolling over must be minimised by means such as the following:

- stabilising the work equipment
- having a structure that does no more than fall on its side
- having sufficient clearance if it overturns further
- providing a suitable restraining system if there is a crushing risk.

Compliance is not required where overall risks would increase, where operation of the mobile work equipment would not be reasonably practicable, or, in relation to equipment provided for use before 5 December 1998, if it would not be reasonably practicable.

Overturning of forklift trucks (Regulation 27)
The effect of this regulation is that forklift trucks fitted with a vertical mast or a roll-over protection system (ROPS) must be provided with an operator-restraining system (such as a seat belt), so far as is reasonably practicable.

Self-propelled work equipment (Regulation 28)
Risks from self-propelled work equipment in motion must be controlled by the following:

- facilities preventing its being started by unauthorised persons
- collision minimising measures for rail-mounted work equipment, e.g. buffers
- a braking and stopping device (possibly accompanied by secondary braking systems)
- vision improvement devices where necessary, e.g. mirrors, CCTV
- appropriate lights, etc., for use at night or in dark places
- fire fighting equipment, where appropriate.

Remote-controlled self-propelled work equipment (Regulation 29)
Risks from radio-controlled and similar equipment must be controlled by the following:

- an automatic stop device once it leaves its control range
- other features or devices to prevent crushing or impact (e.g. alarms, flashing lights, hold-to-run controls).

Drive shafts (Regulation 30)

Where seizure of a drive shaft is likely to involve safety risks, means of preventing such a seizure must be provided or other safety measures provided (e.g. guards). Means of safeguarding an uncoupled shaft from soiling or damage by the ground must also be provided, e.g. a cradle or similar support. The practical application of PUWER to workplace vehicles is considered further in paragraph D1.10.

D1.6   Power presses (Regulations 31 to 35)

Power presses are subject to the requirements of PUWER applying to all types of work equipment (see paragraphs D1.3 and D1.4). However, a number of additional requirements, previously contained in the Power Presses Regulations, are continued in force by the regulations in this part of PUWER. The previous regulations only applied to power presses in factories while PUWER applies to all workplaces and thus to any power presses in educational and research premises. There are a few other minor changes to the previous requirements. The regulations deal with the following topics.

Power presses to which Regulations 32 to 35 do not apply (Regulation 31)

The term 'power press' is defined in Regulation 2 of PUWER as 'a press or press brake for the working of metal by means of tools, or for die proving, which is power driven and which embodies a flywheel and clutch'. This definition brings in some equipment to which the regulations are deemed inappropriate. Regulation 31 excludes such equipment (listed in Schedule 2) from Regulations 32 to 35, but not the rest of PUWER. Schedule 2 contains items such as guillotines and riveting and stapling machines.

Thorough examinations of power presses, guards and protection devices (Regulation 32)

Power presses and their guards and protection devices must be thoroughly examined after being put into service for the first time or after assembly at a new site or location, and at prescribed intervals thereafter. Any defects must be remedied before the power press is used again. (Examinations must be carried out by a competent person.)

Inspections of guards and protection devices (Regulation 33)

Guards and protection devices must be inspected and tested by a competent person (or a person undergoing training under the immediate supervision of a competent person):

- after setting, re-setting or adjustment of its tools; and
- within the first four hours of a working period.

The regulation specifies the content of the certificate of inspection and test.

Reports (Regulation 34)

Persons carrying out thorough examinations under Regulation 32 must:

- notify the employer forthwith of any dangerous defect
- make a report as soon as practicable (containing information specified in Schedule 3)
- notify the enforcing authority (HSE) of dangerous defects as soon as practicable.

Persons making inspections and tests under Regulation 33 also must notify the employer forthwith of any dangerous defect.

Keeping of information (Regulation 35)

The regulation stipulates the periods for which reports of thorough examinations and certificates of inspection and test must be kept.

The HSE has published a separate booklet entitled 'Safe Use of Power Presses: PUWER as Applied to Power Presses' (ref. 4) containing an ACOP and detailed guidance on the application of PUWER to power presses. The booklet highlights the importance of adequate supervision and appropriate training of young people in relation to their use of power presses, because of their immaturity, lack of experience and lack of awareness of risks (see paragraphs C1.4 and C1.5).

D1.7 Dangerous parts of machinery

Regulation 11 of PUWER requires measures to be taken to protect people from dangerous parts of machinery. Some of the more common types of dangerous parts are listed below.

In considering what protective measures are appropriate, attention may also need to be given to the hazards referred to in Regulation 12 of PUWER, particularly the possibility of materials being ejected from work equipment. It must also be borne in mind that manually powered equipment can also present considerable dangers, e.g. a treadle-operated guillotine is capable of amputating fingers.

Type	Caused by	Examples
Traps	Operating parts	Press tools Moving parts and fixed objects
	In-running nips	Drums or rollers Conveyor belts and drums Belts and pulley wheels Gear wheels
	Shearing actions	Between moving parts (e.g. rods and linkages) Between moving parts and fixed objects
Entanglement	Rotating parts and materials (particularly those with projections)	Drive shafts Drill spindles and chucks Rotating workpieces (e.g. stock-bars in lathes)
Contact or impact	Direct contact with sharp or abrasive surfaces (causing amputations, cuts or abrasions) Direct impact of moving parts (causing bruises, fractures, etc.)	Saw blades Other cutting tools Grinding wheels or discs Fast moving parts

D1.8   The hierarchy of protection

Paragraphs (1) and (2) of Regulation 11 of PUWER state:

(1) Every employer shall ensure that measures are taken in accordance with paragraph (2) which are effective –
 (a) to prevent access to any dangerous part of machinery or to any rotating stock-bar; or
 (b) to stop the movement of any dangerous part of machinery or rotating stock-bar before any part of a person enters a danger zone.
(2) The measures required by paragraph (1) shall consist of –
 (a) the provision of fixed guards enclosing every dangerous part or rotating stock-bar where and to the extent that it is practicable to do so, but where or to the extent that it is not, then
 (b) the provision of other guards or protection devices where and to the extent that it is practicable to do so, but where or to the extent that it is not, then
 (c) the provision of jigs, holders, push-sticks or similar protection appliances used in conjunction with the machinery where and to the extent that it is practicable to do so,
 and the provision of information, instruction, training and supervision as is necessary.

Paragraphs (3) and (4) contain requirements relating to the standards of guards, protection devices and protection appliances (these are dealt with in paragraph D1.9) while paragraph (5) of the regulation defines two terms used in the regulation:

'danger zone' means any zone in or around machinery in which a person is exposed to a risk to health or safety from contact with a dangerous part of machinery or a rotating stock-bar;
'stock-bar' means any part of a stock-bar which projects beyond the head-stock of a lathe.

The hierarchical requirement relating to protection measures is similar to that contained in the COSHH Regulations (see paragraph C2.5) except that this requirement is qualified by 'to the extent that it is practicable', as opposed to 'reasonably practicable' in the COSHH Regulations. The requirement can be illustrated by considering its application to a bench-mounted circular saw.

Fixed guards

Access to the motor and transmission equipment driving the saw can be prevented by the provision of a fixed guard. A fixed guard can also prevent access to the portion of the saw blade below the bench. However, it is clearly impracticable to use a fixed guard to guard all of the blade above the bench.

Other guards or protection devices

In most joinery workshops it is normal to provide an adjustable guard over the top of the blade, usually fitted with an adjustable nosepiece to extend it as low as practicable at the front of the blade. The rear of the blade is normally protected by the riving knife. However, a small portion of the blade at the front must be left exposed in order to cut wood or other materials.

Protection appliances

Workers' hands can often be kept away from the exposed part of the blade by using a feed table or jig to feed the workpiece to the saw, and use of a push-stick to push timber onto the blade will almost always be practicable.

Information, instruction, training and supervision

Even when all the above precautions have been taken, those using the saw must still be aware of the risks associated with the machine, techniques associated with its use and know how to adjust the top guard, nosepiece and riving knife and how to apply pressure to the push-stick. Information, instruction, training and supervision are all necessary to impart this knowledge and to ensure that it continues to be applied in practice with the necessary skill and concentration.

D1.9   Types of guards and protection devices

Detailed standards for guards and protection devices are contained in BS EN 953:1998, BS EN 294:1992 and BS EN 292-1:1991. The HSE also publishes guidance on many specific types of equipment. Some of the more common types are referred to on page 273. Guards and protection devices can be separated into the following categories.

Fixed guards

Fixed guards are secured in position either permanently (e.g. by welding) or by screws, nuts, etc., so that a tool is required to move or open them. To meet the requirements of the hierarchy of protection (see paragraph D1.8) they must enclose any dangerous parts. Openings are permitted in fixed enclosing guards (and other types of guards), providing that they are not large enough to permit access to a dangerous part of machinery. The sizes of permitted openings relate to safe reach distances – simplistically, a finger-sized opening must be more than a finger's length from a danger point.

Movable guards

Such guards can be opened without the use of tools. They may be:

- power operated
- self-closing (e.g. spring-loaded, as on the front of many cross-cut saws)
- linked to the machine's controls (so that the hazardous functions cannot operate until the guard is closed).

Many movable guards are linked with interlocking devices (see below).

Adjustable guards

Adjustment of such guards should be fairly straightforward and should not necessarily require the use of tools. (The need for tools could prove a disincentive to keeping the guard adjusted correctly.) The top guard of a bench-mounted circular saw would normally be adjustable and such guards are commonly used on other types of woodworking and metalworking machines.

Interlocking guards

It is often practicable to provide movable guards with interlocking devices so that:

- the hazardous machine functions covered by the guard cannot operate until the guard is closed
- if the guard is opened while hazardous machine functions are operating, a stop instruction is given†
- when the guard is closed, the hazardous machine functions covered by the guard can operate, but the closure of the guard does not by itself initiate their operation.*

* A control guard (an interlocking guard with a start function) may be used in certain situations where frequent access is required.

† A locking device may be required to ensure that an interlocking guard remains closed and locked until any risk of injury has passed, e.g. where there is a run-down period after power is removed.

Protection devices

These devices do not prevent access but stop the movement of the dangerous part before contact is made. They include the following:

- mechanical trip devices, e.g. telescopic probes or trip wires (commonly used respectively on radial arm drills and alongside conveyor belts)
- photoelectric beams and proximity devices (often used on press brakes, guillotines, etc.)
- pressure-sensitive mats
- two-hand controls (where two buttons must be depressed to operate a machine – these are potentially dangerous where more than one person may be present).

Standards for guards and protection devices

Paragraph (3) of Regulation 11 of PUWER states:

All guards and protection devices provided under sub-paragraphs (a) or (b) of paragraph (2) shall –

(a) be suitable for the purpose for which they are provided;
(b) be of good construction, sound materials and adequate strength;
(c) be maintained in an efficient state, in efficient working order and in good repair;
(d) not give rise to any increased risk to health or safety;
(e) not be easily bypassed or disabled;
(f) be situated at sufficient distance from the danger zone;
(g) not unduly restrict the view of the operating cycle of the machinery, where such a view is necessary;
(h) be so constructed or adapted that they allow operations necessary to fit or replace parts and for maintenance work, restricting access so that it is allowed only to the area where the work is to be carried out and, if possible, without having to dismantle the guard or protection device.

(Paragraph (4) of the regulation requires protection appliances such as feed tables, jigs and push-sticks to comply with sub-paragraphs (a) to (d) and (g) above.)

The requirements of the Machinery Directive (implemented by virtue of the Supply of Machinery (Safety) Regulations 1992 and Regulation 10 of PUWER) follow similar principles to those described above.

D1.10 Workplace vehicles

While Regulations 25 to 30 of PUWER deal specifically with mobile work equipment, other requirements of the regulations will also be relevant:

- suitability: vehicles must be suitable for the purpose for which they are used
- maintenance: effective maintenance of vehicles will be necessary
- inspection: regular inspections of some vehicles are likely to be required
- information, instruction and training: these will be particularly necessary for those who are to drive workplace vehicles
- conformity with Directives: several aspects of vehicle construction and design will be subject to the requirements of European Directives
- dangerous parts: access to dangerous parts must be prevented
- controls: must satisfy the requirements of the regulations (or relevant Directive)
- warnings: suitable signs or devices may be required.

Practical issues to be considered are: vehicle design and maintenance; selection and supervision of drivers; and working practices:

- *Vehicle design*: e.g. safe access to cabs and other relevant parts, effective brakes, good visibility and provision of appropriate warning devices (horn, lights, reversing lights or alarms, markings, rotating beacons, etc.).
- *Vehicle maintenance*: The level of maintenance necessary will vary according to the type of vehicle and its circumstances of use. Manufacturers' instructions should be followed and in some cases specific guidance is available from the HSE (e.g. forklift trucks). Maintenance might involve basic driver safety checks prior to use, more formal daily or pre-shift checks and regular planned maintenance inspections. Aspects likely to require special attention include brakes, tyres, steering, mirrors, driver visibility, warning devices and other specific safety systems.
- *Selection, training and supervision of drivers*: Drivers should have a suitably mature and reliable attitude. Some testing of drivers' physical capabilities (e.g. vision testing) may be appropriate and in some cases a test of the driver's competence will be necessary. Detailed formal training requirements exist for some vehicles, e.g. forklift trucks. Training may also be necessary in relation to towed equipment and use of attachments, e.g. in agriculture. Driving standards, speed limits, etc., must be effectively monitored and enforced.
- *Working practices*: Safe systems of work should be developed for work involving vehicles. Activities where safe systems are likely to be appropriate include: reversing of vehicles, parking, coupling and decoupling of vehicles, access onto vehicles, loading and unloading of vehicles, sheeting and tipping, use of towed work equipment and certain types of attachments.

Practical guidance on all aspects of workplace transport is available from the HSE (see ref. 18). Requirements relating to traffic routes are contained in the Workplace (Health, Safety and Welfare) Regulations 1992 – see Chapter D2.

D1.11 Woodworking machines

The five year transition period for brakes being fitted to certain types of wood-working machines expired on 5 December 2003. Braking devices must now be fitted to circular saw benches, dimension saws, powered and hand-fed cross-cut saws (unless there is no risk of contact with the blade during rundown), single-end and double-end tenoning machines and combined machines incorporating a circular saw and/or tenoning attachment.

A similar transition period for the fitting of brakes to narrow band saws, re-saws, vertical spindle moulding machines (unless fitted with a manual or foot-operated brake), hand-fed routing machines, thicknessing machines, planning/thickness machines and surface planning machines expires on 5 December 2005.

A separate requirement for the fitting of limited cutter projection tooling (LCPT) to some woodworking machines also required conversion to take place by 5 December 2003. LCPT (also known as 'chip thickness limitation tooling' reduces the severity of injury should an operator's fingers come into contact with the rotating tool. It can be fitted to vertical spindle moulding machines, single-ended tenoning machines and some rotary knife and copying lathes.

Further information is avaliable in HSE woodworking information sheets No. 37 (Selection of tooling for use with hand-fed woodworking machines) and No. 38 (Retrofitting of braking to woodworking machines). The HSE has indi-cated that it is likely to issue improvement notices where employers have not fitted braking devices or LCPT by the deadline dates.

D1.12 Lifting Operations and Lifting Equipment
Regulations 1998 (LOLER)

LOLER applies wherever HASAWA does. Lifting equipment is also subject to PUWER 1998 and European Product Directives requirements including the Supply of Machinery (Safety) Regulations 1992 and the Lifts Regulations 1997. Many old lifting equipment requirements were replaced by LOLER, including Regulations relating to Quarries, Shipbuilding and Shiprepairing, Construction, Hoists and Lifts, Offshore Installations, as well as the Factories Act (ss. 22, 23, 25, 27) and the Mines & Quarries Act (s. 85). The Docks Regulations 1988 were par-tially revoked and amended. The Mines (Shafts and Winding) Regulations 1993 were unchanged, but there is still a duty to comply with LOLER. An HSE booklet, 'Safe Use of Lifting Equipment', contains the regulations, an ACOP and related guidance. A shorter guidance leaflet is also available ('Simple Guide to LOLER').

Several terms are defined in Regulation 2:

- *Lifting equipment*: 'work equipment for lifting and lowering loads and includes its attachments used for anchoring, fixing or supporting it'.
- *Accessory for lifting*: 'lifting equipment for attaching loads to machinery for lifting'.
- *Load*: 'includes a person'.
- *Lifting operation*: defined in Regulation 8(2) as 'an operation concerned with the lifting or lowering of a load'.

HSE guidance indicates that the following would all come within the definition of 'lifting equipment':

- passenger lift
- rope and pulley
- dumb waiter
- vacuum crane
- vehicle hoist
- scissor lift
- climbing rope
- paper roll hoist
- automated storage/ retrieval system

- tractor front end loader
- 'patient' bath hoist
- loader crane (Hiab)
- refuse vehicle loading arm
- cargo elevating transfer vehicle
- vehicle recovery equipment
- vehicle tail lifts.

Many of the above were not covered by previous requirements.

However, the guidance also states that LOLER does *not* apply to:

- equipment whose primary purpose is not lifting or lowering (a three-point linkage on a tractor is specifically excluded)
- escalators (Reg. 19 of the Workplace Regulations applies).

Separate HSE guidance also indicates that some of the thorough examination and inspection requirements contained in Regulation 9 may not apply to some types of equipment coming within the definition of 'lifting equipment'.

As was the case with PUWER (see paragraph D1.2), Regulation 3 of LOLER creates overlapping responsibilities, with the following all having responsibilities under the regulations:

- employers – in relation to lifting equipment provided for use or used by their employees at work
- self-employed persons – in respect of lifting equipment they use at work
- persons who have control to any extent of:
  - lifting equipment
  - a person at work who uses or supervises or manages the use of lifting equipment
  - the way lifting equipment is used.

The HSE booklet 'Safe Use of Lifting Equipment' provides further guidance on responsibilities.

Ships' work equipment is excluded from the requirements of LOLER (whether used on or off the ship) provided that:

- it is used for specified purposes (sub-paragraphs (9) and (10) of Regulation 3 refer)
- merchant shipping requirements are complied with (as defined in Regulation 3).

D1.13 Suitability of lifting equipment

The HSE booklet 'Safe Use of Lifting Equipment' contains several ACOP requirements and considerable guidance on ensuring that lifting equipment complies with the suitability requirements of Regulation 4 of PUWER. This includes sections relating to the following:

- taking account of ergonomic risks (e.g. in operating positions or reach distances)
- material of manufacture (which must be suitable for the conditions of use)

- means of access to any part of the lifting equipment (e.g. to tower cranes or mobile elevating work platforms)
- protection against slips, trips or falls
- operator protection (e.g. from temperature extremes, weather, air contaminants or noise)
- possible effects of high winds.

Regulation 4 of LOLER contains further requirements on 'strength and stability' while Regulation 5 contains requirements on 'lifting equipment used for lifting persons'.

Strength and stability (Regulation 4)
The regulation states that:

Every employer shall ensure that:

(a) lifting equipment is of adequate strength and stability for each load, having regard in particular to the stress induced at its mounting or fixing point;

(b) every part of a load and anything attached to it and used in lifting it is of adequate strength.

'Safe Use of Lifting Equipment' again includes ACOP requirements and detailed guidance on what constitutes adequate strength and stability, including aspects such as the following:

- taking account of all forces on the lifting equipment (e.g. wind loading as well as the weight of the load and associated accessories)
- potential for fracture, wear or fatigue
- strength and stability of the ground surface
- lifting equipment operating on slopes
- possible needs for stabilising arrangements (anchorage systems, counter-balance weights, ballast, outriggers or stabilisers)
- lifting loads out of water or using lifting equipment on floating vessels
- use of mobile lifting equipment
- lifting equipment that is dismantled and reassembled (e.g. tower cranes)
- rail-mounted lifting equipment
- overload indicators and limiters.

Guidance is also provided on means of ensuring that loads and attachments are of adequate strength.

Lifting equipment used for lifting persons (Regulation 5)
The regulation states that:

(1) Every employer shall ensure that lifting equipment for lifting persons –

(a) subject to sub-paragraph (b), is such as to prevent a person using it being crushed, trapped or struck or falling from the carrier;

(b) is such as to prevent so far as reasonably practicable a person using it, while carrying out activities from the carrier, being crushed, trapped or struck or falling from the carrier;

(c) subject to paragraph (2), has suitable devices to prevent the risk of a carrier falling;

267

(d) is such that a person trapped in any carrier is not thereby exposed to danger and can be freed.

(2) Every employer shall ensure that if the risk described in paragraph (1)(c) cannot be prevented for reasons inherent in the site and height differences –
  (a) the carrier has an enhanced safety coefficient suspension rope or chain; and
  (b) the rope or chain is inspected by a competent person every working day.

The term 'carrier' is not defined in the regulations but is stated in the HSE guidance to be a generic term used to describe the device that supports people while being lifted or lowered. It would include the following:

- the car of a passenger lift
- the cage of a construction site hoist
- a platform on a mobile elevating work platform (MEWP)
- a working platform raised by a forklift truck or telescopic handler
- a cradle suspended from a crane
- a bosun's chair or harness (e.g. used by tree surgeons, etc.).

Sub-paragraph (1)(a) of the regulation relates to carriers such as lift cars or hoist cages and is an absolute requirement. Sub-paragraph (1)(b) applies where activities are carried out from the carrier and is qualified by 'reasonably practicable'.

Further guidance and ACOP requirements are contained in 'Safe Use of Lifting Equipment' (ref. 19) on matters such as the following:

- the use of forklift trucks, telescopic handlers and cranes for the lifting of persons
- requirements for lift cars and hoistways
- working from carriers that are not fully enclosed
- devices to prevent carriers from falling in the event of failure of the primary means of support
- safe rescue of persons in the event of malfunction
- enhanced safety coefficients (para. (2) of the regulation).

D1.14 Positioning and installation (Regulation 6)

The regulation states that:

(1) Every employer shall ensure that lifting equipment is positioned or installed in such a way as to reduce as low as reasonably practicable the risk:
  (a) of the equipment or load striking a person; or
  (b) from a load
    (i) drifting;
    (ii) falling freely; or
    (iii) being released unintentionally
and it is otherwise safe.

(2) Every employer shall ensure that there are suitable devices to prevent a person from falling down a shaft or hoistway.

This regulation applies to both permanent and mobile lifting equipment. ACOP requirements and guidance cover matters such as the following:

- minimising lifting loads over people
- positioning lifting equipment to avoid crushing and trapping points
- provision of enclosures, barriers, etc., for loads moving along a fixed path
- adequate widths of passageways, etc., where lifting operations take place
- measures to prevent loads moving out of control or falling freely
- means of preventing risks in cases of power failure (e.g. for pneumatic, hydraulic, vacuum or magnetic equipment)
- use of hooks with safety catches and correctly designed plate clamps
- avoiding collisions between items of lifting equipment or their loads
- use of gates, interlocks, etc., to prevent falls down hoistways or shafts.

D1.15 Marking of lifting equipment (Regulation 7)

The regulation states that:

Every employer shall ensure that:

(a) subject to sub-paragraph (b), machinery and accessories used for lifting loads are clearly marked to indicate their safe working loads;

(b) where the safe working load of machinery for lifting loads depends on its configuration –
  (i) the machinery is clearly marked to indicate its safe working load for each configuration; or
  (ii) information which clearly indicates its safe working load for each configuration is kept with the machinery;

(c) accessories for lifting are also marked in such a way that it is possible to identify the characteristics necessary for their safe use;

(d) lifting equipment which is designed for lifting persons is appropriately and clearly marked to this effect; and

(e) lifting equipment which is not designed for lifting persons but which might be so used in error is appropriately and clearly marked to the effect that it is not designed for lifting persons.

The term 'safe working load' (SWL) was used in previous legislation. Other terms such as 'rated capacity' and 'working load limit' are sometimes used to mean the same thing. Where the SWL cannot be marked directly onto lifting equipment or accessories for lifting (e.g. slings, shackles, eyebolts, etc.) it should be provided via a tag, by use of a colour code system or other effective means.

The ACOP and guidance on this regulation deal with aspects including the following:

- use of rated capacity indicators and rated capacity limiters (e.g. on jib cranes, telescopic jibs, for lifting of people)
- use of readily available indicators, plates, charts, etc., on other lifting machinery whose SWL may vary (e.g. attachments on forklift trucks, lifting beams with multiple lifting points)
- marking of structural elements of lifting equipment which may be dismantled
- marking of lifting accessories with additional information

269

- marking of carriers with the maximum number of persons to be carried and the SWL.

D1.16 Organisation of lifting operations (Regulation 8)

The regulation states that:

(1) Every employer shall ensure that every lifting operation involving lifting equipment is
   (a) properly planned by a competent person;
   (b) appropriately supervised; and
   (c) carried out in a safe manner.
(2) In this regulation 'lifting operation' means an operation concerned with the lifting or lowering of a load.

In effect this regulation requires lifting operations to be carried out in accordance with a 'safe system of work' (see Chapter B8). The safe system would result from a risk assessment:

- a generic one for operations that are carried out regularly
- a dynamic one for individual lifting operations.

The ACOP requirements and guidance relating to this regulation highlight aspects such as the following:

- the qualities of the competent person planning lifting operations
- matters to be taken into account in planning lifting operations
- a requirement for a written plan for operations involving two or more items of lifting equipment simultaneously ('tandem' lifts)
- selection of lifting equipment and accessories
- what constitutes 'appropriate supervision' of lifting operations
- avoiding carrying or suspending loads over persons, where practicable
- visibility of the load path to operators of lifting equipment (use of CCTV, visual markers, banksmen, communication systems, etc.)
- attaching, detaching and securing loads
- possible impact of weather conditions
- ensuring adequate headroom, including for access and egress of lifting equipment
- precautions to prevent lifting equipment overturning, tilting, moving or slipping
- risks from nearby objects (e.g. power lines and other services) and equipment
- possible need to de-rate the SWL of lifting equipment
- additional precautions necessary for lifting of persons in some circumstances
- conducting of pre-use checks on equipment
- correct storage of lifting accessories.

D1.17 Thorough examination and inspection (Regulation 9)

The requirements of Regulation 9 are detailed and complex and consequently they are summarised below rather than reproduced in full.

Before lifting equipment is put into service for the first time it must be thoroughly examined for any defect (unless the employer has received an EC declaration of conformity or a declaration under the Lifts Regulations 1997 not more than twelve months before).

Where the safety of lifting equipment depends on the installation conditions it must be thoroughly examined:

- after installation and before being put into service for the first time
- after assembly and before being put into service at a new site or in a new location

to ensure that it has been installed correctly and is safe to operate.

Lifting equipment that is exposed to conditions causing deterioration which is liable to result in dangerous situations must be thoroughly examined:

- in the case of lifting equipment for lifting persons or an accessory for lifting, at least every six months
- in the case of other lifting equipment, at least every twelve months OR
- in either case, in accordance with an examination scheme AND
- each time exceptional circumstances liable to jeopardise the safety of the lifting equipment have occurred AND
- inspections by a competent person between thorough examinations must be made, if appropriate

to ensure that health and safety conditions are maintained and any deterioration can be detected and remedied in good time.

Every employer must ensure that no lifting equipment

- leaves his undertaking or
- if obtained from the undertaking of another person, is used in his undertaking

unless it is accompanied by physical evidence of its last thorough examination.

This regulation does not apply to winding apparatus to which the Mines (Shafts & Winding) Regulations 1993 apply. It also contains transitionary provisions where thorough examinations were made under previous legislation.

Employers are given a choice under the regulations of either drawing up an examination scheme or carrying out examinations at the default intervals of six months (for lifting equipment for lifting persons and accessories for lifting) or twelve months (for other lifting equipment). However, HSE guidance indicates that some types of lifting equipment may not require thorough examination at all because any deterioration would not be 'liable to result in dangerous situations'. The HSE's leaflet 'LOLER: How the Regulations Apply to Agriculture' states: 'Equipment that does not lift loads over people and where the operators of the equipment are protected does not need to be thoroughly examined.' It provides examples such as 'foreloaders on tractors with safety cabs, telescopic loaders and fork lift trucks with operator protection where no other people are working in the vicinity'. However, employers should be cautious about taking this approach too far.

ACOP requirements and guidance on this regulation deal with matters such as the following:

- the qualities of the competent person carrying out thorough examinations and drawing up examination schemes (usually an engineer from an insurance company or specialist organisation)
- identification of equipment and situations (e.g. accident, modification, repair) requiring thorough examination
- possible needs for testing of lifting equipment
- examination after installation or reconfiguration
- 'in-service' thorough examinations
- contents of examination schemes
- circumstances necessitating regular inspections (in addition to thorough examinations), e.g. correct operation of limiters and indicators, checks on tyre pressures
- keeping evidence of thorough examinations readily available, e.g. carrying copies of certificates in vans or on mobile cranes.

D1.18 Reports and defects (Regulation 10)

This regulation is also summarised below rather than reproduced in full. A person making a thorough examination for an employer under Regulation 9 must:

- notify the employer forthwith of any dangerous defect*
- make a report as specified in Schedule 1 to
  - the employer
  - any person from whom the lifting equipment has been hired or leased
- send a report to the enforcing authority in respect of serious defects.

A person inspecting lifting equipment must:

- notify the employer forthwith of any dangerous defect
- make a written record.

* Lifting equipment must not be used until defects have been rectified. The ACOP requirements and guidance cover the following:

- time scales for reporting defects to employers and submitting normal reports of thorough examinations
- common types of defects in different types of lifting accessories
- defects serious enough to justify sending a report to the enforcing authority.

D1.19 Keeping of information (Regulation 11)

The requirements of Regulation 11 are summarised below. Employers must keep reports, etc., available as follows:

- EC declarations of conformity – Reg. 9(1)
  - as long as they operate the lifting equipment
- initial thorough examination reports – Reg. 9(1)
  - accessories for lifting, for two years after the report is made
  - all other lifting equipment, until ceasing to use the equipment

- installation thorough examination reports – Reg. 9(2)
  - until ceasing to use the lifting equipment at that place
- periodic thorough examination reports – Reg. 9(3)(a)
  - until the next report or two years, whichever is later
- inspection reports – Reg. 9(3)(b)
  - until the next report (this is stipulated by Regulation 6 of PUWER).

Guidance ('Safe Use of Lifting Equipment') is provided for different means of keeping records and their storage locations.

Source materials

1. HSE (2000) 'Safety of Pressure Systems: Pressure Systems Safety Regulations 2000', ACOP, L 122.
2. HSE (1998) 'Buying New Machinery: A Short Guide to the Law', INDG 271.
3. HSE (1999) 'Safe Use of Work Equipment: PUWER 1998', ACOP and Guidance, L 22.
4. HSE (1999) 'Safe Use of Power Presses: PUWER as Applied to Power Presses', ACOP and Guidance, L 112.
5. BS EN 953:1998. Safety of machinery: Guards – general requirements for the design and construction of fixed and movable guards.
6. BS EN 294:1992. Safety of machinery: Safety distances to prevent danger zones being reached by the upper limbs.
7. BS EN 292:1991. Safety of machinery: Basic concepts, general principles for design.
8. HSE (1998) 'Safeguarding Agricultural Machinery: Advice for Designers, Manufacturers, Suppliers and Users', HSG 89.
9. HSE (2000) 'Safety in the Use of Abrasive Wheels', HSG 17.
10. HSE (1998) 'Safety in the Use of Metal Cutting Guillotines and Shears', HSG 42.
11. HSE (1999) 'Health and Safety in Engineering Workshops', HSG 129.
12. HSE (1986) 'Pie and Tart Machines', HSG 31.
13. HSE (1998) 'Safe Use of Woodworking Machinery', L 114.
14. HSE (1997) 'Health and Safety in Sawmilling', HSG 172.
15. HSE (2000) 'Chainsaws at Work', INDG 317.
16. HSE (1995) 'Workplace Transport Safety', HSG 136.
17. HSE (1998) 'Safe Use of Lifting Equipment: LOLER 1998', ACOP and Guidance, L 113.
18. HSE (1999) 'Simple Guide to LOLER', INDG 290.
19. HSE (1998) 'LOLER: How the Regulations Apply to Agriculture', AIS 28.

The Workplace

D2.1   Introduction

This chapter summarises the requirements of the two main sets of regulations of relevance to the workplace:

- the Workplace (Health, Safety and Welfare) Regulations 1992; and
- the Construction (Health, Safety and Welfare) Regulations 1996 (CHSW).

It also deals with two safety aspects of particular importance in all workplaces but especially in construction activities:

- work at heights (new regulations on this topic are expected imminently)
- vehicle routes.

At the end of the chapter, information is provided on the new

- duty to manage asbestos in non-domestic premises.

D2.2   Workplace (Health, Safety and Welfare) Regulations 1992

Although physical working conditions were previously covered by sections of the Factories Act and Offices, Shops and Railway Premises Act, those Acts only applied to a limited range of workplaces. Many important sectors (e.g. hospitals, educational establishments, leisure centres) were largely outside their scope. The Workplace Regulations apply to most workplaces, replacing large sections of these two Acts and many other older laws. There are several exceptions from application of the regulations, mainly contained in Regulation 3:

- mines, quarries and other mineral extraction sites (subject to other legislation)
- construction sites (subject to CHSW)
- means of transport (apart from Reg. 13 which applies to aircraft, trains and road vehicles when stationary in a workplace)
- outdoor farming and forestry workplaces away from main buildings (apart from limited application of Regs 20 to 22)
- domestic premises, i.e. private dwellings (excluded by Regulation 2).

On temporary worksites, the welfare requirements (Regs 20 to 25) only apply 'so far as is reasonably practicable'.

Interpretation (Regulation 2)

In the regulation, 'traffic route' is defined as a route for pedestrian traffic, vehicles or both and includes any stairs, staircase, fixed ladder, doorway, gateway, loading bay or ramp. 'Workplace' means any premises or part of premises that are not domestic premises and are made available to any person as a place of work, and includes:

- any place within the premises to which such person has access while at work; and
- any room, lobby, corridor, staircase, road or other place used as a means of access to or egress from the workplace or where facilities are provided for use in connection with the workplace other than a public road.

Responsibilities (Regulation 4)

Regulation 4 places duties on employers in respect of workplaces under their control and where any of their employees work. Other persons (e.g. owners, landlords, managing agents) who have, to any extent, control over workplaces are responsible for requirements of the regulations relating to matters within their control.

Typically in multi-occupied premises, owners, landlords or managing agents would be responsible for common parts, facilities, services and means of access, while individual tenant employers would be responsible for their own areas. While employers have no responsibility under these regulations for those workplaces where their employees work but which are not under their control, their general duties under HASAWA (relating to their employees' health, safety and welfare) will still apply to such workplaces.

The requirements of the regulations have been divided below into three broad sections dealing with health, safety and welfare together with an additional section dealing with maintenance. The regulations are available in full, together with an ACOP and associated guidance, in a single HSE booklet: 'Workplace Health, Safety and Welfare'.

Disabled persons (Regulation 25A)

Parts of the workplace (particularly doors, passageways, stairs, showers, wash-basins, lavatories and workstations) used or occupied by disabled persons at work must be organised to take account of such persons.

D2.3 Health

Ventilation (Regulation 6)

Paragraph (1) of the regulation states that: 'Effective and suitable provision shall be made to ensure that every enclosed workplace is ventilated by a sufficient quantity of fresh or purified air.' Windows or other openings normally provide sufficient ventilation to replace stale, hot or humid air and minimise unpleasant smells. However, in some circumstances mechanical ventilation systems may be necessary, possibly accompanied by air conditioning equipment. Mechanical systems must be regularly cleaned, tested and maintained, particularly where there is a possibility of legionella bacteria developing. (Local exhaust ventilation or general ventilation may also be required in order to comply with the COSHH Regulations – see Chapter C2.)

An additional HSE ACOP and guidance is available on the prevention of legionella and there is also general guidance on ventilation.

Temperature in indoor workplaces (Regulation 7)

Paragraph (1) of the regulation states that: 'During working hours, the temperature in all workplaces inside buildings shall be reasonable.' The ACOP refers to a temperature of 'normally' at least 168C (138C if the work involves severe physical effort). Where hot or cold processes or storage requirements make maintenance of a comfortable temperature impractical, all reasonable steps should be taken to achieve a temperature as close as possible to comfortable. In some circumstances use of local heating or cooling, protective clothing or rest facilities may be necessary.

Workplaces must be adequately thermally insulated having regard to the type of work carried out and the physical activity of the persons carrying out the work. Excessive effects of sunlight on the temperature in workplaces must be avoided.

Heating or cooling methods must not result in the escape into the work-place of injurious or offensive fumes, gas or vapour and sufficient thermometers should be provided for persons at work to determine the temperature. An HSE guidance booklet is available on the subject.

Heat stress

Although there are minimum legal temperatures, described above, at which individuals should not have to work, there is no upper limit covering working in very high temperatures. The TUC and human resources professionals are call-ing for a maximum limit to be set and the HSE is currently developing detailed guidance for employers on how to manage and assess the risk of heat stress in their workplace. In the interim, however, the HSE has issued an information sheet regarding the issue of high temperatures and offers guidance on the way to avoid heat stress (GEIS 1).

Heat stress occurs when the body's mechanism for controlling internal tem-perature fails. Factors that contribute to this failure include air temperature, work rate and the clothing worn.

Heat stress can occur, for instance, in hot and humid conditions where someone is wearing protective clothing and performing heavy physical work. An environment where the work operation increases heat levels, such as that found in bakeries, foundries and laundries, or where work is conducted in restricted spaces, poses the risk of heat stress.

Some of the effects of heat stress are:

- lack of concentration
- heat rash
- muscle cramps
- fatigue and giddiness
- heat stroke – hot, dry skin, confusion, convulsions, loss of consciousness and death if the problem is not detected at an early stage.

Identifying workers and operations that may be susceptible can reduce the risk of heat stress. This can be achieved by carrying out a risk assessment. It is important to recognise that one worker may be acclimatised to the prevailing temperature levels while another may not, possibly as a result of the person's age or the pres-ence of a medical condition. It is therefore important to look at the circum-stances of each worker who might be exposed to high temperatures.

Some of the ways in which the risk of heat stress can be reduced include the following:

- Temperature control – the use of fans or air conditioning, or a change in the processes that generate the higher temperatures.
- Reduce the work rate or provide mechanical aids to reduce the physical effort involved.

- Ensure that the length of time to which workers are exposed to high temperatures is regulated – this can be achieved through breaks.
- Prevent dehydration by encouraging workers to drink water – workers should be able to obtain this easily. It will not, however, be possible to provide workers dealing with the removal of asbestos with this facility.
- Permit workers to enter the working area only when temperature is at a set rate.
- Allow workers to acclimatise to their work environment and consult them regarding their strategies for dealing with the heat.

Lighting (Regulation 8)

Paragraph (1) of the regulation states that: 'Every workplace shall have suitable and sufficient lighting', while paragraph (2) requires that this lighting 'shall, so far as is reasonably practicable, be by natural light'.

General lighting may need to be augmented by local lighting at individual workstations. Emergency lighting is required by paragraph (3) if sudden failure of artificial light would create a risk (e.g. from falls, dangerous machinery or molten materials). An HSE booklet, 'Lighting at Work', provides detailed guidance on lighting.

Cleanliness and waste materials (Regulation 9)

Paragraph (1) of the regulation states that: 'Every workplace and the furniture, furnishings and fittings therein shall be kept clean.' Paragraph (2) requires that surfaces of internal floors, walls and ceilings are capable of being sufficiently clean while paragraph (3) states that, so far as is reasonably practicable, waste materials should not be allowed to accumulate, except in suitable receptacles. The ACOP states that standards of cleanliness can vary according to the use to which the workplace is put – an area where meals are taken should be cleaner than a factory floor, which in turn should be cleaner than an animal house. Both the ACOP and guidance refer to frequencies of cleaning and standards of surface treatments necessary in order to maintain cleanliness.

Room dimensions and space (Regulation 10)

The regulation requires that: 'Every room where persons work shall have sufficient floor area, height and unoccupied space for purposes of health, safety and welfare.' The ACOP sets a minimum standard of 11 cubic metres per person, ignoring heights above 3 metres. More space may be required, depending on the room layout, the presence of furniture, etc., and the nature of the work. (The standard of 11 cubic metres is not intended to apply to kiosks, shelters, control cabs, etc., or to rooms used for lectures, meetings and similar purposes.)

Workstations and seating (Regulation 11)

The regulation requires workstations to be arranged so as to be suitable both for the people using them and for the work likely to be done. They must also be arranged so that:

- weather protection is provided, so far as is reasonably practicable
- persons can leave swiftly in the event of an emergency
- persons are not likely to slip or fall.

Where work can or must be done sitting, suitable seating must be provided, together with a footrest where necessary. Seating should, where possible, provide adequate support for the lower back. HSE guidance is available on both seating and ergonomic matters generally (refs 6, 7 and 8). Display screen equipment workstations are dealt with in Chapter C5.

D2.4   Safety requirements

Condition of floors and traffic routes (Regulation 12)
This regulation sets down several requirements relating to floors and the surfaces of both pedestrian and vehicle traffic routes. It requires them to:

- be of suitable construction
- be free from holes, slopes or uneven or slippery surfaces, which create risks (see below)
- have effective means of drainage where necessary
- be kept free from obstructions and from any article or substance that may cause slips, trips or falls, so far as is reasonably practicable.

Apart from the obvious need for floors and traffic routes to be strong enough for the loads placed upon them and the traffic passing over them, their surfaces must be such as to minimise risks of:

- slips, trips or falls
- persons dropping or losing control of loads they are lifting or carrying
- instability or loss of control of vehicles and/or their loads.

The ACOP refers to the need for precautions such as:

- guarding or marking temporary holes, etc.
- ensuring that slopes are no steeper than necessary (hand rails may be necessary, especially for people with disabilities)
- slip-resistant coatings
- gritting and snow clearance to minimise risks from snow and ice
- marking temporary obstructions or slippery areas, e.g. by cones
- fencing open sides of staircases (an upper rail at least 900 mm high and a lower rail)
- providing staircases with at least one hand rail (unless this would cause an obstruction).

Precautions relating specifically to vehicle routes are also referred to in paragraph D2.9. HSE guidance is available on the prevention of slips and trips.

Falls or falling objects (Regulation 13)
The regulation requires suitable and effective measures to be taken, so far as is reasonably practicable, to prevent any person:

- falling a distance likely to cause personal injury
- being struck by a falling object likely to cause personal injury.

These measures must, so far as is reasonably practicable, be measures other than the provision of PPE or information, instruction, training or supervision. Areas where such risks exist should be clearly indicated where appropriate, e.g. by signs.

In addition the regulation requires every tank, pit or structure, where there is a risk of a person falling into dangerous substances within, to be securely covered or fenced, so far as is practicable. Traffic routes over, across or in such tanks, pits or structures must be securely fenced. (The definition of 'dangerous substance' contained in paragraph (7) of the regulations includes substances likely to scald or burn; poisonous or corrosive substances; fume, gas or vapour likely to overcome a person; or granular or free-flowing substances, or viscous substances.)

The ACOP states that secure fencing should be provided:

- wherever possible at any place where a person might fall 2 metres or more
- where a traffic route passes close to an edge
- where large numbers of people are present
- where a fall may be onto a sharp or dangerous surface or into the path of a vehicle.

This regulation applies to means of transport, therefore falls from vehicles during loading or unloading must be considered. Practical means of complying with the regulation are examined in greater detail in paragraph D2.8.

Most of this regulation is expected to be replaced soon by the requirements of the proposed Work at Height Regulations. Further information on those regulations and their implications is contained in Paragraph D2.8.

Windows, and transparent or translucent doors, gates or walls (Regulation 14)
The regulation states that:

(1) Every window or other transparent or translucent surface in a wall or partition and every transparent or translucent surface in a door or gate shall, where necessary for reasons of health and safety:
   (a) be of safety material or be protected against breakage of the transparent or translucent material; and
   (b) be appropriately marked or incorporate features so as, in either case, to make it apparent.

The ACOP states that safety materials or adequate protection against breakage (e.g. a screen or barrier) must be provided:

- in doors and gates, and any side panels, where any part of the transparent or translucent surface is at shoulder level or below
- in windows, walls and partitions, where any part of the transparent or translucent surface is at waist level or below (except in greenhouses).

(This requirement does not apply to narrow panes up to 250mm wide.)

Safety materials are defined in the ACOP as:

- materials such as polycarbonates or glass blocks
- glass that, if it breaks, breaks safely
- glass meeting thickness criteria in a table in the ACOP.

Large glazed areas (windows, doors or partitions) coming down to ground level or thereabouts should be marked to prevent accidental collisions – coloured lines, patterns or stickers are acceptable providing that they are conspicuous enough.

Windows, skylights and ventilators (Regulation 15)
The regulation states that:

(1) No window, skylight or ventilator which is capable of being opened shall be likely to be opened, closed or adjusted in a manner which exposes any person performing such operation to a risk to his health or safety.
(2) No window, skylight or ventilator shall be in a position when open which is likely to expose any person in the workplace to a risk to his health or safety.

As well as the need to provide safe means of opening windows, etc.:

- devices should be provided to prevent windows opening too far (where there is a danger of falling from a height)
- open windows, etc., should not project into areas where persons are likely to collide with them (e.g. walkways).

Ability to clean windows, etc., safely (Regulation 16)
The regulation states that:

(1) All windows and skylights in a workplace shall be of a design or be so constructed that they may be cleaned safely.
(2) In considering whether a window or skylight is of a design or so constructed as to comply with paragraph (1), account may be taken of equipment used in conjunction with the window or skylight or of devices fitted to the building.

The ACOP refers to suitable provision for safe cleaning of windows as including the following:

- pivoted windows, which can be cleaned from the inside
- suspended cradles, travelling ladders and similar access equipment
- firm level surfaces for ladders up to 9 metres long on which to stand (with anchor points for securing ladders over 6 metres long)
- suitable anchorage points for safety harnesses.

Organisation, etc., of traffic routes (Regulation 17)
The regulation creates an overall requirement that:

(1) Every workplace shall be organised in such a way that pedestrians and vehicles can circulate in a safe manner.
(2) Traffic routes in a workplace shall be suitable for the persons or vehicles using them, sufficient in number, in suitable positions and of sufficient size.

Paragraphs (3) and (4) of the regulation require suitable measures to be taken to ensure that:

- pedestrians or vehicles do not cause danger to persons working near traffic routes
- there is sufficient separation of vehicle routes from doors, gates and pedestrian routes leading onto them
- vehicles and pedestrians using the same routes are sufficiently separated
- suitable markings and signs are provided.

The ACOP and guidance go into further detail on these requirements. (Some parts of the ACOP are not applicable to routes in existence prior to 1 January 1993.) Further practical guidance is provided in paragraph D2.9.

Doors and gates (Regulation 18)
Paragraph (1) of the regulation states: 'Doors and gates shall be suitably constructed (including being fitted with any necessary safety devices).'

Paragraph (2) contains requirements for the following:

- devices to prevent sliding doors or gates coming off their track
- devices to prevent upward opening doors or gates from falling back
- features to prevent powered doors or gates trapping persons (e.g. sensitive edges linked to trip devices)
- manual overrides or automatic opening if the power supply of powered doors or gates fails
- a clear view through doors or gates, which can be pushed from either side.

The ACOP and guidance go into further detail on these points.

Escalators and moving walkways (Regulation 19)
The regulation requires that:

Escalators and moving walkways shall:

- function safely;
- be equipped with any necessary safety devices;
- be fitted with one or more emergency stop control which is easily identifiable and readily accessible.

D2.5   Welfare requirements

Sanitary conveniences and washing facilities (Regulations 20 and 21)
These regulations require the provision of suitable and sufficient sanitary conveniences and washing facilities at 'readily accessible places'. Showers must be provided 'if required by the nature of the work or for health reasons'. There are also requirements for the following:

- adequate ventilation and lighting of conveniences and facilities
- keeping conveniences and facilities 'in a clean and orderly condition'
- separate conveniences and showers for men and women (apart from single conveniences and showers in rooms that can be secured from the inside).

Washing facilities must include the following:

- a supply of clean hot and cold, or warm water (running water where practicable)
- soap or other suitable means of cleaning
- towels or other suitable means of drying.

The ACOP goes into considerable detail on the minimum numbers of facilities and also contains requirements relating to the following:

- provision for workers with disabilities
- possible use of facilities outside the workplace (e.g. in multi-occupancy premises)
- standards for water closets (drainage, flushing, provision of toilet paper, coat hooks, means for disposing of sanitary towels)
- circumstances where showers (or baths) should be provided

- prevention of scalding in showers (use of thermostatic mixer valves)
- means of ensuring privacy
- facilities for remote workplaces or temporary worksites.

Drinking water (Regulation 22)

Paragraph (1) of the regulation requires that: 'An adequate supply of wholesome drinking water shall be provided for all persons at work in the workplace.' Paragraphs (2) and (3) require the supply of drinking water to be:

- readily accessible at suitable places
- conspicuously marked by an appropriate sign where necessary (e.g. if some cold water supplies may be contaminated)
- provided with suitable cups or vessels (or in an upward drinking jet).

Where water cannot be obtained from a mains supply, refillable enclosed containers may be used. These should be refilled daily (unless used with chilled dispensers and returned to the supplier for refilling).

Accommodation for clothing (Regulation 23)

Paragraph (1) of the regulation requires 'suitable and sufficient accommodation' to be provided for workers' own clothing that is not worn during working hours and for special work clothing that they do not take home. Paragraph (2) requires such accommodation to:

- provide suitable security for work clothing
- be separate for work clothing and other clothing where necessary to avoid risks to health or damage to clothing
- include drying facilities, so far as is reasonably practicable
- be in a suitable location.

The ACOP states that accommodation may vary, according to circumstances, from a coat hook to individual lockable lockers. Separate 'clean' and 'dirty' lockers may be required in some cases. The COSHH Regulations and regulations dealing with asbestos, lead and ionising radiations contain additional requirements on clothing, and compliance with food hygiene legislation may also be necessary.

Facilities for changing clothing (Regulation 24)

The regulation requires 'suitable and sufficient facilities' to be provided for persons at work to change clothing where they have to wear special clothing for work purposes and cannot, 'for reasons of health or propriety, be expected to change in another room'. (This requirement relates to work uniforms as well as protective clothing.) Separate changing rooms are particularly necessary where workers' own clothing may be contaminated in the workplace.

Amendments to the regulation introduced in 2002 required changing facilities to be easily accessible, of sufficient capacity and provided with seating. Suitable arrangements must be made to ensure privacy and convenient access from clothing accommodation and showers or baths will often be necessary.

Facilities for rest and to eat meals (Regulation 25)

Paragraph (1) of the regulation states that: 'Suitable and sufficient rest facilities shall be provided at readily accessible places.' Paragraphs (2) and (3) require rest facilities to include the following:

- suitable facilities to eat meals where food eaten in the workplace may become contaminated
- suitable arrangements to protect non-smokers from discomfort caused by tobacco smoke.

Paragraph (5) requires suitable and sufficient facilities to be provided for persons at work to eat meals where meals are regularly eaten in the workplace.

The ACOP stresses that in offices and other reasonably clean workplaces, both resting and eating in work areas is acceptable, providing that workers will not be subject to excessive disturbance during breaks, e.g. by members of the public. The ACOP describes the standards required for rest facilities and eating facilities. The latter should include a facility for preparing or obtaining hot drinks (kettle, vending machine or canteen) and, if hot food is not obtainable nearby, means for heating food. Good standards of hygiene must be maintained. Discomfort from tobacco smoke should be prevented either by providing separate areas or rooms for smokers and non-smokers or by prohibiting smoking altogether.

Paragraph (4) of the regulations states that: 'Suitable facilities shall be provided for any person at work who is a pregnant women or nursing mother to rest.' The ACOP requires such facilities to be convenient for sanitary facilities and, where necessary, to include the facility to lie down.

D2.6 Maintenance

Regulation 5 contains a requirement for the workplace in general and for specific equipment, devices and systems to 'be maintained (including cleaned as appropriate) in an efficient state, in efficient working order and in good repair'. This applies to equipment and devices, a fault in which is liable to result in a failure to comply with any of the regulations, and to all mechanical ventilation systems. Where appropriate such equipment, devices and systems must be subject to a system of maintenance.

Effective maintenance arrangements need to be in place for heating and ventilation systems, sanitary conveniences, washing facilities, etc. The ACOP provides a list of examples of equipment and devices requiring a system of maintenance, which includes the following:

- emergency lighting
- fencing
- fixed equipment used for window cleaning
- anchorage points for safety harnesses
- devices to limit the opening of windows
- powered doors
- escalators and moving walkways.

It also contains details of what a suitable system of maintenance should involve.

D2.7 Construction (Health, Safety and Welfare) Regulations 1996

The regulations apply to any construction work carried out by a person at work. The definition of construction work (set out in detail in paragraph B7.5) is extremely wide and includes many activities traditionally regarded as repair or maintenance work. While the regulations contain many requirements of specific relevance to construction, other parts represent the construction equivalents of the requirements contained in the Workplace Regulations or the Fire Precautions (Workplace) Regulations. A summary of the regulations is provided below and some key areas are examined in further detail in paragraphs D2.8 and D2.9. Detailed practical guidance on the regulations is available in the HSE booklet 'Health and Safety in Construction'.

Parts of these regulations, in particular regulations 6, 7 and 8, will be replaced by the proposed Work at Height Regulations. Further details are contained in Paragraph D2.8.

Regulation 4 sets out those who have duties under the regulations:

- employers and the self-employed
- persons controlling the way construction work is carried out
- employees.

Every person carrying out construction work must:

- cooperate with other persons on health and safety matters
- report dangerous defects to those in control.

Safe places of work (Regulation 5)

The regulation requires a safe place of work and safe means of access to and from every place of work to be provided and maintained, so far as reasonably practicable. This duty applies whether at ground level, at a height or below ground level.

Precautions against falls (Regulations 6 and 7)

Regulation 6 contains a general obligation that: 'Suitable and sufficient steps shall be taken to prevent, so far as is reasonably practicable, any person falling.' Where persons may fall more than 2 metres, a hierarchical approach must be followed, with the order of preference being as follows:

- physical safeguards (guard rails and toe boards or their equivalent)
- personal suspension equipment (e.g. rope access, bosun's chairs)
- fall arrest equipment (e.g. safety harnesses, nets).

Regulation 6 also includes requirements relating to the use of ladders, while Regulation 7 specifies the steps that must be taken to prevent falls through fragile materials.

Schedules 1 to 5 of the Regulations set out detailed standards for guard rails and toe boards, working platforms, personal suspension equipment, means for arresting falls and ladders. (Work at heights is dealt with in more detail in paragraph D2.8.)

Falling objects (Regulation 8)

Steps must be taken to prevent dangers to workers and others from falling materials or objects. Preferably this should be by providing guard rails, toe boards, barriers, etc., to stop items falling but, where this is not reasonably practicable, protective screens or covered walkways may be necessary to prevent falling items from hitting people. Materials and objects must not be thrown or tipped from heights if they could strike someone, and materials and equipment must be stored to prevent dangers from their collapse, overturning or unintentional movement.

Stability of structures (Regulation 9)

All practicable steps must be taken to prevent any new or existing structure collapsing and structures must not be overloaded. Temporary supports, etc., must be erected and dismantled under the supervision of a competent person.

Demolition or dismantling (Regulation 10)
Demolition or dismantling of any structure must be planned and carried out in a safe manner under the supervision of a competent person.

Explosives (Regulation 11)
Steps must be taken to prevent injuries from explosions or associated projected or flying material.

Excavations (Regulation 12)
Steps must be taken to prevent dangers from the following:

- collapse of excavations and the ground above
- persons being trapped by falling or dislodged material
- persons, vehicles, plant or equipment falling into or causing the collapse of excavations
- underground services or underground cables.

HSE booklets provide specific guidance on excavations (see page 295 for details).

Cofferdams and caissons (Regulation 13)
These must be properly designed, constructed and maintained.

Prevention of drowning (Regulation 14)
Steps must be taken to prevent persons drowning in water (or other liquids) by:

- preventing falls, so far as is reasonably practicable (by means such as those referred to in Regulation 6)
- minimising the risk of drowning in the event of a fall (e.g. by use of life-jackets, lines, etc.)
- providing, maintaining and using suitable rescue equipment (e.g. lifebelts, lines, etc.)
- providing safe transport by water, controlled by a competent person.

Traffic routes, doors and gates (Regulations 15 and 16)
These requirements are similar to those contained in the Workplace Regulations (see paragraph D2.4). (Further information on vehicle routes is contained in paragraph D2.9.)

Vehicles (Regulation 17)
The regulation contains requirements relating to the following:

- steps to prevent or control unintended movements of vehicles
- giving warning of possible dangerous movements of vehicles (e.g. reversing)
- safe driving, operation, towing or loading of vehicles
- prevention of riding in unsafe positions in vehicles
- precautions during loading or unloading of loose materials
- prevention of vehicles falling into excavations, pits or water or overrunning edges of embankments or earthworks
- dealing with derailed rail vehicles.

Emergency precautions (Regulations 18, 19, 20 and 21)
Steps must be taken to prevent risks from the following:

- fire or explosion
- flooding
- any substance liable to cause asphyxiation.

Specific requirements relate to the following:

- provision and maintenance of emergency routes and exits
- preparation and implementation of emergency procedures
- provision and maintenance of fire fighting equipment, fire detectors and alarm systems.

These follow the principles set out in paragraph B3.14 (Emergency procedures) and Chapter C7 (Fire). The HSE has enforcement responsibilities for means of escape and other fire precautions on most construction sites. HSE guidance is available on fire safety in construction work (refs 14 and 15).

Welfare facilities (Regulation 22)

Paragraph (1) of the regulation gives persons in control of construction sites an overall duty to ensure that suitable welfare facilities are provided (even though they may not themselves provide them), in addition to placing responsibilities on individual employers and self-employed persons. There are specific requirements for the following:

- sanitary conveniences and washing facilities
- drinking water
- facilities for changing and storing clothing
- rest facilities (including arrangements for preparing and eating meals and means for boiling water).

Schedule 6 contains additional requirements which closely match those for other types of workplaces (see paragraph D2.5).

General site requirements (Regulations 23, 24, 25 and 26)

These regulations contain requirements for the following:

- ensuring that sufficient fresh or purified air is available
- ensuring reasonable temperatures in indoor workplaces
- providing weather protection in outdoor workplaces
- workplace and traffic route lighting (including emergency lighting)
- maintaining sites in good order and reasonable cleanliness
- identifying site perimeters with suitable signs
- avoiding risks from projecting nails.

Most of these requirements are qualified by the phrase 'so far as is reasonably practicable'.

(Regulation 27 relating to plant and equipment was revoked by PUWER 1998, which applies to all construction work equipment see Chapter D1.)

Training (Regulation 28)

Persons carrying out activities involving construction work must have the necessary training, technical knowledge or experience to reduce risks, or be under appropriate supervision.

Inspection and reports (Regulations 29 and 30)

Regulations 29 and 30 together with Schedules 7 and 8 contain requirements for regular inspections of the following to be carried out by a competent person:

- working platforms where a person could fall 2 metres or more (see also Reg. 6)
- personal suspension equipment (see Reg. 6)
- supported or battered back excavations (see Reg. 12)
- cofferdams or caissons (see Reg. 13).

(A sample inspection report form and notes on the carrying out of inspections are contained in Figures B5.7 and B5.8 in Chapter B5.)

D2.8   Work at heights

New regulations relating to Work at Height were scheduled to come into operation during 2005 after a lengthy period of consultation. These regulations were to replace parts of the Workplace Regulations and the Construction (Health, Safety and Welfare) Regulations dealing with work at height. The content of this section is based on the expected requirements of the regulations.

Definition of work at height

A key part of the consultation process was the definition of 'work at height'. The consultative document used a definition which included access or egress (except by a staircase in a permanent workplace) and work at or below ground level in a place from which ... 'a person could fall a distance liable to cause personal injury.'

This definition provoked considerable comment as it moved away from the '2 metres rule' contained within previous regulations. Even under 2 metres, precautions still had to be taken when other risks were present e.g. of falling onto sharp or dangerous surfaces, into the path of vehicles or into water. Further consultation on this point closed at the end of 2004 and the definition to be incorporated into the final version of the regulations was awaited with interest.

Management of work at height

Work at height must be properly planned and appropriately supervised. Obviously the degree of planning and supervision necessary for a major construction project would be far greater than that required for a routine piece of maintenance work, but these aspects must not be neglected in any type of work at height.

Planning must include the selection of equipment to be used and take into account possible emergency and rescue situations e.g. rescuing people left suspended from fall-arrest harnesses or caught in safety nets. The competence of those involved in all aspects of work at height is also important – whether they are organising, planning or supervising work or using particular pieces of work equipment.

Avoidance of risks from work at height

A hierarchical approach is at the core of avoiding risks from work at height. The following options must be taken in order of preference, so far as is reasonably practicable:

1. Avoid carrying out work at height.
   e.g. assemble structures at ground level and lift into position using a crane.
2. Take suitable and sufficient measures to prevent falling a distance likely to cause personal injury (including using a fixed place of work or fixed means of access or egress).
3. Take measures to minimise the distance and consequences of a fall e.g. personal or collective fall-arrest equipment.

287

In identifying the measures required, a risk assessment must be carried out. Collective measures (e.g. nets, airbags) should have priority over personal protection (e.g. harnesses).

Work equipment for work at height

A number of basic principles should be taken into account in selecting equipment for use at height. These include:

- working conditions and risks at the place of use
- the distance to be negotiated (for access and egress)
- the distance and consequences of a fall
- the duration of the work and frequency of use of the equipment
- the need for emergency evacuation and rescue
- additional risks from the use, installation and removal of the equipment.

Requirements for specific types of work equipment

The Work at Height Regulations were expected to contain specific requirements on different types of equipment for work at height. These requirements and HSE guidance on their interpretation should be studied in detail. Specific provisions were expected for:

- guard rails, toe boards, barriers, etc.
  (the minimum height for guard rails was expected to rise to 950 mm)
- all work platforms
- scaffolding
- nets, airbags and other collective fall-arrest equipment
- personal fall protection systems (e.g. harnesses)
- work positioning systems
- rope access and positioning techniques
- fall arrest systems
- work restraint systems
- ladders (see later in this section).

Fragile surfaces

Steps must be taken to prevent persons falling through fragile surfaces. Again a hierarchical approach must be taken – avoiding access across or work on fragile surfaces so far as is reasonably practicable. Where this is not the case precautions necessary are likely to include:

- using platforms, coverings or similar means of support
- covering or railing off fragile surfaces near to access routes or work positions
- warning notices at approaches to fragile surfaces
- means for arresting falls where a risk of falling still remains.

Falling objects

Steps must be taken to prevent the fall of objects, so far as is reasonably practicable. These might be by incorporating toeboards, brick guards etc. into scaffolding or other access equipment, or by using tool belts or other means of restraint. However, it may be necessary to take other precautions to prevent persons being struck by falling objects e.g. protective screens.

Materials or objects must not be thrown from height where this is liable to cause injury and items must be stored at height in such a way as to prevent risks from their collapse, overturning or unintended movement (e.g. as a result of strong winds).

Inspections of work equipment

Where necessary for safety purposes, equipment must be inspected prior to use after its initial installation or assembly. All equipment must be inspected at suitable intervals and after exceptional circumstances jeopardising safety.

The previous requirement for scaffolding to be inspected every 7 days will be included in the new regulations, but employees must also consider what frequencies and types of inspection are necessary for other types of equipment e.g. mobile elevating work platforms, ladders, rope access equipment, safety nets, air bags and fall arrest harnesses. Some specific requirements for inspections (and inspection reports) are likely to be incorporated into the Work at Height Regulations or the accompanying HSE guidance.

Ladders

Contrary to statements in some quarters, the new regulations will not prohibit work from being carried out from ladders or stepladders. However, a risk assessment approach must be taken prior to their use.

The consultative document on the regulations stated:

> HSE believes that ladders should only be used as work equipment for access, egress or as a place to work from if a risk assessment has shown that the use of other, more suitable, work equipment is not necessary because of low risk, short duration tasks or topography of the work location. However, we must accept the practicalities of the use of ladders for work at height, and the fact that they are commonly used in a wide variety of situations.

Various factors must be considered in determining whether or not ladders are suitable.

These include:

- working for long durations
- handling heavy or bulky equipment
- risks to (and from) passers-by
- persons working alone
- the availability of a secure handhold and secure support
- avoiding stretching or overreaching
- whether the ladder can be secured to prevent slipping.

It must also not be forgotten that the use of alternative types of access equipment might introduce alternative risks e.g. from manual handling of components or from work at height being necessary to install it. Extensive guidance is already available on the safe use of ladders and more is expected to accompany the new regulations.

D2.9   Vehicle routes

Both the Workplace Regulations and the CHSW Regulations contain requirements relating to the following:

- safe circulation of vehicles and pedestrians
- separation of vehicles from pedestrians
- suitability of routes for the vehicles using them
- indication of traffic routes by road markings or signs
- surfaces of traffic routes (their construction and maintenance).

Some of the more important practical issues affecting vehicle routes are summarised below.

Route design

Routes must be wide enough for the safe movement of the largest vehicles likely to use them, including vehicles making deliveries to or collecting items from the workplace. This includes not just roadways but any gateways and entrances into buildings through which vehicles are likely to pass.

Vehicle routes close to steep unprotected drops or with sharp or blind bends should be avoided where possible, as should routes passing close to high-risk items such as fuel or chemical tanks and associated pipework or dispensing pumps. (Alternatively physical barriers may be necessary.) The need for vehicles to reverse should also be minimised.

In order to ensure that adequate width is available, it may be necessary to introduce one-way systems or impose parking restrictions. Appropriate speed limits must be established and these may need to be supported by the use of 'sleeping policemen' or other traffic calming measures. Care should be taken not to introduce additional risks road humps may be dangerous in areas where forklift trucks circulate. Appropriate use should be made of signs and road markings (see below).

Pedestrian protection

The measures necessary to protect pedestrians from vehicles will depend to a certain extent on the width of the traffic routes, the density of both the pedestrian and vehicle traffic, the types of vehicles involved and the nature of the activities they are performing. Precautions to be considered include the following:

- provision of separate routes for pedestrians
- physical separation of all or part of pedestrian routes (by barriers, guard rails, bollards, etc.)
- separate building entrances and exits for pedestrians and vehicles
- clear marking of pedestrian and vehicle routes on floors and roadways
- designated pedestrian crossing points (with good lighting and clear visibility)
- prohibiting vehicle access at high-risk times (e.g. shift changeovers).

The need to separate pedestrians from vehicles is particularly great where members of the public are likely to visit the premises regularly.

Parking

Adequate provision must be made for parking all types of vehicle:

- delivery vehicles
- vehicles used in construction work or industrial premises
- private vehicles used by workers
- buses and coaches used for transporting workers and others.

Parking areas should have firm, even surfaces and adequate standards of lighting. The need for drivers to cross potentially dangerous roads or work areas should be avoided and safe access with clear visibility must be provided between parking areas and roadways.

The uncontrolled parking of private vehicles around workplaces (especially construction sites) can cause considerable danger to other road users as well as nuisance to local residents.

Loading areas

Loading bays and other areas where vehicles are loaded and unloaded should be appropriately designed, with the following points of particular importance:

- the manoeuvring of vehicles into position (minimising reversing)
- at least one pedestrian exit point or refuge at lower levels of loading bays (reducing the risk of crushing)
- good levels of lighting
- clear marking of the edges of loading bays (with guard rails where this is practical)
- suitable arrangements for sheeting up vehicles
- the possible need for weather protection.

The use of banksmen to assist vehicles in loading areas may be necessary – they and others in the vicinity should wear high visibility clothing.

Lighting, signs and road markings

Adequate lighting should be provided on all roadways and in other areas where vehicles circulate regularly in hours of darkness. Particular attention should be given to the following:

- pedestrian crossing points
- other areas heavily used by pedestrians
- junctions
- loading areas.

Road signs and road markings (of the types used on the public highway) should be provided, where necessary, on workplace vehicle routes. They are particularly likely to be necessary to indicate the following:

- one-way systems
- no-entry areas
- speed limits
- ramps and other traffic calming measures
- road junctions
- sharp bends and blind corners
- pedestrian crossing points
- steep gradients
- low headroom (possibly supported by 'goalpost' protection)
- rail crossings
- roadworks and other temporary obstructions.

Light-controlled crossings may be necessary at:

- major road junctions
- heavily used pedestrian crossing points
- rail crossing points.

Clear directional signs will also aid in the safe flow of traffic.

Maintenance arrangements

Suitable arrangements must be in place to maintain both vehicle and pedestrian routes. These may need to include the following:

- prompt clearance or treatment of spillages and obstructions
- gritting or sanding of icy or snowy surfaces (or even snow removal)
- repairs to the road surface (particularly in respect of temporary road-ways)
- renewal of road markings
- cleaning, repair or replacement of signs and lights.

The HSE booklet 'Workplace Transport Safety' provides detailed guidance not only on vehicle routes but also on vehicle design and good working practices. Specific guidance is also available on transport in the construction industry (refs 22, 23 and 24).

D2.10 Duty to manage asbestos

The Control of Asbestos at Work Regulations 2002, which came into operation on 21 November 2002, introduced a new requirement to manage asbestos in non-domestic premises. This new duty (the main change to the Regulations) contained in Regulation 4 came into effect on 21 May 2004.

Domestic premises are private dwellings in which people live. Non-domestic premises include almost all workplaces as well as common parts of housing developments and flats such as boilerhouses, foyers, corridors, lifts and lift shafts, etc.

Dutyholders

Paragraph (1) of Regulation 4 defines the 'dutyholder' under the Regulation as:

(a) every person who has, by virtue of a contract or tenancy, an obligation of any extent in relation to the maintenance or repair of non-domestic premises or any means of access thereto or egress therefrom; or

(b) in relation to any part of non-domestic premises where there is no such contract or tenancy, every person who has, to any extent, control of that part of those non-domestic premises or any means of access thereto or egress therefrom, and where there is more than one dutyholder, the relative contribution to be made by each such person in complying with the requirements of this regulation will be determined by the nature of the maintenance and repair obligation owed by that person.

Consequently the duty to manage asbestos will fall primarily on either the owner or the occupier depending on the terms of the lease. Owners will normally have responsibility for the common parts of multi-occupancy premises and also for unoccupied premises belonging to them. The ACOP to the Regulations ('The Management of Asbestos in Non-domestic Premises') goes into some detail on the respective responsibilities of owners and occupiers, providing some illustrative examples. A key requirement of Regulation 4 is paragraph (2) which states: 'Every person shall co-operate with the dutyholder so far as is necessary to enable the dutyholder to comply with his duties under the regulation.' Thus, even where the occupier is the dutyholder, the owner must still provide relevant information to enable the occupier to determine whether asbestos-containing materials (ACMs) are present. Similarly, occupiers vacating premises should provide the owner (who becomes the dutyholder for empty premises) with relevant information.

The duties summarised

Dutyholders must:

- take reasonable steps to find ACMs and check their condition
- presume that materials contain asbestos unless there is strong evidence to the contrary
- keep a written record of the location and condition of actual and pre-sumed ACMs and keep it up to date
- assess the risk of people being exposed to asbestos
- prepare and put into effect a plan to manage that risk.

Finding ACMs and assessing their condition

Asbestos was once commonly used in construction materials, particularly dur-ing the 1960s and 1970s. As its role in causing asbestosis, lung cancer and mesothelioma was increasingly recognised, legislation restricting its use was progressively introduced. The use of materials containing blue and brown asbestos was banned in 1985 but building materials containing white asbestos were not banned until 1999.

Consequently asbestos is likely to be present in many work premises, where it may be found as:

- loose asbestos packing (used as fire protection in ceiling voids)
- moulded or preformed lagging (as thermal insulation on pipes and vessels)
- sprayed asbestos providing fire protection (on panels, partitions, ceiling panels, structural steelwork, ductwork, etc.)
- insulating boards (for fire protection and thermal insulation)
- some types of ceiling tiles
- products used for insulation of electrical equipment or as gaskets in pipework systems
- asbestos cement products (roofing, wall cladding, gutters, drainpipes, water tanks)
- some types of textured coatings
- bitumen roofing materials
- vinyl or thermoplastic floor tiles.

Information about possible ACMs should be gathered together from:

- plans and specifications
- information from builders, installation engineers and material suppliers
- previous asbestos survey work.

An inspection of every part of the premises must then be carried out and mate-rials categorised as follows:

- not containing asbestos materials (some are obviously wood, glass, metal, brick, stone, etc.)
- materials known to be ACMs
- materials presumed to be ACMs until there is evidence to the contrary.

Some materials may be inaccessible, e.g. covered by wallpaper, contained within walls or fire doors or in positions that could be dangerous to access such as roof voids. In some cases it may be appropriate to make further investigations as to their composition or it may be acceptable just to presume that they are ACMs for the time being (see below). The locations and condition of actual and presumed ACMs must be clearly recorded on drawings and, where necessary, in support documentation. The HSE booklets 'The Management of Asbestos in Non-domestic Premises' and 'A Comprehensive Guide to Managing Asbestos in Premises' contain further information on the records required.

Who should carry out the assessment?

The dutyholder must ensure that anyone carrying out work as a result of the Regulation is competent for the purpose. HSE guidance document MDHS 100, 'Surveying, Sampling and Assessment of Asbestos-containing Materials', sets out three types of survey that can be carried out, each requiring a different level of competence. If some or all of the work is to be contracted out, the dutyholder should check:

- for evidence of staff training and experience in this work
- for evidence of suitable insurance
- whether the survey will follow the standards in MDHS 100.

Sampling and subsequent analysis of suspected ACMs must also be carried out by suitably trained and competent people. Guidance on selecting and managing contractors is provided in Chapter B7, and more specific guidance is available in the ACOP booklet ('The Management of Asbestos in Non-domestic Premises') and a free HSE leaflet, 'A Short Guide to Managing Asbestos in Premises'.

Managing the risk

Actions identified through the assessment as being necessary to manage the risk from ACMs might include:

- clear identification of actual and presumed ACMs (with signs, colour coding, etc.)
- restriction of access to or maintenance work in ACM areas (through keeping doors locked, use of permits to work or other control systems)
- analysis of suspect material prior to its disturbance
- treatment of ACMs which may release fibres (sealing or removal*)
- protection of ACMs from damage (e.g. from vehicles, trolleys)
- regular inspections of ACMs which may deteriorate or be disturbed or damaged (including damage by vandalism).

*Removal of ACM in many cases can only be carried out by a licensed contractor.

While dutyholders had to ensure that they complied with these requirements by 21 May 2004, they should also try to avoid unnecessary activity and expense (there are always those who try to cash in on new legislation). Dutyholders should avoid:

- removing ACMs unnecessarily (the ACOP states: 'if the materials are in good condition and are unlikely to be damaged or disturbed, then it is better to leave them in place and to introduce a system of management')
- analysing materials when it is clearly unnecessary (either the material obviously does not contain asbestos, or is already known to contain it)
- purchasing expensive record-keeping systems that are inappropriate for their needs (ample guidance is available from the HSE – see 'The Management of Asbestos in Non-domestic Premises' and 'A Comprehensive Guide to Managing Asbestos in Premises').

Source materials

1. HSE (1992) 'Workplace Health, Safety and Welfare: Regulations, ACOP and Guidance', L 24.
2. HSE (2000) 'Legionnaire's Disease: The Control of Legionella Bacteria in Water Systems', L 8.
3. HSE (2000) 'General Ventilation in the Workplace', HSG 202.
4. HSE (1999) 'Thermal Comfort in the Workplace', HSG 194.
5. HSE (2003) 'Heat Stress in the Workplace: What You Need to Know as an Employer', GEIS 1.
6. HSE (1998) 'Lighting at Work', HSG 38.
7. HSE (1998) 'Seating at Work', HSG 57.
8. HSE (1994) 'A Pain in your Workplace? Ergonomic Problems and Solutions', HSG 121.
9 HSE (2003) 'Understanding ergonomics at work', INDG 90.
10. HSE (1996) 'Slips and Trips: Guidance for Employers on Identifying and Controlling Risks', HSG 155.
11. HSE (1996) 'Slips and Trips: Guidance for the Food Processing Industry', HSG 156.
12. HSE (2001) 'Health and Safety in Construction', HSG 150.
13. HSE (1999) 'Health and Safety in Excavations', HSG 185.
14. HSE (2000) 'Avoiding Danger from Underground Services', HSG 47.
15. HSE (1997) 'Fire Safety in Construction Work', HSG 168.
16. HSE (1997) 'Construction Fire Safety', CIS 51.
17. BS EN 1263-1. Safety nets.
18. HSE (1998) 'Health and Safety in Roof Work', HSG 33.
19. HSE (1997) 'Tower Scaffolds', CIS 10.
20. HSE (1997) 'General Access Scaffolds and Ladders', CIS 49.
21. HSE (1999) 'Working on Roofs', INDG 284.
22. HSE (1995) 'Workplace Transport Safety', HSG 136.
23. HSE (1998) 'Safe Use of Vehicles on Construction Sites', HSG 144.
24. HSE (2001) 'Safety of Construction Transport', SIR 58.
25. HSE (1999) 'Construction Site Transport Safety: Safe Use of Compact Dumpers', CIS 52.
26. HSE (2002) 'The Management of Asbestos in Non-domestic Premises', ACOP and Guidance, L127.
27. HSE (2001) 'Surveying, Sampling and Assessment of Asbestos-containing Materials', MDHS 100.
28. HSE (2002) 'A Short Guide to Managing Asbestos in Premises', INDG 223.
29. HSE (2002) 'A Comprehensive Guide to Managing Asbestos in Premises', HSG 227.

Electrical Safety

D3.1 Introduction

The Electricity at Work Regulations 1989 which apply to all those 'at work' contain requirements on the construction and maintenance of electrical systems and work activities on or near them. They cover all 'electrical equipment' – everything from a battery-powered torch to a high voltage overhead line. The purpose of the regulations is to prevent 'danger', defined as death or personal injury from electric shock, electric burn, fires of electrical origin, electric arcing or explosions initiated or caused by electricity.

In the case of electric shock, the effects will depend on its voltage, frequency and duration and the impedance and route of the current path. Because of the variability of conditions it is impossible to specify a 'safe' voltage but the normal mains supply of 240V and 50Hz must always be considered potentially fatal. Quite low currents (of the order of only a few milliamps) may cause a fatal electric shock when passing through the body.

Employers and the self-employed, together with managers of mines and quarries, have the majority of the duties under the regulations although employees also have duties to cooperate with their employers and to comply with the regulations so far as they relate to matters within their own control. The HSE has published a 'Memorandum of Guidance' on the regulations.

This chapter looks at the various requirements contained in the regulations (and to a certain extent elsewhere) for the following:

- standards of electrical systems
- maintenance of electrical equipment
- electrical working practices
- persons engaged in electrical work.

D3.2 Standards of electrical systems

System construction (Regulation 4)

Regulation 4(1) states that 'all systems shall at all times be of such construction as to prevent, so far as is reasonably practicable, danger'. This includes the design both of the system and of the equipment comprising it.

The principal reference work on standards of electrical installation is the Institution of Electrical Engineers Regulations for Electrical Installations (known as the 'IEE Wiring Regulations') – strictly speaking, this is a code of practice which is now in its sixteenth edition. Various British Standards also contain requirements for the standards of electrical equipment. The Electrical Equipment (Safety) Regulations 1994 (which result from an EC Directive) are primarily directed towards consumer protection but also contain requirements

of relevance to the safety of electrical equipment in the workplace. In addition
to the general requirements of Regulation 4(1), several other regulations relate
to the standards of electrical equipment.

Strength and capability (Regulation 5)

Regulation 5 requires electrical equipment to have the strength and capability
to withstand the thermal, electromagnetic and electrochemical effects of
expected currents. This includes transient overloads, fault currents, etc.

This relates to the current carrying capacity of conductors and also to the
standard of insulation. Normally compliance with this regulation should be
achieved by using equipment in accordance with the manufacturer's rating and
instructions.

Adverse or hazardous environments (Regulation 6)

Electrical equipment must be constructed to prevent danger arising from reasonably foreseeable exposure to the following:

- mechanical damage
- the effects of weather, natural hazards, temperature or pressure
- wet, dirty, dusty or corrosive conditions
- flammable or explosive substances (including dusts, vapours or gases).

The IEE Wiring Regulations contain requirements relating to the standards of
protection for electrical equipment in these types of environments. Additional
requirements are contained in the Equipment and Protective Systems Intended
for Use in Potentially Explosive Atmospheres Regulations 1996 and certification
standards are imposed by the Electrical Equipment for Explosives
(Certification) Regulations 1990.

Insulation, protection and placing of conductors (Regulation 7)

Electrical conductors are required to be insulated and further protected (e.g.
by secondary insulation, conduit, armoured cables) as necessary to prevent
danger. The requirement for further protection is primarily to prevent mechanical damage to the insulation but may also involve providing protection against
the risks referred to in Regulation 6. The IEE Wiring Regulations contain
detailed guidance on these matters.

In some situations it may not be reasonably practicable to insulate conductors, e.g. overhead power lines, conductors supplying power for overhead travelling cranes, electrolytic or electrothermal processes. In these cases other precautions must be taken to prevent danger, such as:

- siting such equipment in segregated areas
- providing appropriate warning notices
- use of earth-free areas or insulated working platforms
- separation of conductors at different potentials
- controlling access to equipment (e.g. by permit-to-work procedures)
- establishing other relevant safe systems of work
- restricting work in relevant areas to competent trained staff (see paragraph D3.5).

(See paragraph D3.4 for details of appropriate working practices.)

Earthing or other suitable precautions (Regulation 8)

Regulation 8 relates to metal casings and other conductive parts of electrical equipment which may become charged as a result of a fault condition or the induction effects of an electrostatic or electromagnetic field. It requires precautions to be taken (by either earthing or other means) in order to prevent danger. The most common means of achieving this are by earthing accessible conductors or by using double-insulated equipment. Other means of achieving compliance with the regulation include the use of safe voltages and of 'earth-free' areas. These and other alternatives are described in the HSE booklet on the regulations.

Integrity of reference conductors (Regulation 9)

This regulation is concerned with the basic design of electrical systems and is intended to prevent earthed or other referenced conductors reaching significantly different potentials and causing danger. Technical guidance is available from HSE booklet 'Memorandum of Guidance on the Electricity at Work Regulations 1989'.

Connections (Regulation 10)

The regulation states that: 'Where necessary to prevent danger, every joint and connection in a system shall be mechanically and electrically suitable for use.' This applies to all means of connecting both circuit and protective conductors (whether temporary or permanent). They must both have adequate mechanical strength and provide sufficient electrical conductance. Plugs and sockets must be designed to prevent accidental connection of conductors intended to operate at different voltages. With earthed portable equipment, earthing should be automatically achieved by insertion of the plug.

Means for protecting from excess of current (Regulation 11)

The regulation requires that: 'Efficient means, suitably located, shall be provided for protecting from excess of current every part of a system as may be necessary to prevent danger.' Such protection is usually provided by fuses and residual current devices or circuit breakers. Detailed guidance is provided in the IEE Wiring Regulations.

Means for cutting off the supply and for isolation (Regulation 12)

This regulation requires that:

> suitable means (including, where appropriate, methods of identifying circuits) shall be available for –
> (a) cutting off the supply of electrical energy to any electrical equipment; and
> (b) the isolation of any electrical equipment.

'Isolation' is defined as 'the disconnection and separation of the electrical equipment from every source of electrical energy in such a way that this disconnection is secure' and is distinct from switching off equipment, although switching off and isolation may in some cases be achieved by the same action.

Means for cutting off and isolating the supply should be:

- accessible at a suitable location
- clear and unobstructed
- clearly marked as to the equipment it controls.

Further guidance on design is contained in the HSE memorandum and the IEE Wiring Regulations. The use of isolation switches is dealt with in paragraph D3.4.

D3.3   Maintenance of electrical equipment

Regulation 4(2) requires electrical systems to be maintained so as to prevent danger, so far as is reasonably practicable. Maintenance is required for fixed installations as well as portable appliances and might involve the following:

- visual checks by the user
- formal visual inspections
- combined inspections and tests
- physical attention, e.g. cleaning, component replacement or repair.

The frequency and nature of inspections and tests is a matter for judgement based on manufacturers' guidance, HSE recommendations (see refs 2 and 4 on page 303) and the employer's experience. There is no standard requirement for the annual testing of portable appliances – some equipment and work environments may justify more frequent attention, others much less frequent.

Visual inspections

Visual inspections can be incorporated within general health and safety inspection programmes and should be seeking to identify faults such as:

- damaged or taped cables
- outer cable insulation not secured inside plugs or connectors
- damaged plugs, sockets or switches
- loose or faulty joints
- loose or damaged conduit or trunking
- unsecured panel doors.

(Users should be encouraged to look for similar faults during their visual checks.)

Inspections and tests

More detailed inspections and tests might include the following:

- checks on fuse ratings
- checks on internal connections
- earth continuity tests
- insulation tests.

Those carrying out such tests and inspections must be competent for the purpose but need not necessarily be fully trained electricians. Records should be kept of inspections, tests and maintenance work – these can be very useful in helping to determine whether the frequency and type of maintenance is correct.

HSE recommendations on intervals for combined inspection and testing of portable and transportable equipment are as follows:

- equipment hire: before issue
- construction: three months
- industrial: six to twelve months
- premises used by the public, e.g. hotels: one to two years
- offices and other low-risk environments: up to five years.

299

Inspection and testing of double-insulated equipment, battery-operated equipment (less than 20V) and equipment operating at less than 50V AC is not usually required.

An HSE booklet, 'Maintaining Portable and Transportable Electrical Equipment', provides detailed guidance and other HSE guidance material is available ('Electrical Safety on Construction Sites').

Fixed installations

Inspection and testing of fixed installations should also be carried out periodically. Advice on frequencies and methods is contained in the IEE Wiring Regulations.

D3.4   Electrical working practices

Regulation 4(3) requires all work activities on or near electrical systems to be carried out so as not to give rise to danger, so far as is reasonably practicable. This includes excavations near live cables or work near overhead power lines. Several other regulations also deal with working practices.

Work on or near live conductors (Regulation 14)

The regulation states that:

> No person shall be engaged in any work activity on or near any live conductor (other than one suitably covered with insulating material so as to prevent danger) that danger may arise unless –
> (a) it is unreasonable in all the circumstances for it to be dead; and
> (b) it is reasonable in all the circumstances for him to be at work on or near it while it is live; and
> (c) suitable precautions (including where necessary the provision of suitable protective equipment) are taken to prevent injury.

Note should be taken of the use of the word 'unreasonable' in (a) as opposed to 'inconvenient'. Examples of such situations are given in the HSE 'Memorandum' and include the following:

- cable jointing in the electrical supply industry
- some work on electrical railways
- telephone network connections
- diagnostic work on control circuitry (separated from power circuits)
- some work in continuous process plants.

In such cases (and providing it is reasonable for work to be done live) suitable precautions must be taken (see below) and the persons performing it must be sufficiently competent (see paragraph D3.5). However, the clear preference is for equipment to be made dead.

Precautions for work on equipment made dead (Regulation 13)

This regulation states that:

> Adequate precautions shall be taken to prevent electrical equipment, which has been made dead in order to prevent danger while work is carried out on or near that equipment, from becoming electrically charged during that work if danger may thereby arise.

This will normally involve use of the means for cutting off the supply and isolating the equipment, which is required by Regulation 12 (see paragraph D3.2). The isolator should then be locked off to prevent re-energisation. If this is not possible, then fuses or other links must be removed and kept in a secure place. Such measures can be made more secure by taping over fuse holders and using warning signs.

In some situations equipment may become charged for other reasons, e.g. because of electromagnetic induction, mutual capacitance or stored electrical energy. In these cases temporary earths will need to be applied at appropriate points in the system.

Isolation of equipment must always be proved before work starts. The test method used should assume that the equipment may still be live (see ref. 5). (There have been many instances where the wrong equipment has been isolated – because of mistakes or misunderstandings, or due to incorrect labelling of isolator switches.) The test equipment used should also be proved, immediately before and immediately after the test. Permit-to-work systems can be used as an additional means of ensuring effective isolation (see paragraphs B8.9 to B8.12).

Working space, access and lighting (Regulation 16)

This regulation is partially concerned with working practices but also relates to the standards required when designing electrical systems (see paragraph D3.2). It states:

> For the purpose of enabling injury to be prevented, adequate working space, adequate means of access and adequate lighting shall be provided at all electrical equipment on which or near which work is being done in circumstances which may give rise to danger.

Where there are dangerous exposed live conductors within reach, there should be sufficient space:

- to allow persons to pull back away from the conductors
- for persons to pass one another with ease and without hazard.

Appendix 3 of the 'Memorandum' provides guidance on dimensions of switchboard passageways.

Access around the equipment should be kept clear so that there is no risk of persons tripping and coming into contact with live conductors. Adequate lighting must be provided, preferably from natural light or from permanent artificial lighting. However, in some circumstances temporary lighting rigs, hand lamps or torches will be the only practicable solution.

Precautions in live work

Where live work is carried out as permitted by Regulation 14 (see above) suitable precautions must be taken. These must relate to the level and the nature of the risk and should include, as appropriate, the following:

- using trained and competent people for the work (see paragraph D3.5 below)
- providing adequate information about the equipment and its risks
- using suitable insulated tools, equipment and protective clothing
- using suitable insulated barriers or screens
- using suitable instruments and test probes

- providing an accompanying person (see paragraph D3.5 below)
- carrying out live testing in specially protected areas
- controlling access to the work area.

The use of permit-to-work systems (see paragraphs B8.9 to B8.12) can also provide a systematic method of ensuring that suitable precautions are being taken for live working.

The 'Memorandum' to the regulations provides further guidance on live working while the HSE booklet 'Electricity at Work: Safe Working Practices' (ref. 6) gives guidance on assessment procedures and working practices for both dead and live working, while other HSE publications provide guidance for specific work sectors or types of electrical equipment (see page 303).

D3.5   Persons engaged in electrical work

Regulation 16 states that:

> No person shall be engaged in any work activity where technical knowledge or experience is necessary to prevent danger or, where appropriate, injury, unless he possesses such knowledge or experience, or is under such degree of supervision as may be appropriate having regard to the nature of the work.

The nature of the work

The degree of knowledge and experience necessary will depend on the nature of the work and the dangers that may result from it. Such work may involve the following:

- isolating electrical supplies
- electrical work on or close to live conductors (e.g. fault finding)
- non-electrical work close to live conductors, e.g. a mobile crane driver operating near overhead power lines
- work in electrolytic or electrothermal processes.

Technical knowledge or experience

The scope of the knowledge or experience necessary will depend on the circumstances and may include the following:

- adequate knowledge of electricity
- adequate experience of electrical work
- adequate understanding and practical experience of the type of electrical equipment or system involved
- understanding of the hazards that may arise, and the precautions necessary
- the ability to recognise whether it is safe to continue.

Where persons do not themselves have the necessary knowledge or experience they may still carry out work under an appropriate degree of supervision. Again this degree of supervision may be exercised through the operation of a permit-to-work system.

Accompaniment

A person should be accompanied during live working if an accompanying person can substantially contribute towards the implementation of safe working practice or, if necessary, render assistance in the event of an emergency. Such a person may contribute by:

- assisting in planning work methods
- helping identify live conductors from dead conductors
- keeping other people away
- isolating the supply in an emergency
- providing prompt first aid treatment.

In high-risk situations use of an accompanying person with considerable electrical expertise may be appropriate, but someone without any specialist knowledge should be able to keep others away or take appropriate emergency action. Accompaniment is unnecessary for low-risk live work, e.g. fault finding in low voltage control circuits, where power circuits are screened effectively.

Source materials

1. HSE (1989) 'Memorandum of Guidance on the Electricity at Work Regulations 1989', HSR 25.
2. HSE (1994) 'Maintaining Portable and Transportable Electrical Equipment', HSG 107.
3. HSE (1995) 'Electrical Safety on Construction Sites', HSG 141.
4. HSE (1996) 'Maintaining Portable Electrical Equipment in Offices and Other Low-risk Environments', INDG 236.
5. HSE (1995) 'Electrical Test Equipment for Use by Electricians', GS 38.
6. HSE (2003) 'Electricity at Work: Safe Working Practices', HSG 85.
7. HSE (1997) 'Avoidance of Danger from Overhead Electrical Lines', GS 6.
8. HSE (1997) 'Electrical Safety at Places of Entertainment', GS 50.
9. HSE (1994) 'Electrical Safety in Arc Welding', HSG 118.
10. HSE (1997) 'Do You Use a Steam/water Pressure Cleaner? You Could be in for a Shock', INDG 68.
11. HSE (1993) 'Electrical Storage Batteries: Safe Charging and Use', INDG 139.
12. HSE (1997) 'Electrical Safety for Entertainers', INDG 247.

Occupational Health and First Aid

D4.1 Introduction

Occupational health has been defined as the process of 'keeping the work environment fit for the workers, and keeping the workers fit for the work environment'. It encompasses a variety of topics, some of which are dealt with in greater detail in other chapters:

- hazardous substances (see Chapter C2)
- asbestos
- noise (see Chapter C3)
- vibration (hand-arm and whole body)
- radiation
- ergonomics
- manual handling (see Chapter C4)
- display screen equipment (see Chapter C5)
- stress
- smoking
- health education, including drink, drugs and lifestyle issues.

Even where these issues are not covered by specific sets of regulations, employers still have their general duties under the Health and Safety at Work Act, particularly the duty to ensure, so far as is reasonably practicable, the health, safety and welfare at work of all their employees.

The general obligation to carry out risk assessments (see Chapter C1) means that employers must still systematically assess occupational health risks not covered by specific regulations and both identify and implement appropriate precautions.

This chapter deals with the following:

- the principles of occupational health
- the role of an occupational health service
- health screening
- screening techniques
- the role of the occupational hygienist
- exposure monitoring techniques
- first aid.

D4.2 Principles of occupational health

UK regulations dealing with occupational health issues all require the same basic approach to be taken. This approach can also be applied to those health matters not covered by regulations.

Assess the risk

Potential health hazards must be identified and assessed:

- What types of harm might be caused and in what way?
- How might workers and others be exposed to the hazard?
- How great is their exposure and for how long?

Elimination or substitution

It is always best if exposure to the health hazard can be prevented and there have been many successful examples of that occurring:

- Asbestos has been replaced by non-hazardous or less hazardous substances.
- Water-based paints have been substituted for solvent-based ones.
- Quieter processes have replaced noisy ones.
- Smoking is not permitted in many workplaces.

However, elimination or substitution of hazards is not always 'reasonably practicable'.

Control

Risks must be controlled to an acceptable level through the application of control measures. Control of the risk at source, e.g. by enclosure or through local exhaust ventilation, should always be preferred to the use of PPE. Controls can be divided into three categories:

1. Equipment-based, e.g. enclosure or ventilation.
2. Procedural, e.g. specified work methods or job rotation.
3. People-based, e.g. PPE or use of skilled techniques.

Monitoring

The effectiveness of the control measures can be confirmed by monitoring:

- the workers (through health screening – see paragraphs D4.4 and D4.5)
- the work environment (this type of monitoring is covered in paragraphs D4.6 and D4.7).

Maintenance

The control measures must be maintained effectively. Both occupational hygiene and health screening activity can provide ongoing confirmation that this is the case.

Information, instruction, training and supervision

Employees must be aware of the health risks present in the workplace and know what control measures should be used. Precautions are more likely to be taken if the reasons for them are fully appreciated. Training in the use of some precautions may be necessary. Employers must also provide reasonable supervision to ensure that the precautions are actually taken.

D4.3   The role of an occupational health service

Services provided

Occupational health services are usually involved in the following activities:

- health screening of new and transferred employees
- regular screening of workers exposed to specific health risks
- provision of health education and advice
- rehabilitation of workers (following injuries or illness)
- provision of first aid treatment (including the organisation and training of first aiders and appointed persons) – see paragraph D4.8.

The nature of the service required and the method of providing it will depend on the numbers employed and the risks to which workers are exposed. In small workplaces a limited range of the above activities will often be performed by human resources or personnel staff, possibly supported by external specialists. Larger workplaces are more likely to have their own specialist staff working on a full-time or part-time basis.

Although providing occupational health services will inevitably involve additional overheads, these can be offset by reducing employee absence due to ill health and injury (whether work related or not) and by lessening the possibility of successful civil claims for health-related problems.

Occupational health specialists

Most occupational health services rely heavily upon the full-time or part-time services of an occupational health nurse who should preferably hold a relevant specialist qualification such as the Occupational Health Nursing Certificate or Diploma.

Medical practitioners are usually engaged on a part-time basis and they too should preferably have formal qualifications in occupational medicine or have attended relevant courses. Retaining a local general practitioner with no specialist training and a poorly defined role will usually be a poor investment.

Use of specialist services from physiotherapists or chiropodists can help considerably both in preventing employees going absent from work and in rehabilitating them more quickly following injury or illness. Other specialists such as ergonomists or toxicologists may need to be involved in dealing with specific health-related problems and specialist assistance may be required to operate audiometric and other health screening equipment (see paragraph D4.5).

Some organisations offer a full range of occupational health services to employers on either a contractual or fee-paying basis and this may well be the best way for a small employer with significant health-related risks to obtain the services of a professional team. There may also be the possibility of sharing services with neighbouring employers or larger companies within the same group.

The Employment Medical Advisory Service (EMAS) is attached to the HSE and can be accessed by both employers and workers needing advice on occupational health matters. Disability Service teams within the Employment Service may also be able to provide assistance on work-related matters involving persons with disabilities.

D4.4   Health screening

Pre-placement screening

Screening of new employees or existing employees being transferred into new jobs is desirable from several standpoints. It can avoid putting people into situations where their own health and safety or that of others is affected, e.g. by not placing asthmatics in areas where certain chemicals are used, or not employing crane drivers with defective vision.

Employment legislation forbids discrimination without good reason. The existence of a health problem may provide good reason not to employ a person in particular types of work but employers should not place blanket exclusions on those with health problems or disabilities.

Baselines can also be established against which future comparisons can be made. This can help identify the early stages of occupational health problems and also can protect the employer from unjustified claims for health damage that in fact occurred from previous employment or other causes.

Regular screening

The regular screening of employees' health is desirable in its own right, but some regulations require screening to be carried out on a regular basis in order to confirm that control measures are proving effective and that workers' health is not being endangered. In such cases screening is compulsory, whereas general health screening may be optional (although companies may choose to make it a condition of employment). The health surveillance requirements of the COSHH Regulations are summarised in paragraph C2.14.

Detailed guidance is available in the HSE booklet 'Health Surveillance at Work'.

D4.5   Screening techniques

Pre-placement questionnaires

These can be used to ask new employees (or internal transferees) about health conditions or disabilities that could affect their work. Questionnaires may be completed by prospective employees themselves, although it is preferable for them to be used by an occupational health specialist or other trained member of staff, who can determine the relevance of answers given or probe a little deeper in cases of apparent hesitancy.

Where possible issues are identified, these must be investigated further by appropriate staff and it may be necessary to consult with the person's own general practitioner or even to refer them to a relevant specialist.

General physical examination

Physical examinations are often carried out by medical practitioners in order to determine whether employees have adequate strength, mobility and general physical fitness to carry out their work safely. Such examinations will usually be accompanied by measurement of weight, height, pulse rate, blood pressure, etc.

Lung function tests

The condition of workers' lungs can be monitored by the use of instruments to measure the forced vital capacity (the capacity of the lungs), the forced expiratory volume in one second (an indication of the elasticity of the lungs) or the peak respiratory flow. These readings can be compared to previous results for the same worker or the norms for people of the same age, size and gender.

Vision tests

Vision screening equipment can be used to test visual acuity at near and long distances (or at the normal distance of a display screen), and also peripheral vision (important in driving or operating other mobile equipment), depth perception (important for many crane operators) and colour vision (important for electricians and in recognising colour signals).

Audiometric tests

Audiometric testing equipment is used to test the sensitivity of hearing in each ear to a range of noise frequencies. If correctly used it can identify those whose hearing has already been affected by occupational exposure or for other reasons, and also workers whose hearing is beginning to deteriorate.

Testing should usually take place inside a soundproof booth and at a time when the worker has not been exposed to significant noise for some time (in order to avoid the short-term threshold shift which is caused by recent exposure).

The operation of audiometric equipment and interpretation of results must be carried out by suitably trained staff and the equipment must be calibrated and maintained regularly. (The same applies to lung function and vision testing equipment.)

Blood and urine tests

Analytical techniques are available to test for the presence of hazardous substances (or their metabolites) in the blood or urine, with the former generally producing the more reliable results. In some work sectors (particularly public transport) employees may be subjected to random tests in relation to their use of alcohol or drugs. Testing may also be carried out in relation to general health matters, e.g. to measure levels of cholesterol.

Specific questionnaires or examinations

Where workers are subjected to specific risks (e.g. materials that may cause skin cancer or dermatitis), they may be asked to complete questionnaires periodically about the condition of their skin. Physical examinations may also be carried out regularly or used as a follow-up where possible problems are indicated on questionnaires. A similar approach could be taken to other health problems, e.g. those relating to the lungs.

D4.6   The role of the occupational hygienist

Occupational hygiene has been described as the recognition, evaluation and control of risks to health at work. While occupational health specialists are primarily concerned with monitoring the workers exposed to potentially hazardous work environments, the occupational hygienist is more directly involved with the work environment itself. Cooperation between the two types of specialist is obviously essential.

While staff possessing general health and safety qualifications (e.g. the NEBOSH Certificate or Diploma) may be capable of carrying out a limited range of occupational hygiene work, there are many specialist courses and qualifications, particularly in relation to risks from airborne contaminants, noise and radiation.

Recognition

Preferably occupational hygienists should be able to recognise potential health risks before they are actually created in the workplace. They should be able to anticipate that major modifications to a structure built in the 1970s may involve exposure to asbestos-containing materials or that introduction of a particular type of equipment may create noise risks. However, the hygienist must also operate in a reactive way to problems observed in the workplace or reported by others, to complaints from workers, or to actual cases of ill health (reported by workers or identified by occupational health specialists).

Evaluation

In many cases the hygienist will be able to evaluate whether a significant risk exists simply through observations of the workplace and discussions with relevant workers and their supervisors. However, other risks may necessitate further enquiries (e.g. about the potential hazards associated with a particular substance and the recommendations from the supplier or others as to its control). Some situations may require measurement or sampling to be carried out in the workplace to determine the extent of the risk (see paragraph D4.7 below). The hygienist must, of course, be able to interpret the results of the measurement or sampling activity.

Control

Reference is made at paragraph D4.2 to the different types of control measures, which can be applied to occupational health risks. The occupational hygienist should be aware of or be able to find out which control measures are available for the risk in question, and to help determine what will be appropriate. Tackling risks at source is often the best approach but this may not always be 'reasonably practicable'. However, people-based controls such as PPE may be attractive initially as a low-cost option, but may prove more expensive in the long run, and difficult to enforce.

The occupational hygienist should play a big part in the decision-making process and may also need to devise a monitoring strategy in order to ensure that the chosen control method continues to prove effective.

D4.7   Exposure monitoring techniques

Measurement of noise and the instruments available to measure it were covered in paragraph C3.4. Some of the more common means of monitoring airborne contaminants are outlined below and equipment is also available to monitor other types of health risks such as radiation.

Chemical indicator tubes

Indicator tubes are available from several commercial suppliers. These can be used to measure a wide range of airborne substances, particularly gases, vapours and aerosols. A hand bellows or a mechanical pump is used to draw a measured volume of air through the indicator tube. The contaminant reacts with the chemicals contained in the tube to produce a colour change. The concentration of the contaminant can be measured by calibrated markings showing how far the colour change has penetrated into the tube or alternatively by comparing the intensity of the colour change with a calibrated chart. Different types of tubes can be used to measure concentrations over short-term and long-term sampling periods.

Direct reading instruments

Direct measurements of the concentrations of gases, vapours or dust present in the atmosphere can be provided by direct reading instruments. Many instruments are portable (making them particularly useful for measuring contaminants in confined spaces). Some instruments are designed to be used in fixed positions, sometimes as part of detection networks. As well as providing instant readings of concentrations, some instruments can be set to provide alarms when specified concentrations are reached, e.g. a given percentage of the MEL or the OES. This can be a particularly useful means of ensuring that the MEL is never exceeded.

Sampling pumps and filter heads

These utilise small battery-operated pumps to draw air through filter heads containing filter papers or another absorbent medium. The samples collected can then be weighed, chemically analysed or counted under microscopes (as is the case for asbestos fibres).

The concentration can be determined from the weight of the sample (or the number of fibres counted) and the total volume of air drawn through by the pump over the sampling period. For 'personal sampling' the filter head can be fixed in the worker's breathing zone (often on the lapel) with the pump clipped onto a belt.

Monitoring equipment can be used either to monitor the concentration of a contaminant in a given part of a workplace or for personal sampling to measure the concentration in the breathing zone of an individual worker. Care should be taken in interpreting results – the conditions measured may not be representative of those normally prevailing and allowances must be made for the accuracy of the sampling method. Some occupational hygienists deliberately sample what they believe to be the worst conditions first. If these circumstances produce concentrations that are comfortably within the relevant MEL or OES then the need for further monitoring may be greatly reduced or eliminated. Note should be taken of the requirement to retain all exposure monitoring records for five years and those relating to the personal exposure of identifiable individuals for at least forty years (see paragraph B9.8).

HSE guidance is available on monitoring strategies in general and also on methods for the determination of many individual hazardous substances, via their MDHS series of publications.

D4.8   First aid

The Health and Safety (First Aid) Regulations 1981 require the provision of adequate and appropriate equipment, facilities and personnel to give aid in the event of injury or illness. The regulations, together with an ACOP and detailed guidance, are available in an HSE booklet but a good summary of the requirements is available in a free HSE booklet. A further free leaflet also provides basic advice on first aid techniques (see page 312 for details).

The HSE have recently been reviewing the effectiveness of these regulations and a consultation document about proposed changes to them may be published during 2005.

Definition of first aid

First aid is defined in Regulation 2 as meaning the following:

- In cases where a person will need help from a medical practitioner or nurse, treatment for the purpose of preserving life and minimising the consequences of injury and illness until help is obtained.
- Treatment of minor injuries, which would otherwise receive no treatment or which do not need treatment by a medical practitioner or nurse.

General duties of employers

Regulation 3 places general duties on employers to make provision for first aid, which must include the following:

- adequate and appropriate equipment and facilities
- such number of suitable persons as is adequate and appropriate.

The ACOP and HSE guidance set out what is 'adequate and appropriate' for particular situations and who constitutes a 'suitable person' and these requirements are summarised below. (Regulation 5 requires self-employed persons to ensure that adequate and appropriate equipment is provided for them to render first aid to themselves.)

An absolute *minimum* requirement for any work site is:

- a suitably stocked first aid box (see below)
- an 'appointed person' to take charge of first aid arrangements.

Factors to consider

There are many aspects to consider in determining what constitutes 'adequate and appropriate' provision of first aid, including the following:

- work involving specific risks, e.g. hazardous substances, dangerous equipment
- experience of specific accidents or cases of ill health
- the number of persons employed
- the presence of inexperienced workers or of staff with disabilities or health problems
- multi-storey buildings or spread-out sites
- shift work or out-of-hours working
- the distance from emergency medical services
- employees who travel a lot or work alone
- employees working on other people's sites
- the presence of members of the public (although there is no legal responsibility in respect of non-employees).

First aid equipment

The minimum requirements recommended by the HSE for a first aid box are as follows:

- a leaflet giving general guidance on first aid ('Basic Advice on First Aid at Work')
- twenty individually wrapped sterile adhesive dressings (assorted sizes)
- two sterile eye pads
- four individually wrapped triangular bandages (preferably sterile)
- six safety pins
- individually wrapped sterile unmedicated wound dressings: six medium-sized (approx. 12 cm x 12 cm) and two large (approx. 18 cm x 18 cm)
- one pair of disposable gloves.

(Tablets or medicines must not be kept in the first aid box.)

Depending on the factors referred to immediately above, additional equipment or arrangements may be required, such as the following:

- additional equipment for dealing with specific risks, e.g. sterile water, stretchers, resuscitation equipment
- additional contents for first aid boxes likely to be heavily used
- a wide spread of first aid boxes throughout larger workplaces
- ensuring that some equipment is accessible at all times, e.g. in a security office
- travelling first aid kits to go out with some workers
- special liaison arrangements with local medical services or emergency services
- arrangements with other site occupiers to provide first aid equipment.

First aiders and appointed persons

A 'first aider' is someone who has undergone an HSE-approved first aid course and holds a current First Aid At Work certificate. Courses normally involve attendance at a four-day course and certificates must be kept current by attending a two-day refresher course every three years.

An 'appointed person' is someone chosen by the employer to:

- take charge when someone is injured or falls ill (including calling an ambulance if required)
- look after first aid equipment (including restocking boxes).

Category of risk	Numbers employed at any location	Suggested number of first aid personnel
Lower risk e.g. shops and offices, libraries	Fewer than 50 50–100 More than 100	At least one appointed person At least one first aider One additional first aider for every 100 employed
Medium risk e.g. light engineering and assembly work, food processing, warehousing	Fewer than 20 20–100 More than 100	At least one appointed person At least one first aider for every 50 employed (or part thereof) One additional first aider for every 100 employed
Higher risk e.g. most construction, slaughter-houses, chemical manufacture, extensive work with dangerous machinery or sharp instruments	Fewer than five 5–50 More than 50	At least one appointed person At least one first aider One additional first aider for every 50 employed

It is preferable that appointed persons attend a short emergency first aid training course so that they can provide basic first aid where required.

The numbers of first aiders and/or appointed persons required will depend on the numbers employed and the level and nature of risks. At least one appointed person should be available whenever people are at work, so allowance must be made for holiday and sickness absence as well as shifts and out-of-hours working. Out-of-hours situations can often be covered by designating security staff as appointed persons.

The HSE's 'First Aid at Work: Your Questions Answered' contains the above table of suggested numbers of first aid personnel. However, it also states that additional provision may be necessary for remote locations, spread-out sites or shift work.

First aiders and, where appropriate, appointed persons may need additional training when they are expected to use special first aid equipment or techniques to deal with specific types of accident that may occur in the workplace.

Information about first aid

Employers are required by Regulation 4 to inform employees about their arrangements for first aid, including the location of equipment, facilities and personnel. This should be part of induction training for new or transferred employees but at least one conspicuous notice should also be provided. Account may also need to be taken of employees with language or reading difficulties.

Source materials

1. HSE (2000) 'An Introduction to the Employment Medical Advisory Service', HSE 5.
2. HSE (1999) 'Health Surveillance at Work', HSG 61.
3. HSE (1997) 'Monitoring Strategies for Toxic Substances', HSG 173.
4. HSE (1997) 'First Aid at Work: Health and Safety (First Aid) Regulations 1981', ACOP and guidance, L74.
5. HSE (1997) 'First Aid at Work: Your Questions Answered' (free booklet), INDG 214.
6. HSE (2002) 'Basic Advice on First Aid at Work' (free leaflet), INDG 347.

Occupational Stress

D5.1   Introduction

Work-related stress and its physical and mental effects are never far from the headlines. The issue has recently been brought into sharper focus by news that the HSE has issued enforcement proceedings against West Dorset Hospitals NHS Trust. The Trust has been given just three months to improve the ways in which it identifies and addresses stress in the workplace. If it fails to comply with the enforcement notice it will run the risk of an unlimited fine under the Health and Safety at Work Act.

The HSE took the unprecedented decision to issue the enforcement notice following a complaint that it received from a former member of staff concerning a culture of long hours and alleged bullying. The action, which is a first in relation to work-related stress, together with the HSE's development of a new set of stress management standards, signals a shift in the way that the issue is to be tackled.

The HSE's tough stance on the issue is not that surprising considering that workplace stress has very serious repercussions on the lives of workers and the costs to industry. The 2001/02 survey of Self-reported Work-related Illness (SW101/02) indicated that over half a million individuals in Britain believed they were experiencing work-related stress at a level that was making them ill. Another survey, the Stress and Health at Work Study (SHAW), indicated that one in five working individuals thought that their job was extremely stressful.

Every day 250,000 workers report absent due to stress-related illness and this costs their employers an estimated £5 billion. The extent of the problem and its damaging consequences – absenteeism, poor performance and accidents – are such that employers should attempt to identify and, where possible, eliminate its causes.

D5.2   What is workplace stress?

There is no single agreed definition of stress. One definition describes it as 'a psychological state resulting from an imbalance between job demand and the inability of the individual to cope with that demand'. The HSE proposed its own definition in its discussion document 'Managing Stress at Work', where it described stress as 'the reaction people have to excessive pressures or other types of demand placed on them'. This definition makes an important distinction between pressure that can be a positive state if managed correctly, and stress that can be detrimental to health.

Usually stress can be fairly short-lived and its symptoms removed once the particular pressure has been eliminated or reduced. Pressure in itself is not necessarily bad, as some individuals thrive on it. However, there is a limit to the amount of pressure that individuals can cope with and ill health can result from excessive pressure. In some instances, the effects of stress can be permanent, causing substantial and long-term physical or mental impairment.

D5.3   Stress-related conditions

In itself stress is not a disease, but it can lead to mental and physical ill health such as high blood pressure, heart disease, ulcers and depression. An employee suffering from severe stress might be classed as disabled under the Disability Discrimination Act 1995 (DDA). In order to fall within the DDA definition, an employee would have to establish the following:

> A mental impairment, which had a substantial adverse effect on his ability to carry out normal day-to-day activities, and the impairment must have lasted or be likely to last for twelve months, or be likely to recur.

The DDA makes it unlawful to discriminate against a person with a disability and obliges the employer to make reasonable adjustments to working conditions and practices so that the individual with the disability is not at a disadvantage. Such adjustments might involve, for instance, reducing that person's workload.

It is worth noting that the dismissal of an employee who is absent due to a stress-related illness might result in a disability discrimination claim. In such circumstances there is no limit on the amount that may be awarded to the claimant.

D5.4   Contributory factors

In dealing with stress in the workplace, it is important that employers are able to ascertain some of the possible causes that would contribute to it. Some contributory factors are listed below:

- Physical aspects of work environment, e.g. too noisy, cold, dark.
- Work capacity: there can be a work overload, i.e. too much to achieve in the available time.
- Introduction of new technology.
- Working conditions: long hours, shift work, lack of flexibility.
- Design of work: this may be overly complex or organised in a way that makes it difficult to carry out.
- Poor communication: management may not communicate its objectives or issue instructions clearly. Relations within the company between employees and managers may be poor.

D5.5   Identifying stress in the workplace

The following steps may help to identify the potential occurrences of stress:

- Regular risk assessments.
- Education and training of managers to gain an understanding of the problem.
- A greater reliance on staff appraisals as a means of identifying employees who are experiencing excessive levels of pressure and stress.
- Periodic anonymous staff surveys may bring to light any problem areas.
- The collation of absence or sickness statistics, taken from sickness notes and certificates, will help to show any trends and patterns.

The HSE has published a model stress policy on its website (www.hse.gov.uk). This provides a systematic approach to dealing with stress. It could usefully be incorporated in an organisation's health and safety policy.

D5.6 The legal framework

There is no legal provision that specifically covers stress. However, employers do have statutory and common law duties imposed on them in this area. HASAWA 1974 states: 'it shall be the duty of every employer to ensure, so far as is reasonably practicable, the health, safety and welfare at work of all his employees'. It is clear from section 53 of the Act that this duty applies to both the mental as well as the physical health of employees. It defines personal injury as 'any disease and any impairment of a person's physical or mental condition'. The general duty to ensure a safe working environment (HASAWA 1974, s. 2) therefore clearly extends to the prevention of stress-related illness.

It should be noted that employees are themselves under a statutory duty to exercise care regarding their own physical and mental health and safety and must inform their employer if they believe that there is an imminent and serious threat to health and safety.

D5.7 Assessing the risk

Under the Management of Health and Safety Regulations 1999, employers have a duty to carry out a risk assessment (see paragraph B2.2) with a view to identifying preventative and protective measures that it needs to take in order to comply with HASAWA 1974, s. 2. It must also put into practice the necessary procedures to follow on from the findings of the risk assessment. These cover the planning, organisation, control monitoring and review of the protective measures (see paragraph B1.8).

HSE guidance makes it clear that ill health from stress at work should be treated in the same way as ill health caused by other physical factors present in the workplace. Employers therefore have a legal duty to ensure that reasonable care is taken to ensure that health is not placed at risk through excessive and sustained levels of stress. Employers should therefore keep stress in mind when they are assessing possible workplace health hazards.

D5.8 Duty of care

In addition to health and safety legislation, employers have a common law duty to take reasonable care to ensure the safety of their employees. This duty arises from the contract of employment and also from the law of negligence. Here the law does not recognise a difference between physical and psychiatric injury.

Because the employer's duty of care is personal and owed to each employee as an individual, employers should take into account any particular susceptibilities of an individual or group of individuals of which it is aware or should be aware, in order to prevent reasonably foreseeable risks as a result of stress.

Employers also have an implied contractual term, in every contract of employment, that they will take reasonable steps to ensure the health and safety of their workforce. The employer has a duty to support the employee in order to prevent a stress-related illness arising. Failure to take reasonable steps can result in a claim that the employee has been constructively dismissed.

D5.9   Stress management standards

Following the publication of draft standards in 2003 and a period of consultation, the HSE launched their stress management standards in November 2004. The standards are available via the HSE website (www.hse.gov.uk) and are based on a continuous improvement model. They feature a benchmarking tool to help gauge stress levels, make comparisons with other organisations and work with employees to identify solutions.

The Minister for Work stated that specific legislation was not necessary but that the standards would still be enforced by the HSE – presumably using the general requirements of HASAWA and the Management Regulations, particularly the duty to carry out a risk assessment (as in the case referred to in Paragraph D5.1)

Source materials

1.   ICSA Publishing, Employment Law and Practice. This includes detailed coverage of employment law aspects of occupational stress.
2.   HSE (2001) 'Tackling Work-related Stress: A Managers' Guide to Improving and Maintaining Employee Health and Well-being', HSG218.
3.   HSE (2003) 'Beacons of Excellence in Stress Prevention', RR133.
4.   HSE (2001) 'Work-related stress. A short guide', INDG 281.

The HSE website – www.hse.gov.uk – publishes survey and research results and fact sheets dealing with occupational stress.

Work-related driving

D6.1   Introduction

The HSE leaflet 'Driving at work' published in 2003 estimates that up to a third of all road traffic accidents involve someone who is at work at the time. These may account for over 20 fatalities and 250 serious injuries every week. That equates to an annual death toll of 1000 – three to four times the number of fatalities from other work-related accidents.

Whilst this area is primarily dealt with by road traffic legislation, much of which is enforced by the police, health and safety laws still apply to work activities on the road. The more pro-active application of such laws was a key recommendation of the 'Work-related Road Safety Task Group' which was established by the Government and the HSC. As a result the HSE has embarked on a programme of work focusing on:

- developing and promoting best practice
- awareness raising
- intelligence and data collection
- clarifying investigation and enforcement arrangements (between the police, the HSE and local authorities)
- research and guidance.

D6.2   Relevant legislation

Much information is available elsewhere on road traffic legislation (see Source Materials at the end of this chapter). The general requirements of the Health and Safety at Work Act (see paragraphs A 2.1 to A 2.10) apply to everyone who is 'at work', whether on the roads or elsewhere. These not only protect those who are at work but also others who may be at risk e.g. passengers, pedestrians and other road users.

However, employers and the self-employed also have duties under the Management of Health and Safety at Work Regulations, including the duty to carry out risk assessments (see Chapter C1). In conducting a risk assessment of work-related driving there are a number of aspects which must be taken into account including the driver, the vehicle and systems of work routes, scheduling etc.

The risk assessment must also take into account the wide range of people who may be involved in work-related driving:

- those driving large goods vehicles (LGVs) or passenger service vehicles (PSVs)
- drivers of small and medium vans, often providing delivery services
- sales and technical representatives, frequently travelling long distances
- directors and managers, sometimes making lengthy journeys after a full day spent working elsewhere.

Some regulations also contain specific risk assessment requirements of relevance to drivers on the road, in respect of manual handling (See D6.6 and Chapter C4) and the use of display screen equipment (See D 6.7 and Chapter C5)

As is the case for all risk assessments, appropriate steps must be taken to ensure that effective management systems are in place to deliver the precautions required (see paragraphs B3.3 to B3.8 on the management cycle).

D6.3   The driver

Many aspects to be considered during the risk assessment centre around the drivers themselves. These could include general competence, general and specific training and fitness to drive.

General competence

The following formal checks on driving licences and insurance are likely to be necessary:

- possession of a valid current driving licence
- compliance with specific LGV/PSV licence requirements
- possession of insurance for business use and a current MOT certificate (where private vehicles are used)
- capability checks (by reference to previous employers or driving trials).

A form to use in checking drivers' own documentation is shown below at Figure D6.1.

General training

Many of the following aspects could be included in induction training for those driving regularly for work purposes.

- communication of the organisation's requirements and policies on driving
- guidance on avoiding fatigue
- how to carry out routine checks on vehicles (lights, tyres, screenwash etc.)
- how drivers should ensure their safety in the event of a breakdown
- basic repair work and adjustments e.g. wheel changing, headlamp beams, head restraints
- provision of additional training for high risk drivers.

FIND THIS ON CD

Figure D6.1   Driver documentation check

(To be carried out annually)

Driver's name	Driving licence	Business Insurance	MOT certificate	Date Checked	Signature

Specific training

The need for certain types of training will vary according to the vehicles involved. The following areas could be considered:

- maximum permitted loads and safe load distribution
- how to secure loads
- safe sheeting of loads
- vehicle heights (laden and empty)
- use of ancillary equipment e.g. tail-lifts, vehicle-mounted cranes, fork lift trucks
- what to do in emergencies or if the vehicle is considered unsafe
- use of anti-lock brakes
- availability of handbooks and other relevant guidance.

Fitness to drive

There are specific medical requirements for drivers of heavy lorries. There may also be a need for regular medicals and/or eyesight tests for other drivers. Drivers should also be aware of the possible effects of any medication they are taking.

D6.4   The vehicle

Vehicles must be suitable for their purpose, properly equipped and maintained in a safe condition. Aspects to consider during the risk assessment include:

Suitability of vehicles

- arrangements for selecting vehicles (preferably with driver input)
- arrangements for hiring in additional vehicles
- suitability of vehicles for the loads to be carried (weights, configuration etc.)
- suitability for loading and unloading, including headroom (see also paragraph D6.6 relating to manual handling)
- the ergonomic suitability of seats and availability of adjustment mechanisms.

Equipment

- the availability of suitable tools for basic repair work and adjustments
- the availability of equipment and handling aids for loading and unloading (see also paragraph D6.6)
- means for securing the loads being carried
- suitable sheeting and covers
- means for storing securely tools and equipment accompanying the vehicle.

Maintenance

- arrangements for planned and preventive maintenance
- compliance with MOT certification requirements
- arrangements for regular safety checks, in addition to those carried out by drivers (lights, tyres, screenwash, wiper blades, seatbelts, head restraints etc.)
- monitoring mechanisms to ensure maintenance, repairs and safety checks are carried out to an acceptable standard.

D6.5   Systems of work

It is a requirement of section 2 of HASAWA that employers provide and maintain safe systems of work for their employees. In establishing safe systems of work for driving, the following aspects are important and must be considered during the risk assessment process:

Routes

- routes to be used must be suitable for the type of vehicle involved (motorways are preferable, particularly for larger vehicles)
- overhead restrictions (e.g. those on bridges or tunnels) must be considered for some vehicles
- level crossings should be avoided for long vehicles
- areas of traffic congestion should be avoided, particularly at peak times
- weather conditions may justify alternative routes e.g. avoiding high ground in snow, using lit routes in night-time fog
- the direction of approach to premises may need to be considered e.g. turning into restricted openings or reversing into entrances.

Scheduling

Scheduling of driving is important not just for LGV/PSV drivers (whose driving hours are regulated by road traffic legislation), but also for representatives and managers who may set themselves (or be set) unrealistic programmes. The two key issues are the potential for fatigue and the temptation to drive excessively fast.

- scheduling and arrival times must be realistic, allowing for driving within speed limits, breaks, traffic and weather conditions and time spent loading and unloading
- breaks should be taken regularly on long journeys (the Highway Code recommends a 15 minute break every two hours)
- drivers should be given guidance on what to do if they feel tired e.g. stop overnight rather than driving at the end of a long day
- is it possible to avoid driving when drivers are most likely to feel sleepy? (sleep-related accidents are most likely to occur between 2 am and 6 am, and also 2 pm and 4 pm).

Weather

Weather conditions are a big factor in the day-to-day planning of driving activities. Information should be readily available on weather conditions in areas in which staff will be driving. Snow and fog present obvious problems as do high winds, particularly for high-sided vehicles. Routes and schedules should be modified to take account of conditions, with drivers having the discretion to modify or even abandon their programmes should conditions dictate.

Whilst employers must establish information and communication systems to take account of weather, a degree of reliance must be placed on driver's own dynamic risk assessments based on information available and prevailing conditions. The prevailing culture should be 'don't drive if the weather makes it dangerous'.

Mobile phones

Mobile phones are now an essential tool for those driving regularly, particularly through remote areas. They are particularly useful in the case of bad weather, breakdowns, unexpected delays or driver fatigue. Problems can be identified and alternative approaches discussed between the driver and those responsible for his or her management. However, the use of hand-held phones by drivers in their vehicles is now prohibited whenever the engine is running. That means that it will be an offence to use a hand-held mobile telephone even when the user is stuck in traffic. The prohibition covers:

- speaking or listening to a phone call;
- using a device interactively for accessing any kind of data , including the internet; or
- sending or receiving text messages or other images.

Contravention of this rule may result in a three-point penalty on the driving licence of the offender and a £30 fixed-penalty; or up to £1000 fine, £2500 for drivers of goods vehicles or vehicles manufactured to carry nine or more passengers.

It is an offence to 'cause or permit' employees to lose proper control of a vehicle, and even though employers can continue to make calls to the mobile telephones of employees, it is incumbent upon them to advise personnel that they are not obliged to and should not answer calls whilst their engine is running. Such guidance should be reflected in health and safety and risk management policies. Failure to do this may make them solely liable if an offence is committed.

Whilst use of hands-free equipment is still permitted, it is safer to make and receive calls whilst stationary rather than on the move. The introduction of regular phone stops and effective voicemail should be considered when reviewing safe working practices for mobile workers

Alternatives

The possible use of alternative transport for goods is likely to be a strategic policy decision for most businesses. However, in the case of personal travel it will usually be safer to use public transport (trains or planes) where available, rather than drive long distances, particularly at the end of a hard day. Company transport policies should reflect this. Other options which should be conveyed to staff include:

- making overnight stops e.g. in the case of fatigue or bad weather
- modifying or abandoning programmes where necessary
- arranging taxis or lifts home for tired LGV/PSV drivers.

Managers should monitor whether appropriate systems of work are being followed e.g. by checking drivers' tachographs or quizzing staff about their driving practices and their awareness of the organisation's policies on driving.

D6.6 Manual handling

Chapter C4 provides much more detail on factors to be taken into account when assessing manual handling operations generally. However, handling activities carried out away from the employer's base often do not receive as much attention as they should. Employees at risk from manual handling are not just those involved in delivering or collecting goods but also sales representatives (loading and unloading samples), technical staff (who may need to handle instruments and other equipment) and employees involved in setting up displays at conferences and exhibitions. Consideration must be given to vehicle design, handling equipment and training.

Vehicle design

- ensuring that there is enough space to gain safe access to loads
- ensuring loads can be moved safely out of and into the vehicle (estate cars and vans are safer than the boots of saloon cars)

- providing adequate internal headroom for the handling of larger loads
- ensuring sufficient space is available to use handling equipment and aids.

Handling equipment and aids

Equipment may need to be carried with the vehicle to use when handling loads:

- powered equipment such as tail-lifts, vehicle-mounted cranes, fork lift trucks
- handling aids e.g. trolleys, wheeled platforms, skates, jacks.

As stated in paragraph D6.3 staff must be trained in the use of such equipment.

Assistance

Drivers may need to be provided with assistance in the loading and unloading of their vehicles. Such assistance may come from customers or suppliers (possibly using their own handling equipment or aids) or from staff at conference or exhibition venues. Where loads cannot be handled safely by the driver alone, employers should ensure that potential sources of assistance are identified in advance, or provide assistance of their own.

Training

Drivers who are expected to carry out manual handling activities regularly or where a significant amount of risk is involved must be provided with appropriate training. Paragraph C4.9 provides guidance on the content of manual handling training.

D6.7   Display screen equipment

Many staff, particularly sales and technical representatives, use their vehicles as mobile offices. This is likely to involve the use of laptop computers, possibly for extended periods. Whether this is sufficient to bring such staff within the 'user' definition of the regulations (see paragraph C5.5) is to some extent immaterial as employers still have legal obligations under the general sections of HASAWA.
    Practical issues to be considered during assessments are:

- the ergonomic suitability of vehicle seats for laptop use
- the adequacy of internal lighting if night time keyboard use is involved
- availability of sun shields (to prevent the sun shining onto laptop screens).

Use of the passenger seat for laptop work may be preferable in order to avoid space restrictions from the steering wheel and to use the glove compartment door as an additional support surface.

Source materials

1. HSE (2003) 'Driving at Work. Managing work-related road safety'. INDG 382
2. The Stationery Office (2001) 'The Highway Code'.
3. The Stationery Office (2002) 'Code of practice. Safety of loads on vehicles'.

The Royal Society for the Prevention of Accidents (ROSPA) and Brake Road Safety have a wide range of publications available on work-related driving. These can be found via their websites www.rospa.com and www.brake.org.uk.

Homeworking

D7.1   Introduction

The Office of National Statistics estimates that there are more than 2 million homeworkers in the UK. Working from home can have many advantages from the individual's point of view, such as eliminating commuting time or making work more compatible with domestic duties such as childcare.

Much of the recent increase in homeworking can be attributed to teleworking, where professional or administrative services can be provided effectively from home using modern IT. However, many workers also use their home as a base from which to travel to carry out professional, technical or sales functions. This is often accompanied by some work carried out at home e.g. administrative duties, repair of equipment or sorting of stock. Some home work also involves the more traditional type of home-based activities – the manufacture, assembly or packing of items, often as part of a much larger manufacturing operation.

D7.2   Relevant Legislation

Most health and safety legislation is still relevant to those who work at or from home. Employers still have duties to their employees and also to others who may be affected by their activities (such as other family members or visitors), by virtue of sections 2 and 3 of HASAWA. Even where a homeworker is self-employed for tax purposes, they will still often be regarded as an employee under health and safety legislation. Should this not be the case, the self-employed are still afforded protection from employers engaging their services through section 3(1) of HASAWA.

Homeworkers are also covered by the Management of Health and Safety at Work Regulations 1999, particularly in respect of the need to carry out risk assessments (see D7.3 below). Other regulations of particular relevance to homeworking are those dealing with work equipment (PUWER), display screen equipment, manual handling, dangerous or hazardous substances (DSEAR and COSHH) and personal protective equipment. Each of these topics is dealt with in more detail later in the chapter.

D7.3   Risk assessment of homeworkers

Where employers employ or engage the services of homeworkers, such work must be included in their risk assessments and the significant findings of these risk assessments must be recorded (if they have five or more employees). Chapter C1 provides further details of general risk assessment requirements with information on several more specific types of risk assessment also contained in section C.

The risk assessment process must take account of others present in the home who may be put at risk by the activities of the home worker. Children, the elderly or those with special needs are all particularly vulnerable, especially if the work involves the use of potentially dangerous equipment, or dangerous or hazardous substances.

Special consideration must also be given during the risk assessment to homeworkers who become pregnant or have recently given birth (new or expectant mothers) and also any homeworkers under the age of eighteen (young people). Guidance on assessing risks to these groups is contained in paragraphs C1.4 to C1.7.

Some of the more common topics likely to be of relevance during risk assessments of homeworkers are dealt with below, although employers must always be alert to other potential risks. A proforma checklist is also provided (see Figure D7.1) to aid in the risk assessment process.

D7.4 Work equipment

The requirements of the Provision and Use of Work Equipment Regulations 1998 (PUWER) and of the Electricity at Work Regulations 1989 are of relevance to equipment supplied for use by homeworkers. These regulations are explained in some detail in chapters D1 and D3 respectively.

Equipment must be suitable for its intended purpose and also suitable for use in the home environment. Homeworkers must receive appropriate information, instruction and training both in the purpose for which the equipment is intended and in how to use it safely. Any dangerous parts of machinery must be guarded or protected in other ways (see D1.7 to D1.9) and arrangements established for any necessary maintenance of equipment.

Electrical equipment must also meet appropriate standards and arrangements made for its inspection and testing. Homeworkers must be trained so as to be capable of carrying out simple visual inspections of electrical equipment (see the checklist in D3.3). Arrangements must also be in place for combined inspection and testing of electrical equipment to take place (see D3.3 also). This may involve homeworkers bringing equipment in to the employer's premises for inspection and testing, or for this to be carried out by a competent person visiting workers' homes.

Whilst employers cannot be held responsible for the standards of the fixed electrical installation in workers' homes, they may consider it prudent to supply certain types of electrical equipment with residual current devices. Homeworkers should also be made aware of risks caused by overloading power sockets and obstructing cooling air ducts.

Arrangements must be in place for homeworkers to report any defects in equipment and for the prompt and effective repair of such defects (or the temporary replacement of defective equipment). Homeworkers must be discouraged from attempting their own repairs.

D7.5 Display screen equipment (DSE)

Many employees working at home will be categorised as 'users' under the regulations dealing with DSE and entitled to the full protection of those regulations as set out in chapter C5. This will include the carrying out of an assessment on their workstation. The self-assessment checklist and associated guidance provided in Figures C5.1 and C5.2 will be particularly useful for homeworkers.

However, it must be made clear to homeworkers what must be done with the completed assessment checklist and how problems identified during the self-assessment will be rectified. This may involve arranging for a competent person to visit their home to carry out a more detailed assessment. Homeworkers who are 'users' will be entitled to eye tests and, where necessary, to the provision of corrective spectacles. They must also be provided with adequate training and information on DSE work (see C5.7).

Many of those working from a home base are likely to use laptop computers and they must be given instruction on the particular importance of ensuring good posture and taking adequate breaks away from the screen. Where laptops may be used for extended periods at the home base, it may be necessary to provide their users with suitable stands or docking stations and separate detachable keyboards.

Practical issues particularly likely to be important for DSE work at home include:

- availability of a suitable chair with adjustable seat and back
- adequate levels of suitable artificial lighting
- means of screening out strong sunlight (curtains or blinds).

'Users' who are self-employed are categorised as 'operators' under the regulations and an employer has a duty to carry out an assessment of 'those workstations which have been provided by him and are used for the purposes of his undertaking by operators'. In practice where employers do supply equipment to the self-employed, the workstation is likely to consist of a mixture of equipment supplied by the two parties. It would be prudent for the employer to insist on the type of self-assessment referred to above, but there would only be a duty to rectify problems relating to equipment supplied by the employer.

D7.6   Manual handling

Manual handling by homeworkers might involve moving stationery or records, components being manufactured or assembled, stored products or samples, work equipment or items used in displays or exhibitions. Chapter C4 details what is required in the assessment of manual handling activities carried out by employees. Aspects likely to be of relevance in respect of homeworkers are:

- storing materials and records in manageable-sized containers
- ensuring adequate space is available for safe storage
- providing suitable racking or shelving where appropriate
- ensuring heavier items are stored between knee and shoulder height
- maintaining clear access to racks and shelves
- providing handling aids, where necessary
- ensuring that assistance is available where team handling is required.

Manual handling into and out of vehicles is dealt with in D6.6. Some homeworkers may require special training in manual handling techniques. Guidance on training is provided in C4.9.

D7.7   Dangerous and hazardous substances

Guidance on the requirements of DSEAR (which applies to flammable gases and liquids, explosive dusts etc.) is provided in C7.4 while the COSHH Regulations are covered in detail in chapter C2. Both of those regulations contain important principles of relevance to homeworkers:

- substituting the substance by a less dangerous or hazardous one (or using an alternative work method), where reasonably practicable
- minimising the quantities of substances present in the workplace (in this case the home).

Where dangerous or hazardous substances do have to be used by homeworkers, then this use must be subject to the normal processes of risk assessment and appropriate control measures provided e.g.:

- suitable containers and dispensers
- local exhaust ventilation, where necessary
- personal protective equipment.

Further details on potential controls are provided in C7.4, C2.7 and C2.8. Controls of particular importance in homeworking relate to the protection of others (especially children and other vulnerable persons). These are likely to include:

- provision of suitable locked storage facilities for substances
- clear labelling of all containers
- close supervision over substances whilst not in locked storage.

D7.8   Personal protective equipment

The personal protective equipment (PPE) needs of homeworkers should be assessed as for other categories of employees (see chapter C6). In the home these are likely to relate primarily to the use of hazardous substances e.g. gloves, eye protection, although some manual handling activities may require the use of gloves or safety footwear.

However, account must also be taken of work carried out away from the home base which may involve visits to or work at construction sites or industrial premises. Employees involved in such activities may need to be provided with a much wider range of PPE together with guidance on when and where it should be used.

D7.9   Personal safety issues

Personal safety of staff working away from their normal base is an important topic in its own right and many aspects are of even greater relevance to those working from home. Issues to be addressed include:

- provision of details of expected visit locations and approximate timings
- follow-up actions when staff do not return or cannot be contacted
- visits outside normal working hours
- activities involving increased levels of risk (encounters with aggrieved persons, home visits, work in remote locations etc.)
- availability and awareness of appropriate precautions (e.g. mobile phones, attack alarms, choice of work locations, accompaniment)
- what staff should do if encountering problems.

Guidance on risks associated with work-related driving is provided in chapter C6. Where homeworkers are expected to meet customers, clients or others in the workers' own homes, the risks associated with such encounters must be assessed and appropriate precautions introduced e.g. restricting visits to persons already known to the organisation, requesting incoming telephone calls to check that everything is satisfactory.

D7.10 Accidents and first aid

HSE guidance on homeworking refers to the need to ensure that adequate first aid arrangements are made for employees working at home. This is particularly relevant for those using potentially dangerous work equipment, carrying out significant manual handling or using hazardous substances. Although more specialist risks may require specific first aid equipment, provision of a small travelling first aid kit should suffice in most cases. Basic guidance on first aid should also be provided – the HSE leaflet INDG 347 'Basic Advice on First Aid at Work' is particularly useful in this respect.

Homeworkers must also be made aware of the importance of reporting accidents and incidents associated with their work (including accidents to others). RIDDOR requirements apply to homeworkers as does the need to enter all work-related accidents in the HSE Accident Book (BI 510). Further details of these requirements are contained in chapter B4 which also emphasises the importance of accident and incident investigation as an accident prevention tool.

D7.11 Supervision, communication and consultation

Many employees are allowed to work at or from home because employers feel that they do not need close supervision. However, some degree of supervision is still necessary and effective communications with homeworkers are essential. Issues to be addressed include:

- mechanisms for obtaining additional equipment (including PPE)
- arrangements for reporting and rectifying defects in equipment
- arrangements for DSE eye tests and workstation assessments
- procedures for reporting accidents and near-miss incidents
- systems for communicating the workers' whereabouts when away from home
- contact points for advice or to express health and safety concerns.

Several of these have related training needs, some of which were referred to earlier in the chapter.

Employers also have duties to consult with employees on health and safety matters. These are explained in more detail in chapter B6. It is important that this consultation effectively includes homeworkers. This may involve identifying a specific employee representative drawn from homeworkers' ranks or alternatively ensuring that other employee representatives are effectively representing their interests.

Source Materials

1. HSE (2003) *Homeworking. Guidance for employers and employees on health and safety* (free leaflet) INDG 226
2. HSE (2002) *Basic Advice on First Aid at Work* (free leaflet) INDG 347.

| | | FIND THIS ON CD | Figure D7.1 | Homeworker risk assessment checklist |

Name of homeworker
Brief summary of work activities etc.

	Topic	Details – state what or how (or not applicable)
1.1	Work equipment supplied	
1.2	Arrangements for electrical inspection and test	
1.3	Any other maintenance required?	
1.4	Relevant training provided	
2.1	DSE eye test offered	
2.2	DSE self-assessment carried out	
2.3	Problems identified rectified	
2.4	Relevant training provided	
3.1	Manual handling – risks assessed	
3.2	Relevant training provided	
3.3	Handling equipment, racking etc. provided	
4.1	Dangerous/hazardous substances – risks assessed	
4.2	Control measures provided	
4.3	Secure storage provided	
4.4	Relevant training provided	
5.1	PPE supplied	
5.2	Guidance on use provided	
6.1	Personal safety guidance provided	
6.2	Mobile phone / attack alarm supplied	
7.1	First aid kit supplied	
7.2	Accident reporting arrangements explained	
8.1	Main contact for reporting defects/problems	
8.2	Name of employee representative	
9	Any other risks identified or precautions required	

I confirm that the equipment referred to above has been supplied and that the necessary training, guidance etc. referred to has been provided.

Employer's representative name   Signature   Date

Homeworker name   Signature   Date

Sources of Information

Sources of Information and Advice

E1.1   Introduction

The range of information available on health and safety matters has increased considerably in recent years. Much of this is due to the efforts of the HSE and this chapter provides a guide to the information and advice available from the HSE in printed publications, via their website and in other forms. However, the HSE is not alone in providing information and advice. Much is also available from commercial publishers, employers' organisations, trade unions, professional bodies, etc.

It should also be noted that employers have a duty under the Management Regulations to appoint someone to provide competent health and safety assistance (see paragraphs B3.9 to B3.13). While the regulations express a preference for this person to be a member of the workforce, assistance may also be required from external sources. The HSE leaflet 'Need Help on Health and Safety?' (ref. 1) provides useful guidance on such sources.

E1.2   HSE information sources

The HSE is constantly seeking new ways of providing information to employers and workers. Its current information sources (mostly available from HSE Books) include the following:

The services available from these sources are described in paragraphs E1.3 to E1.8 below.

- HSE printed publications
- HSE videos
- subscription services
- the HSE website
- HSE Infoline.

E1.3   HSE Books

HSE Books is the publishing arm for HSE publications and videos and for HSE's subscription services. Orders can be placed by:

- telephone: 01787 881165
- fax: 01787 313995
- post: HSE Books, PO Box 1999, Sudbury, Suffolk CO10 2WA
- email: hsebooks@prolog.uk.com
- internet: http://www.hsebooks.co.uk

Payment can be made by cheque (payable to HSE Books) or via credit card (American Express, MasterCard, Visa). Approved credit account customers may pay via BACS.

HSE Books normally produces a catalogue twice a year detailing the various publications, videos, etc., currently available and will supply this free of charge on request. Details of these publications (including recent additions) are also available via the main HSE website (see paragraph E1.7).

E1.4   HSE printed publications

HSE's printed publications are available directly from HSE Books and also from a network of bookshops situated in major cities and towns. A visit to such a bookshop can provide an opportunity not only to see the range and quality of the publications available but also to check whether an individual publication (if available in stock) provides the information required.

Some publications are priced whereas others are available free of charge, although charges are made for multiple copies. The most common categories of publications are as detailed below.

Legal (L) series priced

An L series booklet normally accompanies any new set of regulations of any significance. It usually contains the regulations themselves, guidance on interpretation and how to achieve compliance and, where relevant, details of any Approved Code of Practice (ACOP).

Health and Safety Guidance (HSG) series priced

These booklets provide guidance on the practical application of legislation. They deal with a wide range of issues such as:

- types of equipment, e.g. lift trucks, agricultural machinery
- specific hazards, e.g. compressed air, electricity, violence
- specific activities, e.g. meat preparation, roof work, welding
- types of precautions, e.g. local exhaust ventilation, lighting
- specific substances, e.g. flammable liquids, chlorine
- health and safety management techniques.

Of particular use to some readers will be the booklets in the series that provide comprehensive guidance on health and safety in a particular work sector. These include the following:

- HSG 62: tyre and exhaust fitting premises
- HSG 67: motor vehicle repair
- HSG 79: golf course management and maintenance
- HSG 112: motor sport events
- HSG 129: engineering workshops
- HSG 150: construction
- HSG 172: sawmilling
- HSG 179: swimming pools
- HSG 192: charity and voluntary workers
- HSG 220: care homes.

Guidance notes priced

These are sub-divided into several separate series providing guidance on specific topics or pieces of equipment:

- chemical safety (CS), e.g. disposal of waste explosives
- environmental hygiene (EH), e.g. carbon monoxide
- general (GS), e.g. pressure testing, overhead lines
- medical (MS), e.g. colour vision, audiometric testing
- plant and machinery (PM), e.g. freight containers, drilling machines.

INDG leaflets free

Some of these publications provide summaries of the requirements of specific sets of regulations while others deal with other health and safety topics. The majority of the series are aimed at employers but many are also intended to provide information for workers and can be very useful as training aids, justifying the cost of obtaining bulk quantities. Many INDG leaflets can be downloaded via the HSE website.

Information sheets free
These leaflets provide information for specific work sectors. There are several series, including the following:
Other important guidance priced

- agriculture (AIS and AS)
- chemicals (CHIS)
- construction (CIS)
- diving (DVIS)
- docks (DIS)
- education (EDIS)
- engineering (EIS)
- food/catering (CAIS and FIS)

- foundries (FNIS)
- leisure (ETIS)
- paper and board (PBIS)
- plastics (PPS)
- printing (PIS)
- radiation (IRIS)
- textiles (TIS)
- woodworking (WIS).

Three key booklets are not included in any of the above series. These are:

- 'Essentials of Health and Safety at Work': This booklet provides general guidance for small firms on managing health and safety and on a range of specific topics likely to be of relevance to most smaller businesses. It also contains a good reference section.
- 'Fire Safety: An Employer's Guide': Containing guidance on all aspects of fire precautions, this booklet is an essential reference for anyone carrying out fire risk assessments on workplaces of any significant size (see Chapter C7).
- 'Accident Book': This booklet contains individual record sheets which can be removed and stored securely. It also includes guidance on the Reporting of Injuries, Diseases and Dangerous Occurrences Regulations 1995 and the Health and Safety (First Aid) Regulations 1991.

Other series
There are also a number of other types of HSE publications, some of which are no longer being added to but which still contain titles of current relevance:

- COP: previously used for ACOPs and guidance booklets
- HSR: previously used for guidance on sets of regulations.

Other publications include the following:

- forms and posters (including the Health and Safety law poster)
- railway accident/incident reports
- contract research reports (on topics such as stress and musculoskeletal injuries)
- MDHS: Methods for the Determination of Hazardous Substances (analytical methods for use by occupational hygienists see Chapter D4)
- research papers (on topics such as grain dust, health promotion in primary care)
- annual reports and plans of work (of the HSC/HSE and on railway safety).

E1.5   HSE videos

The HSE has a variety of videos available for purchase via HSE Books. Some of these are of general relevance, e.g. preventing slips, vehicle safety or work at height, some relate to specific industries (e.g. 'Health and Safety in Motor Vehicle Repair' and various aspects of farm safety) and some cover specific topics, e.g. hand/arm vibration, indoor climbing walls, dermatitis.

At a price range of 19.50 to 75.00 (plus VAT) these videos are much more economical to purchase than most of those available in the commercial sector, providing that the topics covered are of relevance.

E1.6   Subscription services

Services available from HSE Books include the following:

- HSC Newsletter A bi-monthly newsletter providing information such as: national and international developments (including Euronews); forthcoming legislation and new codes of practice; guidance on a range of processes and hazards; new HSE and HSC publications; reports on accidents. This currently costs 15 per annum with significant discounts for additional subscriptions.
- HSE News Bulletin An annual subscription of 50 provides a weekly compilation of all HSE press releases.
- New Books News A free service providing details of all new HSE and HSC publications as well as statutory instruments.
- Quarry Fact File A twice-yearly publication for a 12.50 annual subscription (1.50 for additional subscriptions).

E1.7   HSE website

The main HSE website is www.hse.gov.uk and this gives access to HSE's publications and several of its other information sources. The free leaflets referred to in paragraph E1.4 above can be downloaded and printed via the website. The format of the website's home page changes regularly but it offers access to the following topics and facilities:

- About HSE – details of the HSC and HSE structure
- Strategies and plans
- Your views – public consultation and discussion
- Careers – details of employment opportunities within HSE
- Publications – access to free HSE leaflets via an alphabetical index
- Statistics – in various forms (causation, occupational area etc.)
- Research – HSE research publications
- Science and innovation
- Small businesses – guidance and publications targeting those starting up
- Workers – employers' and employees' responsibilities, safety representatives, whistle blowing, complaints about HSE
- Getting started – for new businesses
- Enforcement – a database of prosecutions taken and enforcement notices issued, enforcement policy
- Forms – an interactive link
- Selling to HSE
- Other languages (inc. Welsh)
- Contact HSE – making complaints, visiting offices (inc. addresses and maps), out of hours contact
- Feedback – on the website, enquiries, problems, frequently asked questions
- Help
- A–Z index
- Site Map
- What's new
- Press office
- News
- Events

- Campaigns
- Search – by industry or health and safety topics
- Ask an expert – HSE's Info-line (see E1.8 below)

plus a variety of links.

E1.8   HSE Infoline
This service is available by:

- telephone: 08701 545500
- email: hseinformationservices@natbrit.com
- internet: www.hse.gov.uk (see paragraph E1.7).

It can provide information on what HSE publications are available on specific subjects (orders for publications, where known, should be placed with HSE Books see paragraph E1.3) and also deal with general information enquiries. Enquiries are often transferred to Information Centres, the HSE's Field Operations Directorate or HSE Books. (HSE's main offices can also deal with enquiries their telephone numbers are usually shown under 'Health and Safety Executive' in local phone books and details are also available via the HSE website see paragraph E1.7 above.)

E1.9   Employment law and health and safety
Employment Law and Practice, produced by ICSA Publishing, is a looseleaf manual which includes detailed coverage of employment law issues in the area of providing a safe working environment. It also provides numerous case summaries which describe the consequences of injury claims made by employees.

E1.10   Other publishers
Several commercial organisations publish works both on health and safety in general and on specific aspects of the subject. While much of this is in book form, several companies offer subscription services through which the initial publications are kept current through regular updates. As well as revising out-of-date material, the updates also often contain news items (recent accidents, prosecutions, etc.) and details of new HSE publications.

Some subscription services are offered in a loose-leaf format but there is increasing use of CD-ROM systems. The Institute of Chartered Secretaries and Administrators is active in this field, as are Tolley's, Croner and Gee.

E1.11   Important organisations connected with health and safety
Many other organisations can provide information and advice on health and safety and, in some cases, can also offer relevant training. Some of these deal with specific health and safety topics, some relate to specific occupational sectors, while others cover a wide range of issues. Among the more important organisations are the following:

- British Occupational Hygiene Society (tel. 01332 298101): A professional organisation for occupational hygiene specialists, the BOHS produces a number of technical publications on occupational hygiene and health.
- British Safety Council (tel. 020 8741 1231, www.britishsafetycouncil.org): As well as offering a wide range of publications, the BSC offers training courses on many aspects of health and safety.
- British Standards Institution (tel. 020 8996 9000, www.bsi-global.com): In addition to many technical specifications (a number of which are referred to in this book), the BSI offers standards relating to the management of

333

health and safety (BS 8800 and OHSAS 18001 and 18002).

- Chartered Institute of Environmental Health (tel. 020 7928 6006, www.cieh.org.uk): The CIEH is the professional body for Environmental Health Officers. It offers a number of certificated training courses on health and safety as well as on food hygiene matters.
- Chartered Society of Physiotherapy (tel. 020 7306 6666, www.csp.org.uk). Although aimed at practitioner members, this site has useful documents.
- Construction Industry Training Board (tel. 01485 577577, www.citb. co.uk): The CITB offers many publications on health and safety in the construction industry and provides a wide range of training courses and training materials, including a number of videos.
- Fire Protection Association (tel. 020 7902 5300, www.thefpa.co.uk): As well as offering many books, videos and CD-ROMs on fire precautions for a variety of premises and work activities, the FPA also has a limited range of general health and safety materials.
- Institute of Occupational Hygienists (tel. 01332 298087, www.bohs.org): The IOH is another professional body for occupational hygienists and produces a variety of technical publications on the subject.
- Institution of Occupational Safety and Health (tel. 0116 257 3100, www.iosh.co.uk): The IOSH is the main professional body for those actively involved in health and safety and has recently received a Royal Charter. It produces a limited range of publications and provides general training courses (Managing Safely and Working Safely) through a network of accredited centres and also more specialist courses targeted at its own membership. Its monthly magazine, The Safety and Health Practitioner, is widely read and provides the main national forum for those advertising vacancies or seeking posts in health and safety.
- National Examination Board in Occupational Safety and Health (tel. 0116 288 8858): The NEBOSH Certificate and Diploma are the most widely respected qualifications in health and safety (see paragraph B3.12). They also are offered through a network of accredited centres.
- Royal Society for the Prevention of Accidents (tel. 0121 248 2000, www.rospa.co.uk): The RoSPA has a wide range of publications on health and safety at work as well as road safety and other areas of accident prevention. It also offers training courses and materials, including videos.
- Employers' organisations: A number of employers' associations are active in the field of health and safety, with the Engineering Employers Federation (tel. 0207 222 7777, www.eef.org.uk) in particular offering publications, training and consultancy services. The EEF has a network of regional associations.
- Trade unions: The TUC and many individual unions provide publications, training courses and advice for the benefits of their members. The TUC website, www.tuc.org.uk, includes information on new risks, reports and surveys. Users can sign up for free e-mail alerts.
- Manufacturers and suppliers: Manufacturers and suppliers of equipment and materials for use at work can also be useful sources of information and advice, particularly on technical matters. Some are able to provide training on the safe use of their products.

Most of the organisations referred to above have websites providing further information on their publications and services.

Index